Crustal Heat Flow
A Guide to Measurement and Modelling

Crustal Heat Flow: A Guide to Measurement and Modelling is a handbook for geologists and geophysicists who manipulate thermal data, particularly for petroleum exploration.

Heat flow data can provide an indication of past geological events and present economic potential of fossil fuel deposits. In theory and with practical examples the first two parts of the book discuss the sources of heat within the crust, describe how to maximise the accuracy of temperature data, cover the measurement of the thermal properties of rocks, and explain a number of maturity indicators. The last part of the book covers a range of thermodynamic models of the lithosphere and goes on to show how these can be used to reconstruct the thermal history of individual sedimentary basins. The focus remains consistently on practical applications, with worked examples providing a comprehensive guide for data reduction and interpretation.

This text will be attractive to a broad range of earth scientists. Many geologists will be interested in applications for global modelling and structural geology, but the topics covered will also be of specific interest to the oil exploration industry. This book will be regarded as a long-term reference source for professionals and researchers. It will also form the basis of advanced undergraduate and graduate student courses in geology, geophysics, engineering, mining, and environmental science, and will be a valuable text for petroleum industry training schemes.

G. R. Beardsmore is a Research Fellow at the Department of Earth Sciences, Monash University, Australia. Dr Beardsmore received his PhD from Monash in 1996. During his time as a student he was the principal researcher and coordinator of an industry-sponsored programme, *'Heat Flow Compilations and Thermal Maturation for Petroleum Exploration in NW Australia'*. After completing his PhD he spent time working with heat flow researchers in China (Changsha Institute of Geotectonics) and the United States (Southern Methodist University). He has published around a dozen papers on heat flow and tectonic evolution and has contributed to chapters in the volumes *'Applications of Emerging Technologies: Unconventional Methods in Exploration for Petroleum and Natural Gas'* (1997; ISEM, SMU, Dallas TX) and *'Geothermics in Basin Analysis'* (1999; Kluwer Academic/Plenum Publishers). Dr Beardsmore is a Member of the Petroleum Exploration Association of Australia (PESA), the American Geophysical Union (AGU) and the American Association of Petroleum Geologists (AAPG).

J. P. Cull is a Professor of Geophysics at the Department of Earth Sciences, Monash University, Australia. After joining the Australian Bureau of Mineral Resources he received an Australian Public Service scholarship leading to the award of a PhD from Oxford University in 1973–75. He was subsequently awarded a French Government Fellowship for Science and Technology concentrating on geothermal studies at BRGM (Orleans). After returning to Australia he was employed by CRA Exploration Pty Ltd as Principal Geophysicist responsible for the application of electrical and EM methods for base metal exploration. He joined the staff of Monash University in 1986 and was responsible for introducing a full range of geophysical coursework options. Professor Cull continues to conduct research in heat flow, geodynamics, electromagnetics, mineral exploration, and geotechnical studies. He is a Fellow of the Australian Institute of Mining and Metallurgy (F.AusIMM), a Fellow of the Australian Institute of Energy (F.AIE), a Member of the Australian Society of Exploration Geophysics (ASEG) and a Member of the Environmental and Engineering Geophysics Society (EEGS).

Crustal Heat Flow

A Guide to Measurement and Modelling

G. R. BEARDSMORE
Monash University

J. P. CULL
Monash University

CAMBRIDGE
UNIVERSITY PRESS

PUBLISHED BY THE PRESS SYNDICATE OF THE UNIVERSITY OF CAMBRIDGE
The Pitt Building, Trumpington Street, Cambridge, United Kingdom

CAMBRIDGE UNIVERSITY PRESS
The Edinburgh Building, Cambridge CB2 2RU, UK   http://www.cup.cam.ac.uk
40 West 20th Street, New York, NY 10011–4211, USA   http://www.cup.org
10 Stamford Road, Oakleigh, Melbourne 3166, Australia
Ruiz de Alarcón 13, 28014 Madrid, Spain

First published 2001

Printed in the United States of America

Typeface Times 10.75/13pt

*A catalog record for this book is available from
the British Library*

Library of Congress Cataloging-in-Publication Data

Beardsmore, G. R. (Graeme Ross)
   Crustal heat flow: a guide to measurement and modelling / G. R. Beardsmore, J. P. Cull.
      p.   cm.
   Includes bibliographical references and index.
   ISBN 0-521-79289-4 – ISBN 0-521-79703-9 (pb)
      1. Terrestrial heat flow – Measurement.   2. Earth – Crust.   I. Cull, J. P. (James Phillip)   II.
Title.

   QE509.8.B43 2001
   551.1′4 – dc21

00-048634

ISBN 0 521 79289 4 hardback
ISBN 0 521 79703 9 paperback

Contents

Preface   *page* ix

PART ONE. THE THERMAL STATE OF THE EARTH   1

1. Terrestrial Heat   3
  1.1. The Origin of the Earth   4
  1.2. The Present Thermal State of the Earth   9
  1.3. Basic Heat Flow Terms   11
  1.4. Units   18
  1.5. Heat Flow Resources   19
  1.6. Summary   21

2. Heat Generation   23
  2.1. Radiogenic Heat   24
    2.1.1. Heat Flow Provinces   28
    2.1.2. Radiogenic Heat Generation from Well Logs   30
    2.1.3. Seismic Correlations   36
  2.2. Frictional Heating along Faults   39
  2.3. Metamorphic Reactions   41
  2.4. Summary   43

PART TWO. MEASUREMENT TECHNIQUES   45

3. Thermal Gradient   47
  3.1. Direct Measurement Techniques   48
    3.1.1. Precision Temperature Logs   50
    3.1.2. Deep-Water Probes   54
    3.1.3. Borehole Convection   58
    3.1.4. Bottom-Hole Temperatures in Deep Wells   59
  3.2. Indirect Temperature Indicators   67
    3.2.1. Groundwater Geochemistry   67
    3.2.2. Curie Depth   70
    3.2.3. Xenoliths   71
    3.2.4. Upper Mantle Resistivity   72
  3.3. Surface Temperature   73

3.3.1.	Offshore	73
3.3.2.	Onshore	77
3.3.3.	Daily and Seasonal Cycles	78
3.3.4.	Climate Changes	82
3.4.	Data Integration	86
3.5.	Summary	87

4. Thermal Conductivity — 90

4.1.	Heat Transfer Theory	91
4.1.1.	Phonon Conduction	91
4.1.2.	Radiation	94
4.2.	Mixing Laws	97
4.2.1.	Harmonic Mean	97
4.2.2.	Arithmetic Mean	99
4.2.3.	Geometric or Square-Root Mean	100
4.3.	Measurements of Rock Conductivity	102
4.3.1.	The Lithological Column and Porosity Data	102
4.3.2.	Sample Preparation	105
4.3.3.	Steady-State Method	108
4.3.4.	Transient Methods	116
4.3.5.	Compaction Models	127
4.3.6.	Bulk Rock Thermal Conductivity	132
4.3.7.	Temperature Correction	134
4.4.	Well Log Analysis	136
4.5.	Shale Conductivity	142
4.6.	Summary	144

5. Thermal Maturity — 146

5.1.	The Generation of Hydrocarbons from Organic Matter	146
5.1.1.	Kerogen	146
5.1.2.	Kinetics	149
5.2.	Geochemical Indicators of Maximum Palaeotemperature	153
5.2.1.	Vitrinite Reflectance (VR)	155
5.2.2.	Fluorescence Alteration of Multiple Macerals (FAMM®)	169
5.2.3.	Thermal Alteration Index (TAI)	171
5.2.4.	Conodont Alteration Index (CAI)	172
5.2.5.	Clay Mineralogy	174
5.2.6.	Pyrolysis (RockEval)	178
5.2.7.	Fluid Inclusion Microthermometry (FIM)	179
5.2.8.	Molecular Biomarkers	181
5.3.	Fission Track Thermochronology (FTT)	188
5.3.1.	Choice and Preparation of Samples	190
5.3.2.	Analysis and Interpretation	191
5.3.3.	Limitations to FTT	195
5.4.	Deciding which Thermal Maturity Indicators to Use	197
5.5.	Summary	200

PART THREE. MODELLING TECHNIQUES 205

6. Heat Flow 207
  6.1. Product Method 207
  6.2. Bullard Plots 210
  6.3. Non-Linear Bullard Plots 212
    6.3.1. Reasons for Non-Linear Bullard Plots 212
    6.3.2. Climatology 214
    6.3.3. Sedimentation 216
    6.3.4. Erosion 219
    6.3.5. Groundwater Migration 219
    6.3.6. Deep Flow of Hot Fluids 221
  6.4. Non-Vertical Heat Flow 221
    6.4.1. Basement Relief 222
    6.4.2. Topography 224
    6.4.3. Salt Domes 226
  6.5. Heat Flow Correlations 229
    6.5.1. Heat Flow and Heat Production 229
    6.5.2. Heat Flow and Continental Age 230
    6.5.3. Heat Flow and Oceanic Age 231
    6.5.4. Heat Flow and Seismic Data 232
    6.5.5. Heat Flow and Electrical Conductivity 234
  6.6. Summary 235

7. Lithospheric Models 237
  7.1. Stable Lithosphere 238
    7.1.1. Oceanic Lithosphere 238
    7.1.2. Continental Lithosphere 245
  7.2. Hot Spots 248
    7.2.1. Cause of Hot Spots 249
    7.2.2. Thermal Effect of Hot Spots 250
  7.3. Subduction Zones 252
    7.3.1. Driving Forces 252
    7.3.2. Trenches 255
    7.3.3. Island Arcs 255
    7.3.4. Back-Arc Basins 255
    7.3.5. Continental Collision 256
  7.4. Extension 257
    7.4.1. Instantaneous Pure Shear 257
    7.4.2. Constant Pure Shear 261
    7.4.3. Simple Shear 263
    7.4.4. A Pure-Shear/Simple-Shear Coupled Model 265
    7.4.5. The Blanketing Effect of Sediments 269
    7.4.6. Underplating 270
    7.4.7. Passive Margins 272
  7.5. Summary 273

8. Numerical Modelling   275
   8.1. Finite Difference Approximations   275
   8.2. Relaxation   279
      8.2.1. Discretising the Model   279
      8.2.2. Boundary Conditions   284
      8.2.3. Equilibrium Nodal Temperature   285
      8.2.4. Precision   288
   8.3. Errors   290
      8.3.1. Discretisation Error   290
      8.3.2. Round-Off Error   291
   8.4. Summary   291

9. Unravelling the Thermal History of Sedimentary Basins   293
   9.1. Determining Present Heat Flow Distribution   293
   9.2. Understanding the Tectonic History of the Basin   294
   9.3. Determining the Stretching Factor, β 295
   9.4. Determining the Heat Flow Anomaly and Distribution for
          each Time Step   297
   9.5. Solving for Temperature Distribution In Crust   300
   9.6. Summary   301

References   303

Index   321

Preface

The first step is to make sure of the facts themselves. In other words – What is the law regulating this increase of heat? At what rate does the temperature augment? This might be supposed to be a point easily settled. It might be supposed that nothing would be easier than to insert thermometers into the rock at different depths in a mine, and to read off their indications; or to lower them into a borehole for the same purpose. But there are many difficulties to be overcome, and a host of disturbing causes present.

Physics of the Earth's Crust – Rev. Osmond Fisher, 1881, p. 4.

Heat flow data have been used extensively by geoscientists to provide an indication of temperature variations within the Earth. However, there are relatively few experts available with the necessary experience to provide a critique of the central constraints or to obtain new data for second-order models. Terminology relating to thermal conductivity, diffusivity, vitrinite reflectance, xenolith geochemistry, and so on, is widely recognised but poorly understood, particularly by industry professionals.

As an example, the present state of geothermal modelling and interpretation in the petroleum industry is a peculiar one. On the one hand, there are numerous highly sophisticated software applications on the market for solving the complex equations governing the thermal evolution of sedimentary basins and the maturation of organic material. On the other hand, however, the accurate use of these packages requires considerable knowledge of the underlying physical concepts and a wealth of reliable data. Both commodities are in short supply and complex interpretations are often based on an inadequate appreciation of the primary constraints. Consequently, this book grew from a desire to address the problem of education, by covering the fundamental concepts of heat flow from both a practical and a theoretical point of view.

When we first began work on this book, our objectives were relatively simple, with few concerns regarding the scope and range of topics it would eventually cover. The text began its life as a short monograph covering some fundamental aspects of heat flow measurements and geothermal modelling, but the more we progressed into it, the more we discovered the truth of the good Reverend Fisher's words from 1881. The book ended up as a text covering almost every aspect of thermal modelling in at least a basic way. From discussions with practising petroleum geologists we were able to pinpoint some of the 'many difficulties' and 'disturbing causes' and address them directly by way of explanation and example.

In this day and age, print media can be somewhat limited in its abilities to present available information. For this reason, a web site has been set up at

http://www.earth.monash.edu.au/heatflow/

as an ongoing source of support in addition to the information and concepts described and explained in the text. The web site includes colour versions of many of the figures, links to useful web sites elsewhere in the world, an up-to-date list of relevant publications, and small software applications that implement some of the concepts discussed in the book. It is hoped that this book and the associated web site will become useful companions for the general geophysical community, and more specifically the petroleum geologist, struggling to make sense of the sophisticated software tools at his or her disposal.

It is necessary now to acknowledge and thank some of the many people who have directly or indirectly helped in the production of this manuscript. In a way, Barry Goldstein is the godfather of the work, without whom G.R.B. would not have been lured into the field of geothermal research at all. Trevor Graham and Mike Swift, of the then Bureau of Mineral Resources in Canberra, were instrumental in providing G.R.B. with his first geothermal field experience in the active volcanic cauldera of Rabaul Harbor, Papua New Guinea. Such exotic locations are one of the great rewards of this field of work. Deeper in the past, J.P.C. was privileged to enjoy the fundamental supervision of Ron Oxburgh and Steve Richardson at Oxford. They greatly assisted in developing a range of heat flow methods still in current use.

More direct involvement in the manuscript preparation came by way of discussion of content and critical feedback from Neil Sherwood, Gareth Cooper, Frank Maio, Iain Bartholomew, Jason McKenna and particularly David Blackwell. Others who have in some way contributed, through casual discussions, responses to requests for information, or various other means, include Malcolm Altmann, George Asquith, Will Gosnold, Greg Houseman, Mark Lisk, Arthur Mory, Bob Nicoll, Paul O'Sullivan, Walter Pickel, Nigel Russell, Sun Shaohua, Alister Terry and Yoshihiro Ujiié. Our editor, Matt Lloyd, was particularly prompt and helpful whenever we had need of his assistance.

G. R. Beardsmore
J. P. Cull

PART ONE

The Thermal State of the Earth

CHAPTER ONE

Terrestrial Heat

From thence I gathered the creation of the world did fall out upon the 710 year of the Julian Period, by placing its beginning in autumn: but for as much as the first day of the world began with the evening of the first day of the week, I have observed that the Sunday ... happened upon the 23 day of the Julian October; from thence concluded that from the evening preceding that first day of the Julian year, both the first day of the creation and the first motion of time are to be deduced.

The Annals of the World, IV – Archbishop James Ussher, 1658.

Geothermal data provide critical constraints for models of global accretion, core–mantle segregation and crustal evolution. Temperature history is a controlling factor in the maturation of organic compounds into fossil fuels. Geothermal energy is a relatively clean, renewable means of heat and electricity production. Yet, in spite of the number and variety of applications of geothermal research, the means of measuring, assessing and interpreting basic geothermal data are poorly understood by all but a small number of specialists, even among those whose job it is to manipulate such data.

This book aims to bring together many aspects of basic geothermal research in an accessible way. Although some concepts require mathematical treatment, these will be kept to a minimum and, where possible, practical examples will be used as illustrations. Broadly, the book aims to cover geothermal research to the extent that the following general questions can be answered: What are the physical properties of rocks relevant to thermal studies? How are they most accurately measured? How are the results best interpreted? The book is not intended to be an exhaustive text on the current state of geothermal research, but rather a 'how to' guide for geoscientists who need to manipulate geothermal data but lack the necessary knowledge to do so with confidence.

The people at whom this text is mostly focussed are petroleum geologists who have been given the task of discovering new energy reserves. Petroleum exploration requires specialist knowledge in many fields of geology so petroleum geologists are drawn from such varied groups as sedimentologists, structural geologists, seismic processors and interpreters, sequence stratigraphers, geophysicists and geochemists. Many have no prior experience with thermal

3

data and, yet, are required to manipulate and interpret it. This text is intended to be a simple guide for such people.

The general layout of the book is divided into three Parts. Part I contains Chapters 1 and 2, which discuss the origins and present thermal state of the Earth, introduce the basic units and terms used in geothermal studies, and investigate the sources of heat within the Earth. Part II contains Chapters 3, 4 and 5, which look in detail at the measurement techniques for the basic parameters of heat flow – present temperature, thermal conductivity and palaeotemperature. Part III covers modelling and applications in Chapters 6, 7, 8 and 9. Chapter 6 explores the ways in which basic geothermal data are integrated into a coherent picture of present-day heat distribution. Chapter 7 looks at forward modelling, or how we can predict the thermal history of a region given its tectonic history. Chapter 8 introduces the reader to a simple but powerful method of numerical modelling. Chapter 9 integrates components of all the previous chapters into a work plan for the assessment of the thermal history of sedimentary basins.

1.1. The Origin of the Earth

The present thermal structure of the Earth is intimately linked to its formation and composition. Thus, to understand the present thermal structure of the planet, it is useful to have at least a basic understanding of the processes that led to its development. The origin of the Earth has been a topic of hot debate ever since man developed an active curiosity about his home. For much of recorded history, the majority of people have accepted the explanation of religious texts and creation fables. This tradition culminated in the Christian world with James Ussher, Anglican Archbishop of Armagh and Primate of All Ireland, declaring, after careful analysis of the Bible, that the world was brought forth as an Act of God on Sunday 23 October, 4004 BC (Craig and Jones, 1982). Since that time, geoscientists have aspired to equal such precision.

Note: Archbishop Ussher made his calculations using the outdated Julian Calendar. Using our present Gregorian calendar, the world was created on 21 September 4004 BC, and turned 6000 on 21 September 1997 AD.

With the rise and development of scientific methodology, views such as Archbishop Ussher's have lost much of their popular appeal and an explanation for the origin of the Earth has been sought through empirical measurements. One such model that presently commands much support is that the Earth accreted from a cloud of gas and dust some 4550 million years ago (e.g. Press and Siever, 1986). Brown and Mussett (1993) provided a comprehensive review of current thoughts on the formation of the Earth and the reader is referred to their text for details. We present here a short summary of their Chapters 4 and 5.

The solar system began as a rotating cloud of stellar gas, and developed through three distinct phases. First, the cloud of gas flattened gravitationally onto the median plane. Much of the gas collapsed into the centre of the disc, ultimately to form the sun, but a small amount retained enough angular momentum to resist the central pull. Small dust particles began to condense from the gas in compositions determined primarily by temperature, which was in turn related to the distance from the proto-sun. As these dust particles collided, they tended to stick together, decreasing their surface-area-to-mass ratios. The drag of the gas on the dust decreased proportionally, increasing the rate of fall of the dust towards the median plane. Further collisions exponentially increased the size of the solid particles as larger bodies collided with an increasingly greater amount of dust. After several tens of thousands of years there existed a plane of planetesimal bodies encased in the remnants of the initial gas cloud, and orbiting an increasingly active hot ball of gas. At the distance of Earth's current orbit, the planetesimal bodies grew to several metres in diameter and movement was effectively independent of the remaining gas.

At about this time, the new sun entered what is known as the T-Tauri stage of development, which lasted for approximately 10 Ma. During this stage in development, a star produces a strong solar wind that moves radially outwards. Although the planetesimal bodies were mainly unaffected, all remaining stellar gases and microscopic dust particles were swept from the inner solar system, leaving only the disc of solid material.

The second stage of planetary accretion began with a densely populated disc of planetesimal material revolving in haphazard orbits around the sun. Mathematical modelling of such a system predicts a large number of collisions. Some of these collisions result in the destruction of one or both bodies, but others result in the creation of a larger, combined body. The factor dominantly controlling the outcome of a collision was the relative velocity of the two bodies. If it was less than half of the escape velocity (the velocity required for two bodies initially in contact to escape their mutual gravitational attraction by moving directly away from each other) they tended to merge. Escape velocity is proportional to the combined mass of the two bodies, which leads to runaway growth. Within 50,000 years there was probably only one dominant planetary embryo ($\sim 10^{23}$ kg – between 1% and 10% of the Earth's current mass, 6×10^{24} kg) revolving in a roughly circular orbit between 0.99 and 1.01 of Earth's current orbital radius. This embryo was probably accompanied by up to 100 others of similar size orbiting within the region presently occupied by the terrestrial planets (Mercury, Venus, Earth and Mars), and a multitude of significantly smaller bodies ($< 10^{18}$ kg).

The third and final step in planetary accretion was the merging of the planetary embryos and remaining planetesimals into the few planets that we see today. This process is much harder to model statistically because of the small number of bodies involved, but the qualitative argument is as follows.

The planetary embryos were numerous and large enough to affect each other's orbits. Over a significant period of time (~100 Ma), perturbations in adjacent orbits led to further, catastrophic collisions, reducing the overall number of bodies to the present small number. This mechanism accounts for the different sizes and rates of spin of the planets. It also accounts for the formation of the moon – the result of a Mars-sized body colliding with the embryonic Earth about 50 Ma after the formation of the solar system (Lee et al., 1997). However, the model predicts a Mars significantly larger than observed, and a single body occupying the orbit where the asteroid belt is found (between Mars and Jupiter). These discrepancies are attributed to the gravitational influence of Jupiter, which even today exerts a significant force on the orbits of individual asteroids within the belt.

The initial temperature distribution within the Earth is unknown. Brown and Mussett (1993), however, argued that the energy liberated by the immense collisions of planetary embryos was more than sufficient to keep the planet in a mostly molten state for most of the accretionary phase. They argued for the early existence of a deep magma ocean at a temperature of several thousand kelvin, beneath a thin chilled crust. Cameron and Benz (1991) suggested near-surface temperatures in excess of 10,000 K and a global-scale redistribution of mass associated with the development of the moon. These molten conditions facilitated the early (< 50 Ma after the formation of the solar system) gravitational separation of dense metallic material from lighter silicates (Lee et al., 1997). Up to 2×10^{30} J of gravitational potential energy may have been released as large drops of iron settled into the middle of the planet. As the middle of this molten iron core subsequently fused under intense pressure, even more latent heat was released ($\sim 1.5 \times 10^{28}$ J). The temperature at the centre of the early core may have been as high as 7000 K, with a temperature of about 5500 K at the core–mantle boundary. High temperatures prevailed throughout the mantle, and molten magma oceans may even have penetrated to the surface at some periods.

Note: The heat released during the formation of the core is alone sufficient to sustain an average surface heat flow of about 27.5 mW m^{-2} over the entire 4.55 Ga geological history of the planet. This is almost one third of the present average (Pollack, Hurter and Johnson, 1993).

The present thermal state of the Earth has interested scientists for several centuries. Although data have only been collected on a systematic basis since Sir Edward Bullard's measurements in South Africa (1939), earlier investigations of thermal gradient had been conducted at a number of arbitrary locations (e.g. Forbes, 1849; Thomson, 1860; Prestwich, 1886). The significance of these measurements had even been recognised by William Thomson.

> *Note:* William Thomson (1824–1907) was raised to the Peerage as Baron Kelvin of Largs in 1892, and it is as Kelvin that he is most widely remembered. He formulated the first and second laws of thermodynamics and the absolute temperature scale is named after him. From this point on in this text we refer to him only as Kelvin.

Kelvin argued that a positive thermal gradient with depth means that the Earth must be cooling. By assuming the Earth began as a sphere initially at a constant temperature, he was able to calculate when the cooling must have commenced. Using an average gradient of $36.5°C\,km^{-1}$ ($1°F/50$ ft), he calculated the age of the Earth to be 200 Ma if the initial temperature was 5540°C (10,000°F), or only 98 Ma if the initial temperature was 3870°C (7000°F). Kelvin preferred the latter estimate, but included a sizeable error margin, 20–400 Ma, due to the uncertainty in the value of thermal diffusivity for crustal rocks (Thomson, 1862). Given the knowledge of the day, Kelvin's age estimate was not only acceptable, but also virtually unassailable.

> *Note:* Kelvin's age estimate was made using an average thermal diffusivity derived from measurements on only three rocks, all from around Edinburgh! Given that, his value of 1.2×10^{-6} $m^2\,s^{-1}$ ($400\ ft^2\,yr^{-1}$) is remarkably representative of crustal rocks, being approximately that of marble at room temperature.

Based on sound mathematics and solid thermodynamic principles, Kelvin's age estimate baffled geologists and biologists, who required a much older Earth to suit their respective models of land-forming processes and evolution of species. At the time, they were unable to fault his reasoning, but have since been vindicated. We now realise that the Earth is at least an order of magnitude older than Kelvin calculated. So where was the flaw in Kelvin's logic?

The problem was that Kelvin assumed the Earth to be an inert body cooling in space. As such, he inadvertently ignored a major internal source of heat. We can forgive him this oversight, however, because it had yet to be discovered! Soon after Henri Becquerel announced the discovery of radioactivity in 1896, it was realised that here was a source of heat that could account for present thermal gradients, yet allow a much greater age of the Earth than Kelvin's estimate. Kelvin's age estimate thus became a minimum age, applicable only for the case of zero internal heat generation. The age can be lengthened indefinitely with the inclusion of radioactive material in the interior of the Earth.

> *Note:* Kelvin was a thorough and careful scientist and had covered his bases. He wrote, 'It is certain that either the earth is becoming on the whole cooler from age to age, or the heat conducted out is generated in the interior ...' (Thomson, 1862). That

he discounted the latter possibility as 'extremely improbable' and 'contrary to all analogies in nature' should not be held against him.

The heat contribution from a radioactive element is proportional to its absolute abundance, its half-life, and the heat generated from a single decay event of each isotope of the element. Given the present abundance of a number of prominent radioactive isotopes (e.g. 32 ppb uranium in the mantle; Turcotte, 1980), it is possible to estimate the total rate of radiogenic heat generation through time. Figure 1.1 illustrates the historical influence of radiogenic heat generation on surface heat flow for three different models (Lubimova, 1969).

Note: The job of publicly repudiating Kelvin's estimate of the age of the Earth fell to Ernest Rutherford (1871–1937), a Nobel Prize–winning New Zealander and early investigator of radioactivity. In 1904 he gave an address to the Royal Society of London, during which he demonstrated that the Earth is heated by radioactive elements in its rocks, and that the decay rate of

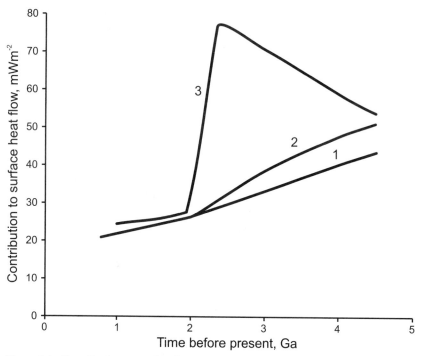

Figure 1.1. Contribution to surface heat flow of radiogenic heat production through time. Lines 1 and 2 assume that radiogenic isotopes fractionated into the crust gradually, with the lines representing two different rates of fractionation. Line 3 assumes a sudden fractionation at about 2 Ga before present. Modified after Lubimova (1969).

these elements provides a means of measuring the age of the planet. He suggested an age much in excess of 100 Ma. Lord Kelvin, the highly respected 80-year-old undisputed elder states-man of British science, was in the audience. The story is told (Brennan, 1997) that although Kelvin slept through much of Rutherford's address, he awoke for the conclusion and glared at the impudent young upstart who was disputing his long-accepted deduction. In a moment of diplomatic brilliance, Rutherford pointed out that Kelvin himself had concluded that his age estimate was not valid if the heat conducted out was generated in the interior, and remarked 'That prophetic utterance refers to what we are considering tonight.' Kelvin was won over, and Rutherford went on to fame and fortune.

Major modifications had to be made to Kelvin's model after it was recog-nised that radiogenic heating acted to significantly extend the cooling time. Rutherford began the job, but it was not until models of the distribution of radioactive isotopes within the Earth were developed that serious age estimates could be put forward (e.g. Lubimova, 1958; MacDonald, 1959). Turcotte (1980) attributed 83% of the present surface heat flow to the decay of radio-active isotopes, and only 17% to the cooling of the Earth. He concluded that the mantle is presently cooling at a rate of $36°C\,Ga^{-1}$, and that three billion years ago it was likely 150°C hotter than at present. The presently accepted age of the Earth is around 4.55 Ga.

1.2. The Present Thermal State of the Earth

That the interior of the Earth is considerably hotter than the surface has never been seriously questioned. That the heat is transported internally to the surface where it is radiated into space is likewise accepted. The manner and rate of heat transfer, however, has been the subject of considerable debate over the years. With our increased understanding of fluid dynamics and the temperature dependence of viscosity, initial purely conductive models have given way to mantle convection as the favoured mechanism of heat loss from the bulk Earth. The exact nature of such convection is still poorly understood, though, and several models are possible that conform to present data sets (Wyllie, 1988; Figure 1.2).

Heat is moved through much of the Earth via convection, but conduction appears to be the dominant means by which it is transported through the crust. Conductive heat flow has been measured over much of the Earth's surface and a recent compilation of 20,201 data points (Pollack et al., 1993) suggested a global mean of $87 \pm 2\,mW\,m^{-2}$, with a total heat loss of $44.2 \pm 1.0 \times 10^{12}$ W. Heat loss is by no means evenly distributed, however, with average heat loss through the oceans ($101 \pm 2.2\,mW\,m^{-2}$) being considerably higher than through the continents ($65 \pm 1.6\,mW\,m^{-2}$). Pollack et al. (1993) broke the

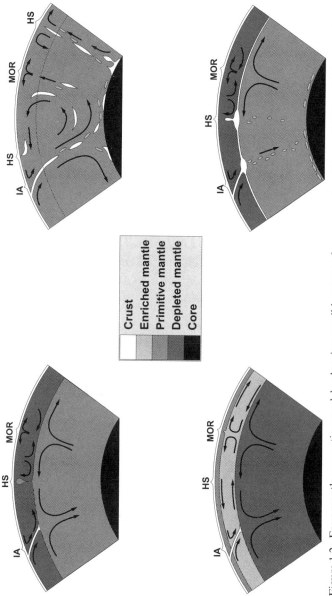

Figure 1.2. Four mantle convection models, showing possible sources for common basalt types. MOR = Mid-ocean ridge, HS = hot spot, IA = island arc. The mantle becomes depleted in certain elements through partial melting, fractionation and separation of relatively enriched magma. Modified after Wyllie (1988).

data down to provide statistical heat flow averages based on age and dominant regional geology. Their data are reproduced in Table 1.1.

The distribution of surface heat flow is closely related to the temperature of the upper mantle, as evidenced by seismic shear-wave velocity data (Pollack et al., 1993). Regions of lower than average shear-wave velocity in the upper mantle coincide with higher than average heat flow at the surface. Lower shear-wave velocity implies higher temperature.

1.3. Basic Heat Flow Terms

Before progressing further into an investigation of geothermal methods, it is useful, and perhaps in many cases necessary, to revise basic thermal physics. This section introduces and defines the terms that will be used extensively throughout the remainder of the text. The basic heat flow equations are derived

Table 1.1. Global Mean Heat Flow as a Function of Surface Geology and Age

Description	Mean Heat Flow $(\text{mW m}^{-2})^a$	Standard Error (mW m^{-2})	Age Range (Ma)
Oceanic			
Undifferentiated Cenozoic	89.3 (125.2)	2.8	0–65
Undifferentiated Mesozoic	44.6 (51.0)	2.8	65–251
Quaternary	139.5 (806.4)	10.1	0–1.8
Pliocene	109.1 (286.0)	5.9	1.8–5.3
Miocene	81.9 (142.2)	2.3	5.3–23.8
Oligocene	62.3 (93.4)	2.2	23.8–33.7
Eocene	61.7 (75.7)	1.6	33.7–54.8
Paleocene	65.1 (65.1)	2.8	54.8–65
Late Cretaceous	61.5 (60.0)	2.1	65–84
Middle Cretaceous	56.3 (53.9)	1.0	84–119
Early Cretaceous	53.0 (50.5)	1.4	119–141
Late Jurassic	51.3 (49.4)	1.2	141–159
Continental			
Undifferentiated Subaqueous	77.7	1.9	
Cenozoic (sed./met.^{b})	63.9	0.9	0–65
Cenozoic (igneous)	97.0	5.6	0–65
Mesozoic (sed./met.)	63.7	1.3	65–251
Mesozoic (igneous)	64.2	3.0	65–251
Paleozoic (sed./met.)	61.0	1.2	251–545
Paleozoic (igneous)	57.7	2.6	251–545
Proterozoic	58.3	1.4	545–2500
Archean	51.5	2.4	2500–3800

[a] Values in (brackets) are from Stein and Stein (1992) and are recommended by Pollack et al. (1993) over actual measured averages.

[b] sed. = Sedimentary, met. = metamorphic.

Source: Pollack et al. (1993).

using examples, and a table is provided for converting units between different systems of measurement.

Human beings are physically able to sense relatively small changes in temperature, and major changes have a direct bearing on personal comfort and welfare. Temperature changes also affect other forms of life, such as food crops, which indirectly impact on human well-being. Consequently, there has been a long history of observation and measurement of air and ocean temperature for improving weather prediction and detecting long-term climate changes.

As well as temperature, human perception also encompasses the concepts of heat transfer and thermal diffusivity. It is considered a matter of common sense that a steel poker will transfer heat from a fire to the hand and cause discomfort. At the same time, it is recognised that the degree of discomfort can be controlled by encasing the grip of the poker in non-metallic insulators such as wood or textiles. It is not necessary to understand the theory of heat conduction to recognise these effects. However, the nomenclature developed in the following paragraphs provides a method for systemising and quantifying these intuitive concepts, and includes definitions of several important parameters relevant to the thermal investigation of the Earth's crust.

Temperature, T (kelvin, K), provides an indication of the free energy available at a location. This energy can be visualised as vibrations within a molecular lattice. The greater the temperature, the greater the magnitude of vibration. Intuitively, if one part of a lattice is set vibrating, the vibrations will disseminate through the lattice in such a way as to produce conditions of greatest harmony (or minimum enthalpy). Put more simply, if one section of a substance is heated, the heat will disseminate (or flow) to regions of lower temperature.

Thermal gradient is defined according to

$$\partial T / \partial z = (T_2 - T_1)/\Delta z \qquad (1.1)$$

where T_1 and T_2 are the temperatures at two points separated by a distance Δz. $\partial T/\partial z$ is expressed in terms of temperature per unit distance (K m^{-1}), and is a vector quantity having both magnitude and direction. By convention, a positive gradient is in the direction of increasing temperature.

Question: A basalt prism 2 cm long is held in a device that maintains a constant temperature of 25°C at one end and 45°C at the other. What is the magnitude of the thermal gradient?

Answer: From Equation (1.1):

$$\partial T/\partial z = (45 - 25)/0.02 = 1000 \text{ K m}^{-1}.$$

The rate of *heat flow*, **Q**, between two points is given by

$$\mathbf{Q} = -\lambda \times \partial T/\partial z \qquad (1.2)$$

where **Q** is expressed in units of power per unit area (watts per square metre, $W\,m^{-2}$) and is a vector quantity. Positive heat flow is conventionally taken to be in the direction of decreasing temperature, which is the opposite of the convention for thermal gradient and the reason for the negative sign in Equation (1.2). λ is a physical property of the medium known as the *thermal conductivity*. It is an operator that relates the heat flow vector to the thermal gradient vector. Thermal conductivity is expressed as power per unit distance per unit temperature ($W\,m^{-1}\,K^{-1}$).

Question: The basalt prism from the previous example has a thermal conductivity $\lambda = 1.8\ W\,m^{-1}K^{-1}$ in the direction of the thermal gradient. At what rate and in what direction is heat flowing through the prism?

Answer: From Equation (1.2):

$$\mathbf{Q} = -1.8 \times 1000 = -1800\ W\,m^{-2}$$

in the direction of increasing temperature.

More conventionally, $\mathbf{Q} = 1.8\ kW\,m^{-2}$ in the direction of decreasing temperature.

Note: λ is a tensor operator, not a constant, because it can be anisotropic, having different values for different directions through the same medium.

Equation (1.2) applies only where we have a constant thermal gradient, attainable only in homogeneous media bounded by fixed temperature conditions. In most practical situations, the temperature of the body is controlled by the magnitude of the heat source, H (watts), and depends on the *heat capacity*, C (energy per unit temperature, $J\,K^{-1}$), of the medium, defined by

$$H \times \Delta t = C \times \Delta T \tag{1.3}$$

If a body with heat capacity C is heated at a rate H, for a time Δt (s), the temperature of the body will rise by ΔT (K). Put simply, heat capacity is the amount of energy required to raise the temperature of the body by 1°. The *relative heat capacity*, or *specific heat*, c ($J\,kg^{-1}K^{-1}$), is the amount of energy required to raise the temperature of one unit mass of the body by 1°. Equation (1.3) can also be written as

$$H = c \times \rho \times V(\partial T/\partial t) \tag{1.4}$$

where ρ is the density ($kg\,m^{-3}$) of the body and V its volume (m^3).

Question: Our basalt prism from previous examples is actually a cylinder with a diameter of 2.5 cm, so it has a volume $V = 9.8 \times 10^{-6}$ m^3. The density of the prism is $\rho = 2850$ kg m^{-3} and relative heat capacity, $c = 800$ J kg^{-1} K^{-1}. What is the total heat capacity of the cylinder? If the basalt is heated at a rate of 10 W for 10 s, by how much does its temperature increase?

Answer: By comparing Equations (1.3) and (1.4), the heat capacity of the cylinder is

$$C = c \times \rho \times V = 800 \times 2850 \times 9.8 \times 10^{-6} = 22.4 \text{ J K}^{-1}$$

Then the temperature increase is found using Equation (1.3):

$$H\Delta t = C\Delta T, \qquad \text{so } 10 \times 10 = 22.4 \times \Delta T,$$

or $\Delta T = 4.47$ K.

Experience tells us that heat does not flow instantaneously through a body, but diffuses over a period. Points further from the heat source take longer to rise in temperature, as any child who has ever held an iron poker in a fire will tell you. Equation (1.3) tells us by how much the temperature of a body will rise due to an applied amount of heat, but it does not indicate the length of time required for equilibrium.

Visualise a long, cylindrical body, like an iron poker, initially at constant temperature, T. All points along the poker are at the same temperature, so from Equation (1.1) the thermal gradient $\partial T/\partial z = 0$. Thus, from Equation (1.2), heat flow along the poker is $\mathbf{Q} = 0$.

One end of the poker is now placed within the hot coals of a fireplace (Figure 1.3). To investigate the effect on the rest of the poker, it is convenient if we reduce the argument to one dimension only. That is, we wish to investigate the effect of heat flowing along the poker, without concerning ourselves with the flow of heat radially through the poker. This is a reasonable assumption if the poker is thin relative to its length, as most pokers are.

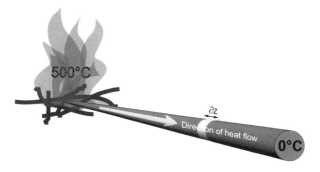

Figure 1.3. The flow of heat along a metal poker. Heat flows from the hot end towards the cooler end. A small amount of heat, $\partial \mathbf{Q}$, is required to raise the temperature of each small length of the poker, ∂z, by an amount, ∂T.

Taking Equation (1.4), we can replace the volume term with a length term:

$$V = z \times A$$

$$H/A = c \times \rho \times z \times (\partial T/\partial t) \tag{1.5}$$

where z is length along the poker and A is cross-sectional area of the poker.

The left-hand side of Equation (1.5) has units of watts per square metre, and is the same as $\mathbf{Q}$ in Equation (1.2). So,

$$\mathbf{Q} = c \times \rho \times z \times (\partial T/\partial t) \tag{1.6}$$

As each small portion of the poker, ∂z, heats up, a small portion of the available heat, $\partial \mathbf{Q}$ is absorbed:

$$\partial \mathbf{Q} = c \times \rho \times \partial z \times (\partial T/\partial t) \tag{1.7}$$

Rearranged, Equation (1.7) now gives us the distribution of heat flow along the poker:

$$(\partial \mathbf{Q}/\partial z) = c \times \rho \times (\partial T/\partial t) \tag{1.8}$$

Recall from Equations (1.1) and (1.2):

$$|\mathbf{Q}| = \lambda \times (\partial T/\partial z) \tag{1.9}$$

so Equation (1.8) becomes

$$(\partial T/\partial t) = \kappa \times (\partial^2 T/\partial z^2) \tag{1.10}$$

where $\kappa = \lambda/(c \times \rho)$ is the *thermal diffusivity* of the poker.

Note: For most minerals and crystalline rocks, the product, $c \times \rho$, lies within 20% of $2.3 \times 10^6 \, \mathrm{J\,m^{-3}\,K^{-1}}$ (Beck, 1988). This simplifies the relationship between diffusivity and conductivity:
$$\kappa \approx \lambda/(2.3 \times 10^6) \, \mathrm{m^2\,s^{-1}}$$
where λ is in $\mathrm{W\,m^{-1}K^{-1}}$.

Thermal diffusivity (area per unit time, $\mathrm{m^2\,s^{-1}}$) is a physical property that controls the rate at which heat dissipates through a material. Equation (1.10) is Fourier's equation for the linear conduction of heat, and is often referred to simply as the *heat flow equation*.

Question: A poker of length $L = 1$ m is initially at a constant temperature $T_0 = 0°\mathrm{C}$. At time $t = 0$ s, one end of the poker ($z = 1$ m) is placed in hot embers ($T_1 = 500°\mathrm{C}$) while the other end ($z = 0$ m) is insulated. After 20 min, the temperature 20 cm from the hot end of the poker is measured at $T = 20°\mathrm{C}$. What is the thermal diffusivity, κ, of the poker?

Answer: The solution to this problem is not trivial. Carslaw and Jaeger (1959, p. 99) derived the following solution for the temperature distribution within the rod, and interested readers are referred to their text for a detailed derivation:

$$\frac{T}{T_1} = 1 - \frac{4}{\pi}\sum_{n=0}^{\infty}\frac{(-1)^n}{2n+1}\exp[-(2n+1)^2\pi^2\tau/4]\cos\frac{(2n+1)\pi z}{2L}$$

where τ is a dimensionless time constant (or Fourier number):

$$\tau = \kappa t/L^2$$

Carslaw and Jaeger (1959, p. 101) also presented a graphical representation of their solution, and for the conditions in which we are interested ($z/L = 0.8/1.0 = 0.8$, $T/T_1 = 20/500 = 0.04$) the solution is

$$\tau \approx 0.005$$

Thermal diffusivity, κ, can now be derived easily from the definition of τ, given that $t = 20$ min $= 1200$ s:

$$\tau = \kappa t/L^2, \qquad \text{so } 0.005 \approx \kappa \times 1200/1.0^2$$

$$\kappa \approx 4 \times 10^{-6}\ \text{m}^2\,\text{s}^{-1}$$

Note: The form of τ indicates that a substance of half the diffusivity will take twice as long to attain the same temperature at the same point.

We generally think of heat as something applied to an object from an outside source. That is, if we wish to increase the temperature of an object we usually place it within or next to a hotter object. Some objects, however, generate heat internally. If present, internal *heat production*, or *heat generation*, A (power per unit volume, $\text{W}\,\text{m}^{-3}$), reduces the amount of external heat required to raise the temperature of an object. Equation (1.10) must now include an extra term:

$$(\partial T/\partial t) = \kappa \times (\partial^2 T/\partial z^2) + A/(\rho c) \tag{1.11}$$

Question: A poker similar to the one in the previous example has thermal conductivity $\lambda = 16.3\ \text{W}\,\text{m}^{-1}\,\text{K}^{-1}$. This poker, however, contains within its structure a certain amount of radioactive material that generates heat through nuclear decay. Both ends of the poker are maintained at 0°C and the rest of its length is allowed to heat up owing to the internal heat production. After 7 h the temperature at the mid-point of the poker is found to have risen to 5°C.

If we assume that no heat is lost through radiation, at what rate is heat being generated within the poker?

Answer: Again, this is not a trivial problem and an exact solution is difficult to derive. Carslaw and Jaeger (1959, p. 130) presented a solution in both mathematical and graphical form:

$$T = \frac{AL^2}{2\lambda} \left\{ 1 - \frac{z^2}{L^2} - \frac{32}{\pi^3} \sum_{n=0}^{\infty} \frac{(-1)^n}{(2n+1)^3} \cos \frac{(2n+1)\pi z}{2L} \exp[-(2n] \right.$$

$$\left. + 1)^2 \pi^2 \tau/4] \right\}$$

We need to redefine a couple of our parameters from the previous example to conform to the way Carslaw and Jaeger defined the problem. Specifically, the total length of the poker is defined as $2L$, so now $L = 0.5$ m. The mid-point, our point of measurement, is defined as $z = 0$. So, at $t = 7\,\text{h} = 25{,}200\,\text{s}$,

$$\tau = \kappa t/L^2 = 4 \times 10^{-6} \times 25{,}200/0.5^2 = 0.4032 \approx 0.4$$

The solution for $\tau = 0.4$ and $z/L = 0$ is

$$2\lambda T/AL^2 \approx 0.62$$

We know $\lambda = 16.3 \ \text{W}\,\text{m}^{-1}\,\text{K}^{-1}$, $T = 5°C$ and $L = 0.5$, so

$$2 \times 16.3 \times 5/(0.5^2 \times A) = 0.62$$

or

$$A \approx 1050 \ \text{W}\,\text{m}^{-1}$$

If left alone, the mid-point of the poker reaches a maximum temperature defined by $2\lambda T/AL^2 = 1$, or $T = AL^2/2\lambda$, which in this particular case is just over $8°C$. Once the steady state is reached, heat is conducted to the ends of the poker at the same rate as it is produced.

An important sub-set of Equation (1.11) is the steady-state equation, where $(\partial T/\partial t) = 0$:

$$0 = \lambda(\partial^2 T/\partial z^2) + A \tag{1.12}$$

Integrating Equation (1.12) with respect to z, and rearranging, we obtain

$$\mathbf{Q}_0 = \lambda_d[\partial T/\partial z]_d + \int A(z)\partial z \tag{1.13}$$

For the purposes of geothermal investigations, $\mathbf{Q}_0$ is the surface heat flow, and the integral is measured from the surface down to a depth d. Three parameters from Equation (1.13) must be quantified in order to characterise completely the thermal conditions within a vertical section. We need to define the distribution of heat generation, $A(z)$, temperature, $T(z)$, and thermal conductivity, $\lambda(z)$. The measurement of these three parameters is covered in detail in Chapters 2, 3 and 4.

A high proportion of geophysical heat flow problems can be reduced to a form of Equation (1.12). Carslaw and Jaeger (1959) demonstrated a wide range of analytical solutions for the equation subject to many and varied geometries and boundary conditions. However, these solutions are highly idealised. Most geothermal problems involve complex geometries and boundary conditions that cannot be modeled with exact solutions. Numerical methods are normally required and such methods for solving Equation (1.12) are introduced in Chapter 8.

1.4. Units

Up until now, we have been using SI units (*Système International d'Unités*) for all parameters, and we will continue to do so throughout this book. The general practice is for modern heat flow data to be published in SI units, with some notable exceptions. The United States oil industry still commonly reports temperature in degrees Fahrenheit and depth in feet, necessitating some familiarity with the older units and conversion factors. Also, many older publications from other parts of the world give heat flow parameters in units other than SI, and a means to convert between measuring systems is necessary to make sense of historical records. Table 1.2 shows the conventional SI units for all common heat flow parameters, and the equivalent values and units for the centimetre-gram-second (cgs) and foot-pound-second (fps) systems.

Question: An old paper lists the thermal conductivity of a basalt specimen as 4.4 $\mathrm{mcal\,cm^{-1}\,^{\circ}C^{-1}\,s^{-1}}$, and the heat flow in the area whence the basalt came as 1.6 $\mathrm{\mu cal\,cm^{-2}\,s^{-1}}$. What are the equivalent values in SI units?

Answer: Using the conversion factors in Table 1.2:
   4.4 $\mathrm{mcal\,cm^{-1}\,^{\circ}C^{-1}\,s^{-1}}$ = SI $\times$ 2.388, so SI = 1.84 $\mathrm{W\,m^{-1}\,K^{-1}}$
   1.6 $\mathrm{\mu cal\,cm^{-2}\,s^{-1}}$ = 0.0016 $\mathrm{mcal\,cm^{-2}\,s^{-1}}$ = SI $\times$ 0.02388,
   so SI = 0.067 $\mathrm{W\,m^{-2}}$

Question: A recent well completion report for a well in America indicates a temperature gradient of 1.6°F/100 ft. What is this in SI units?

Answer: 1.6°F/100 ft = 0.889 K/100 ft = 2.92 K/100 m = 0.029 $\mathrm{K\,m^{-1}}$

Although the advantages of the SI system are widely recognised, allowing direct comparison between different types of energy and power, the basic SI units are inappropriate for many geological applications and smaller divisions of the units are commonly used. In particular, the preferred units for heat flow and heat generation are milliwatts per square metre ($\mathrm{mW\,m^{-2}}$) and microwatts per cubic

Table 1.2. Système International d'Unités (SI) Heat Flow Units and Equivalent Values for the Centimetre–Gram–Second (cgs) and Foot–Pound–Second (FPS) Measurement Systems

Parameter	SI Unit	cgs System	fps System
Time	seconds (s)	s SI	s SI
Temperature	kelvin (K)	Celsius (°C) SI–273.15	Fahrenheit (°F) SI × 1.8–459.67
Energy/heat	joules (J)	calories (cal) SI × 0.2388	BTU[a] SI/1055
Power	watts (W)	cal s^{-1} SI × 0.2388	BTU[a] h^{-1} SI × 3.413
Distance	metres (m)	centimetres (cm) SI × 100	feet (ft) SI × 3.2808
Mass	kilograms (kg)	grams (g) SI × 1000	pounds (lb) SI × 2.2046
Heat flow	W m^{-2}	cal cm^{-2} s^{-1} SI × 2.388 × 10^{-5}	BTU[a] ft^{-2} day^{-1} SI × 7.608
Thermal conductivity	W m^{-1} K^{-1}	cal cm^{-1} C^{-1} s^{-1} SI × 2.388 × 10^{-3}	BTU[a] ft^{-1} F^{-1} h^{-1} SI × 0.578
Thermal diffusivity	m^2 s^{-1}	cm^2 s^{-1} SI × 10^4	ft^2 h^{-1} SI × 3.875 × 10^4
Heat generation	W m^{-3}	cal cm^{-3} s^{-1} SI × 2.388 × 10^{-7}	BTU[a] ft^{-3} day^{-1} SI × 2.319

[a] BTU = British thermal units.

metre (μW m^{-3}), respectively. These smaller units yield values typically in the range 25–100 mW m^{-2} for heat flow and 0–5 μW m^{-3} for heat production. For Equation (1.2) to hold true, thermal gradient must be cited in terms of millikelvin per metre (mK m^{-1}) or the equivalent degrees Celsius per kilometre (°C km^{-1}). Typical geothermal gradients are on the order of 15–50°C km^{-1}.

Some historical geothermal studies used certain abbreviations for convenience. For example, to make cgs units easier to work with, geothermal data were often quoted in heat flow units (1 HFU = 10^{-6} cal cm^{-2} s^{-1} = 41.87 m^{-2}), heat generation units (1 HGU = 10^{-13} cal cm^{-3} s^{-1} = 0.4187 μW m^{-3}) and thermal conductivity units (1 TCU = 10^{-3} cal cm^{-1} °C^{-1} s^{-1} = 0.4187 W m^{-1}K^{-1}). Very rarely, some authors make up their own arbitrary units, such as the Bolderij unit of heat flow (1 BU ~ 77 mW m^{-2}; Houbolt and Wells, 1980).

Note: Wherever possible, it is strongly recommended that all data be converted to SI units.

1.5. Heat Flow Resources

Heat flow data have been acquired around the world on a regular basis dating at least from Sir Edward Bullard's measurements in South Africa in 1939.

However, the distribution of observations has remained highly uneven. Continental data mainly reflect the location of drill holes associated with mining and other activities. Some systematic surveys have been undertaken in regions of tectonic significance, but full funding for heat flow programmes based on deep drill holes has rarely been available. In contrast, marine heat flow probes constructed for rapid observations at sea allow heat flow data to be obtained as a matter of routine in major offshore surveys, at little additional cost. Global heat flow data have been collated and published at irregular intervals throughout the years (e.g. Lee and Uyeda, 1965; Simmons and Horai, 1968; Jessop, Hobart and Sclater, 1976; Cermák, 1979; Pollack et al., 1993). At the time of publication, the latest compilation was available via anonymous ftp over the Internet from

  ftp://ftp.ngdc.noaa.gov/Solid_Earth/Global_Heatflow/

Alternatively, follow the direct link from the web site mentioned in the Preface.

As a first step in any new heat flow study, it is good practice to collate data from previous work. However, locating all such data may not be simple. The compilations listed above may include most published data available at the time, but not all data are published and new data are always being collected. Heat flow measurements taken as part of commercial exploration often remain confidential. For individual countries or regions, geological survey organisations may maintain up-to-date compilations and require companies to make their data available. The United States Geological Survey, for example, sells CD-ROM compilations of heat flow data and thermal gradient data for North America. For other parts of the world, including Australia, compilations are often non-existent or incomplete and data may be found only in global compilations and primary literature sources.

EXAMPLE

Newstead and Beck (1953) published the first estimates of Australian heat flow. Sass, Jaeger and Munroe (1976) and Lilley, Sloane and Sass (1978) subsequently published compilations detailing the results from ninety sites distributed across the country. The most recent global compilation included just 119 sites from onshore Australia. Marine data have not been accumulated on a systematic basis around Australia but a number of Deep-Sea Drilling Project sites have been located in Australian territorial waters. Cull and Denham (1979) considered regional correlations between heat flow and other data sets, and heat flow data quality was reviewed by Cull (1982).

Heat flow data are reported in many journals, sometimes as by-products of other experiments. Specifically measured and interpreted heat flow data can be found in such journals as the *Journal of Geophysical Research, Pure and*

Applied Geophysics (Pageoph), American Association of Petroleum Geologists Bulletin, Geophysics, Geothermics, and the *Proceedings of the Royal Astronomical Society*, often included in papers describing new technology or methodology. Oceanic heat flow data can be found in some reports of the *Deep-Sea Drilling Project*, and the subsequent *Ocean Drilling Project*. Unpublished theses or dissertations are another primary source of heat flow data, but unless these data are referred to or published in some other volume their existence may remain unknown.

Geothermal data are not restricted to heat flow measurements. Other types of data are often sought. These include sub-surface temperature, geothermal gradient, thermal conductivity, thermal diffusivity, heat capacity and heat generation. Locating such data is often more difficult than finding heat flow values, but various sources are worth investigating. For rock properties such as thermal conductivity, thermal diffusivity and heat capacity, there are numerous compilations upon which to draw. Thermal conductivity compilations for different rock types are, arguably, the easiest to find (e.g. Touloukian et al., 1970b; Majorowicz and Jessop, 1981; Reiter and Tovar, 1982), but excellent compilations of other thermal properties also can be found in Touloukian, Liley and Saxena (1970a) and Touloukian et al. (1970b). Specific measurements of thermal conductivity for individual regions or rock formations are sometimes published in journals, but often such measurements are recorded only in confidential reports to companies, or uncirculated open-file reports to geological survey organisations. These may be difficult to trace, and access may be restricted.

Note: When searching for existing geothermal data for an area, an exhaustive search of sources may still turn up no data. Many parts of the Earth have not been the subject of any geothermal study.

Books on the subject of geothermal research are few and most look at detailed applications and case studies rather than basic principles. However, they are useful to gain a feel for the types and range of applications for the principles introduced in this book. For a broader understanding and feel for the subject, see Lee (1965), Gretener (1981), Buntebarth and Stegena (1986), Haenel, Rybach and Stegena (1988), Naeser and McCulloh (1989), Jessop (1990), Cermák and Rybach (1991), Barker (1996), Welte, Horsfield and Baker (1997), Förster and Merriam (1999) and Kutasov (1999).

1.6. Summary

Earth accreted from a hot cloud of gas and dust about 4550 million years ago. Accretion and the separation of the core from the mantle liberated a vast amount of heat, which has since been dissipating through the Earth's surface. However, only part of the present surface heat flow is due to primordial heat –

radiogenic sources within the mantle and crust account for as much as 83% of the total. The crust itself generates approximately 40% of the observed surface heat flow. Heat flows through the Earth primarily by convection in the mantle and conduction and advection in the crust; heat flow is not uniform across the surface. More heat is lost through the oceans than the continental shields.

A self-consistent system of units allows us to quantify many of the concepts encountered in geothermal studies. Use of the SI system of units is widespread and encouraged for the easy communication and interpretation of results from all parts of the world.

CHAPTER TWO

Heat Generation

It is the *surface* of the earth which was heated, by combustion. It is composed in great part of metals, such as sodium and potassium, which catch fire by mere contact with air and water; this took place whenever rain fell, and as the water penetrated the cracks in the earth's crust, further combustion took place, causing explosions and eruptions.

A Journey to the Center of the Earth – Jules Verne, 1864.

The physical properties of the Earth's deep interior are unavoidably obscure. Direct human observation is limited to mines in the top 3 km or less, while deep drilling has provided core samples and instrument access to maximum depths of around only 12 km (Kozlovsky, 1987). Deeper physical evidence, down to several hundred kilometres, is obtained from xenoliths and other volcanic products ejected onto the surface from the mid-crust to upper mantle, but this still represents a very small portion of the 6370 km radius of the Earth. The bulk of our knowledge of the Earth's interior composition and physical properties has been deduced by indirect means. Major internal boundaries and structural features (Figure 2.1) have been delineated using seismic methods and supported by density models deduced from gravity readings (e.g. Ringwood, 1969, 1979; Brown and Mussett, 1993).

One conclusion from velocity modelling is that the interior of the Earth is significantly hotter than the surface. The internal heat is derived both from primordial sources related to the formation of the globe and from secondary processes generating heat internally. The internal heat sources are not distributed uniformly through the globe. Of the heat observed flowing out through the surface of the Earth, 40% actually originates within the thin outer crust (Pollack and Chapman, 1977). The heat of radioactive isotopic decay forms the dominant part, but there are also contributions from the friction of intraplate strain and plate motions, and heat from exothermic metamorphic and diagenetic processes. Still other processes, such as cosmic neutrino interaction with the Earth's mass (Hamza and Beck, 1972) and the distortion of the Earth by gravitational forces, have been suggested as possible internal heat sources, but the magnitudes of such sources are generally considered too small to be significant.

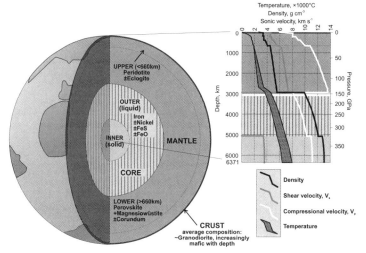

Figure 2.1. The structure, composition and physical properties of the Earth. The graph to the right shows the variation in pressure, density, sonic velocity (V_p and V_s) and temperature with depth. The temperature profile is shown as a range as it remains poorly constrained in deeper sections. Data from Dziewonski and Anderson (1981), Montagner and Anderson (1989), Brown and Mussett (1993).

The generation of heat throughout the crust is also not uniform. Radioactive elements are more concentrated in granitic, compared with basaltic, rocks. Frictional heating is most likely to be significant along plate boundaries and major faults. Metamorphism and diagenesis take place under specific geological conditions. For this reason, an examination of the mechanisms and magnitudes of different heat sources is necessary to help the reader gain an intuitive feel for the possible sources of heat in a specific area.

2.1. Radiogenic Heat

Heat is generated in rocks principally through the radioactive decay of unstable isotopes that release energy in the form of alpha (α), beta (β) and gamma (γ) particles, neutrinos (ν) and antineutrinos ($\bar{\nu}$) (Figure 2.2). Rock is virtually transparent to neutrinos and antineutrinos and most of the energy carried by these particles is lost into space (Hamza and Beck, 1972). However, the surrounding rocks absorb the kinetic energy carried by the other particles, thus generating heat. The rate of radiogenic heat generation within rocks is related to the quantity of radioactive material, the rate of decay and the energy of the emitted particles. The energy emission and rate of decay depend only on the species of radioactive isotope (Table 2.1), so the absolute abundance of individual isotopes in a rock wholly determines the rate of heat production.

Approximately 98% of geothermal radiogenic heat arises from the decay of single isotopes of uranium (^{238}U), thorium (^{232}Th) and potassium (^{40}K). The energy released by the decay of the uranium isotope is considerably greater

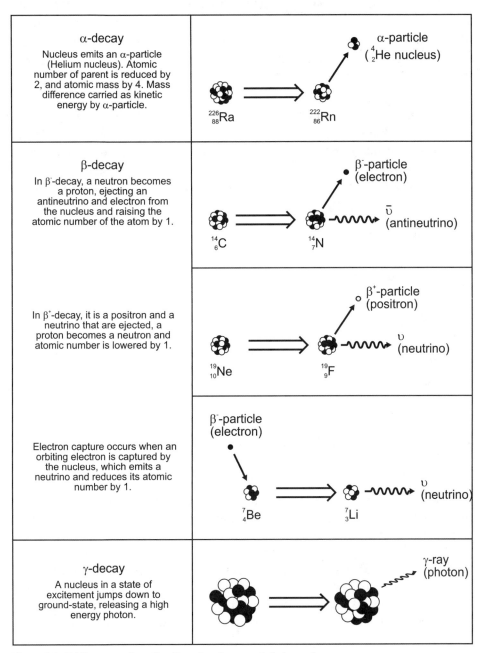

Figure 2.2. Different modes of radioactive decay and their products.

Table 2.1. Major Heat-Generating Elements in Rocks

Element	Heat Generation, A' (μW/kg element)	Average Crustal Abundance, n (weight/weight)	$A' \times n$ (μW/kg rock)
Uranium	96.7	2.3 ppm	2.22×10^{-4}
Thorium	26.3	8.1 ppm	2.13×10^{-4}
Potassium	0.0035	1.84%	6.44×10^{-5}

Sources: Emsley (1989) and Jessop (1990).

than by thorium, which in turn is greater than by potassium. However, the relative contributions of each isotope to the total heat generation are of the same order of magnitude due to their relative abundances in typical crustal rocks (Table 2.1).

Empirically, the decay chains of these isotopes can be reduced to the following:

$$^{238}U \Rightarrow ^{206}Pb + 8\alpha + \gamma \tag{2.1}$$

$$^{232}Th \Rightarrow ^{208}Pb + 6\alpha + \beta + \gamma \tag{2.2}$$

$$^{40}K \Rightarrow ^{40}Ca + \beta^- + \bar{v} + \gamma \tag{2.3}$$

or

$$^{40}K \Rightarrow ^{40}A + v + \gamma \tag{2.4}$$

Note: Potassium decays by one of two paths; 89% decays to calcium and 11% to argon.

Gamma-ray spectrometers provide the most direct method for measuring the abundance of uranium, potassium and thorium in rock. The gamma-ray energy spectrum emitted from a rock is the sum of the individual characteristic spectra of the radiogenic components. The total signal can thus be analysed to determine the proportion of each element. Measurements can be completed under field or laboratory conditions. Laboratory measurements are of higher quality if lead castles are used to screen the system from background radiation. Heat production is found by multiplying the proportion of each component by the activity levels in Table 2.1.

Question: A sample of granite is collected in the field and brought back to the laboratory. Subsequent measurements reveal that the granite has a density of 2.7 g cm^{-3} and contains 4.3 ppm uranium,

12.2 ppm thorium and 5.1% potassium. At what rate is heat generated within the sample?

Answer:

Rate of heat generation = abundance in rock × density × heat generation (from Table 2.1):

For uranium, $A = 4.3 \times 10^{-6} \times 2700 \times 96.7 \times 10^{-6}$
$$= 1.123 \times 10^{-6} \text{ W m}^{-3}$$

For thorium, $A = 12.2 \times 10^{-6} \times 2700 \times 26.3 \times 10^{-6}$
$$= 0.866 \times 10^{-6} \text{ W m}^{-3}$$

For potassium, $A = 5.1 \times 10^{-2} \times 2700 \times 3.5 \times 10^{-9}$
$$= 0.482 10^{-6} \text{ W m}^{-3}$$

Total heat production from the three elements is 2.471 μW m^{-3}.

Cermák et al. (1990) published heat generation data for a number of rock types from West Germany, Switzerland and Italy. Although their findings (Table 2.2) are regionally biased, they clearly indicate greater heat production in acidic, as opposed to basic, rocks.

Table 2.2. Heat Generation, A, velocity, V_p, by Rock Type

Rock Type	No. of Samples	A (μW m^{-3})	V_p (km s^{-1})
Acidic rocks			
Rhyolite	5	2.80 ± 0.28	5.57 ± 0.12
Granite	16	2.82 ± 1.03	5.85 ± 0.16
Granodiorite	11	2.45 ± 1.29	5.88 ± 0.22
		1.0	
Tonalite	8	1.48 ± 0.54	5.89 ± 0.20
Diorite	7	0.88 ± 0.30	6.24 ± 0.25
Basic rocks			
Gabbro	6	0.11 ± 0.13	6.19 ± 0.82
		0.03	
Amphibolite	6	0.37 ± 0.19	6.30 ± 0.30
Ultrabasic rocks			
Hornblendite	3	0.12 ± 0.15	6.91 ± 0.78
Pyroxinite	4	0.06 ± 0.04	7.20 ± 0.33
Peridotite	4	0.01 ± 0.01	7.82 ± 0.12
Serpentinite	6	0.01 ± 0.02	6.49 ± 0.64
Miscellaneous crustal rocks			
Various	23	1.91 ± 1.40	5.78 ± 0.53
Granulite		*0.13*	
Bulk Earth		*0.014*	
Carbonaceous chondrite		*0.01*	
Ordinary chondrite		*0.015*	

Note: Uncertainty in A and V_p = one standard deviation.
Sources: Cermák et al. (1990) and Brown and Mussett (1993).

Note: It is immediately obvious from this that oceanic heat flow
is influenced very little by internal heat generation in comparison
with continental heat flow.

2.1.1. Heat Flow Provinces

Observations of surface heat generation, A_0, and surface heat flow, $\mathbf{Q}$, lead to
the development of crustal models of heat generation. Lachenbruch (1968) and
Birch, Roy and Decker (1968) independently proposed a simple linear relationship:

$$\mathbf{Q} = q + A_0 D \tag{2.1}$$

where $\mathbf{q}$ is a constant component of heat flow from the mantle and D represents
a depth scale for the vertical distribution of heat-producing elements.

Linear relationships obtained for many parts of the world show a remark-
able consistency in parameter D, typically between 10 and 15 km (Drury,
1989). Such observations led Roy, Blackwell and Birch (1968) to define a
heat flow province as a region of common tectonothermal history within
which consistent values of $\mathbf{q}$ and D are obtained (see also Section 6.5.1).
Much effort has been expended in identifying and quantifying heat flow
provinces around the world.

The consistency in D suggests that it relates directly to some physical property of the province. Lachenbruch (1970) pointed out that if the linear relationship is to hold for regions with varying erosion, heat generation must decrease
exponentially with depth. It follows that the parameter D most likely relates to
the exponential decay constant of the heat generation distribution. Jaupart,
Sclater and Simmons (1981) advocated separate exponential trends for
uranium, potassium, and thorium concentrations, resulting in more complicated heat production profiles.

Unfortunately, simple trends of heat production are unstable in nature
because they are readily perturbed by magmatic activity, surface weathering
and fluid circulation (Cermák, Bodri and Rybach, 1991). Apart from the
empirical observations that led to the initial proposal of the theory, no evidence has been found for a systematic decrease in heat generation with depth in
the crust. The most applicable data so far come from the deepest borehole ever
drilled, SG-3 on the Kola Peninsula in the Baltic Shield. The hole penetrates
more than 12 km of Proterozoic rocks and Archean gneiss, which places the
bottom of the hole around the typical depth of the D parameter. Far from
showing an exponentially decreasing distribution of heat generation, there is,
instead, a step-wise distribution, closely related to lithology (Arshavskava et
al., 1987). The average heat generation in the Archean gneiss is about
$1.2~\mu\mathrm{W\,m^{-3}}$, while heat generation in the overlying Proterozoic rocks averages
approximately $0.8~\mu\mathrm{W\,m^{-3}}$. A similar distribution has been reported from the

Kapuskasing Structural Zone of the Canadian Shield, interpreted to be an exposed, oblique section through the top two thirds of the crust (Percival and Card, 1983). Weighted mean values for heat generation are $0.72 \ \mu W \, m^{-3}$ for greenstones in the upper part of the crustal section, $1.37 \ \mu W \, m^{-3}$ for amphibolite facies gneiss from the middle crust, and $0.44 \ \mu W \, m^{-3}$ for granulite assumed to represent the lower crust (Ashwal et al., 1987). These results are in stark contrast to the exponentially decreasing heat generation distribution proposed to explain the observed linear relationship between surface heat flow and surface heat generation.

Drury (1989) suggested that the observed relationship between surface heat flow and surface heat generation could be modelled differently. He proposed a two-layer crust consisting of an upper layer of variable heat generation, approximately 10–15 km thick, underlain by a layer of lower and less variable heat generation about 20 km thick. Such a model is consistent with other geological and geophysical observations and is able to produce $\mathbf{Q} - A_0$ relationships like those observed.

Vigneresse and Cuney (1991) provided an alternative view that heat generation is randomly distributed in near-surface granitic layers, so systematic values for D must relate to the top of a generally depleted zone in the lower crust. In some respects this model resembles the uniform slab theory described above, with well-mixed components providing a near-surface average.

Any simple analytical model for heat production must be considered only a first approximation prior to subsequent measurement and modelling. The exponentially decreasing heat generation model, in particular, appears to have no justification in reality. The heat flow province concept, however, has been remarkably successful in separating the continental landmasses into regions of shared heat flow characteristics, and $\mathbf{q}$ may still prove to have physical relevance.

The true distribution of heat-generating elements affects lower crustal heat flow and temperatures, which, in turn, may drive regional deformation and orogenic processes (e.g. Sandiford, 1999). However, only upper crustal temperatures impact on most emplacement and migration models in geothermal exploration. For most applications, the physical relevance of $\mathbf{q}$ and D are not important, and the heat flow province concept is adequate to characterise shallow temperatures.

Question: In the eastern United States, Roy et al. (1968) identified a linear relationship between $\mathbf{Q}$ and A_0:

$$\mathbf{Q} = q + A_0 D.$$

They found the data are best fit with $q = 33 \pm 1 \ mW \, m^{-2}$ and $D = 7.5 \pm 0.2$ km. Assuming a simple, steady-state, two-layer model of heat production, an average surface temperature of $10°C$, an average thermal conductivity of $2.7 \pm 0.1 \ W \, m^{-1} \, K^{-1}$ and a measured surface heat generation of $3.0 \pm 0.1 \ \mu W \, m^{-3}$, what is the temperature, T, at the base of the heat-generation layer?

Answer: The linear relationship tells us that surface heat flow

$$\mathbf{Q}_0 = 33 + (3.0 \times 7.5) = 55.5 \text{ mW m}^{-2}$$

and the two-layer model tells us that heat flow decreases linearly with depth to $\mathbf{q} = 33$ mW m^{-2} at 7.5 km. Thermal gradient ($\partial T/\partial z = \mathbf{Q}/\lambda$) therefore decreases linearly with depth from $20.56°$C km^{-1} at the surface to $12.22°$C km^{-1} at 7.5 km, or $\partial T/\partial z = 20.56 - 1.11z$. Integrating this with respect to z, and imposing the surface temperature boundary condition, $T = 10 + 20.56z - 0.555z^2$. So at $z = 7.5$ km, and carrying the uncertainties through the calculation, $T = 133 \pm 13°$C.

2.1.2. Radiogenic Heat Generation from Well Logs

Heat generation within sediments is poorly constrained and has not been extensively investigated. Only a small number of papers note individual measurements of sedimentary heat generation (e.g. Epp, Grim and Langseth, 1970; McKenna, Sharp and Lynch, 1995). Alternatively, Issler and Beaumont (1989) derived an empirical relationship between the fractional proportion of quartz in a matrix (FQ) and heat generation measured in sedimentary rocks from the Labrador Shelf:

$$A = -0.96(\text{FQ}) + 1.29 \qquad (2.2)$$

where A is in μW m^{-3}.

This equation yields a range of heat generation values from 0.33 μW m^{-3} for pure quartz lithologies to 1.29 μW m^{-3} for quartz-free rocks. However, it is calibrated for a particular region and its relevance to regions other than that in which the measurements were made is questionable.

Slichter (1941) suggested 0.59 μW m^{-3} as an average heat generation for sedimentary rocks. Keen and Lewis (1982) examined sediments from the continental margin of eastern North America and measured heat generation ranging from 0.3 μW m^{-3} for limestones to 1.4–1.8 μW m^{-3} for shales. Rybach (1986) tabulated average heat-generation values for various sediment types, ranging from a low of 0.012 μW m^{-3} for salt, up to 5.5 μW m^{-3} for black shale. Cenozoic and Mesozoic lacustrine sediments in Chinese basins yielded values in the range 1.02–3.28 μW m^{-3} (Zhang, 1993). McKenna and Sharp (1998) published measurements of heat generation in sediments from several units in the Gulf Coast region of Texas (Table 2.3). They discovered significant heat generation in the region, and recommended that it not be ignored during basin modelling.

Table 2.3. Heat Generation in Texas Gulf Coast Sedimentary Units

Unit	No. of Samples	Range of Heat Generation (μW m^{-3})	Mean (μW m^{-3})
Frio mudrock	18	1.23–2.21	1.72
Wilcox mudrock	52	0.86–1.87	1.50
Frio sandstone	16	0.58–1.53	1.19
Wilcox sandstone	8	0.43–1.27	0.88
Stuart City limestone	6	0.07–1.03	0.38

Source: McKenna and Sharp (1998).

Rybach (1986) developed a simple relationship between heat generation and the natural gamma log, GR [Standard American Petroleum Institute (API) units], for use when electric well logs are available:

$$A = 0.0145(GR - 5.0) \qquad (2.3)$$

where A is in μW m^{-3}. This relationship is based on very limited data, however, and was revised after more data were collected (Bücker and Rybach, 1996):

$$A = 0.0158(GR - 0.8) \qquad (2.4)$$

The database is still derived from a small number of wells, mostly in Europe, and relevance to other regions is arguable. However, the idea of using well logs to estimate heat generation in petroleum exploration regions is good. The best logs for estimating heat generation are derived from the natural gamma spectrometer (NGS) tool. The NGS tool records the total count and frequency distribution of gamma rays and uses the data to estimate the absolute abundance by mass of each gamma-producing element in the rock. The results are generally output as three logs – URAN (ppm uranium), THOR (ppm thorium) and POTA (percentage potassium). Utilising the compensated density log, RHOB, and the data in Table 2.1, the contribution to heat production of the ith element, A_i, is

$A_i =$ (heat production per mass of i) $\times$ (proportion of i in rock) $\times$ (density)

and the total heat production is the sum of the individual contributions:

$$A = (A_{\mathrm{U}} + A_{\mathrm{Th}} + A_{\mathrm{K}}) \qquad (2.5)$$

Note: Care must be taken to ensure that appropriate units are retained throughout the calculation. In particular, close note must be taken of the units of URAN, THOR and POTA. Usually, a correction factor must be applied to ensure that heat production values are determined in μW m^{-3}.

Heat production profiles are easily produced for any well with the RHOB, POTA, THOR and URAN logs (e.g. Figure 2.3). Unfortunately, only a small percentage of wells are logged with an NGS tool. This is especially true for older wells. Other logs are much more common and present other possibilities for log-derived heat production estimates.

The total gamma-ray count (GR) is routinely included in most log suites and is also related to the radioactive decay of uranium, thorium and potassium. It is a record of the total number of γ-rays detected by the tool within a given time. The relative gamma activities (or relative number of γ-rays produced in the same time by the same mass) of the three important elements are shown in Table 2.4.

It follows that if POTA, THOR and URAN are recorded as percentage, ppm and ppm, respectively, then

$$GR \propto POTA + (0.13 \times THOR) + (0.36 \times URAN) \tag{2.6}$$

The proportionality in Equation (2.6) is not constant, but depends on the distance into the rock that the tool is able to detect γ-ray emission. The radius of influence varies with formation density, mud density, hole diameter, gamma-ray energy and tool type (Serra, 1984), but Rider (1991) suggested an average radius of 30 cm. If the radius of influence is the only factor affecting the proportionality, then GR versus [POTA + (0.13 $\times$ THOR) + (0.36 $\times$ URAN)] should be approximately linear for an entire well, with a mean gradient, X:

$$GR = X \times [POTA + (0.13 \times THOR) + (0.36 \times URAN)] \tag{2.7}$$

Then, A is related to GR by combining Equations (2.5) and (2.7):

$$\frac{A}{GR} = \frac{RHOB \times 10^{-3} \times [(35.0 \times POTA) + (26.3 \times THOR) + (96.7 \times URAN)]}{X \cdot [POTA + (0.13 \times THOR) + (0.36 \times URAN)]} \tag{2.8}$$

or:

$$A = \frac{3.5 \times 10^{-2}}{X} \times RHOB \times GR \\ \times \left[\frac{POTA + (0.751 \times THOR) + (2.76 \times URAN)}{POTA + (0.13 \times THOR) + (0.36 \times URAN)} \right] \tag{2.9}$$

Table 2.4. Relative Gamma Activity of Uranium, Thorium and Potassium

γ-Producing element	Relative γ-Activity A	Average Abundance n (%)	$A \times n$
U	3600	0.00023	0.828
Th	1300	0.00081	1.05
K	1	1.84	1.84

Sources: Adams and Weaver (1958) and Emsley (1989).

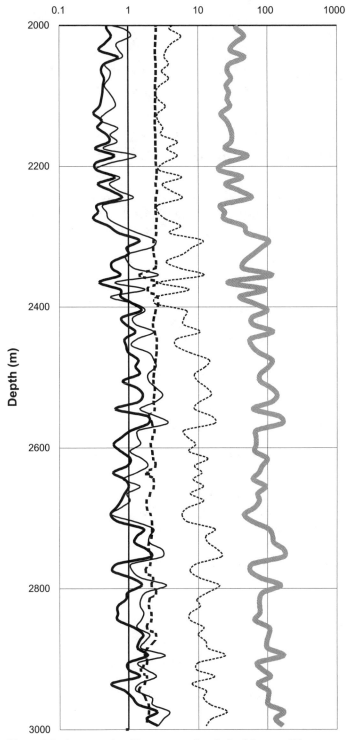

Figure 2.3. An example of heat generation derived from well logs measured in East Yeeda 1 in the Canning Basin, Western Australia. Shown are the logs POTA (%, light solid), THOR (ppm, light dashed), RHOB (g cm^{-3}, heavy dashed) and calculated heat generation (μW m^{-3}, heavy solid). URAN log is excluded for clarity. GR log (API, grey) is shown for comparison.

or

$$A = 3.5 \times 10^{-2} \times \text{RHOB} \times \text{GR} \times \frac{Y}{X} \tag{2.10}$$

where

$$Y = \left[\frac{\text{POTA} + (0.751 \times \text{THOR}) + (2.76 \times \text{URAN})}{\text{POTA} + (0.13 \times \text{THOR}) + (0.36 \times \text{URAN})} \right] \tag{2.11}$$

Note: Remember that this relationship only holds if the proper units for POTA (%), URAN (ppm) and THOR (ppm) are maintained.

It appears that we must still know the absolute abundance of each heat-producing element before we can utilise Equation (2.10). Note, however, that Y is limited to values between 1 (if THOR = URAN = 0, POTA > 0) and about 7.67 (if POTA = THOR = 0, URAN > 0). In addition, in regions where sediment has been derived from a common source through time, the relative proportions of the elements in Equation (2.11), and hence Y, should remain relatively constant with depth. The uncertainty in Equation (2.10) can be derived empirically by observing the scatter in a cross-plot of A versus RHOB $\times$ GR (e.g. Figure 2.4).

EXAMPLE

The well, East Yeeda 1 in the Canning Basin, Western Australia, was logged with an NGS tool at 995–3553 metres depth below kelly bushing (mKB). A plot of A (calculated from the NGS logs) versus RHOB $\times$ GR approximates a straight line of gradient 0.005, with minimal scatter (Figure 2.4). The implication from Equation (2.10) is that $Y/X = 0.143$. So Equation (2.10) can be written for East Yeeda 1 and surrounding regions as

$$A = 0.005 \times \text{RHOB} \times \text{GR}$$

The scatter in the graph suggests that the uncertainty in A is about $\pm 0.5 \ \mu\text{W m}^{-3}$.

Note that the values of A derived for East Yeeda 1 are generally less than those predicted by Equation (2.4).

The gradient of A versus RHOB $\times$ GR is dependent only on the relative proportions of the radioactive elements in the sediment. As well as remaining uniform with depth, the gradient should also be similar over any region containing sediment derived from the same source.

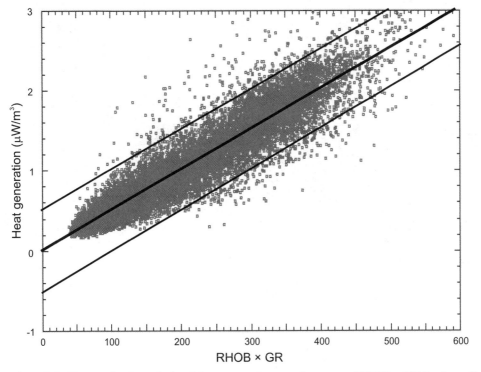

Figure 2.4. Heat production calculated from spectral gamma logs versus RHOB × GR for the well East Yeeda 1 in the Canning Basin (17°37′59″ S, 124°02′50″ E), at depths 995–3553 metres below kelly bushing. Centre line is the best fit by linear regression and outer lines are at one standard deviation.

EXAMPLE

There are at least two wells other than East Yeeda 1 with NGS logs in the Canning Basin. Their cross-plots (Figure 2.5) yield very similar gradients to East Yeeda 1. This lends strong support to the assumption that the relationship holds across regional scales.

Note: Heat generation in the Canning Basin lies mainly between 0 and 2.5 μW m^{-3}, a range similar to that found by McKenna and Sharp (1998) in the Gulf Coast region of Texas. The values are all significantly lower than the 5.5 μW m^{-3} quoted by Rybach (1986) for black shale. These results reinforce the point that reliance on global averages can have serious implications when modelling heat flow in specific regions. Local data should be sought and used whenever possible.

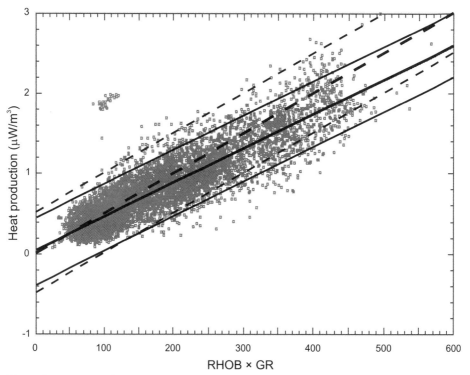

Figure 2.5. Heat production calculated from spectral gamma logs versus RHOB × GR for the wells Blina 2 (17°37′07″ S, 124°29′46″ E) and Sunup 1 (17°37′12″ S, 124°18′58″ E). Solid lines show line of best fit and one standard deviation, dashed lines are from Figure 2.4.

2.1.3. Seismic Correlations

Apart from a small number of measurements on exhumed lower crustal rocks (e.g. Ashwal et al., 1987), no heat generation data have been obtained from below the level of human accessibility. The difficulty in assigning heat generation values to lower levels of crustal models was examined and discussed by Smithson and Decker (1974). Subsequently, Allis (1979) and Rybach (1979) suggested an approach based on correlations between density, seismic velocity, and heat production (Figure 2.6). Their model assumed that gradients in seismic velocity are accompanied by gradients in heat production.

Rybach and Buntebath (1984) explored these correlations in more detail. They suggested that separate estimates are required for heat production in Precambrian and Phanerozoic sections, reflecting variations in crustal evolution and seismic complexity. Separate linear relationships were described for each domain according to the expression

$$\ln(A) = B - 2.17 \, V_p \tag{2.12}$$

where A ($\mu W \, m^{-3}$) is the equivalent heat generation for compressional seismic velocity V_p ($km \, s^{-1}$), and $B = 12.6$ for the Precambrian and 13.7 for the Phanerozoic.

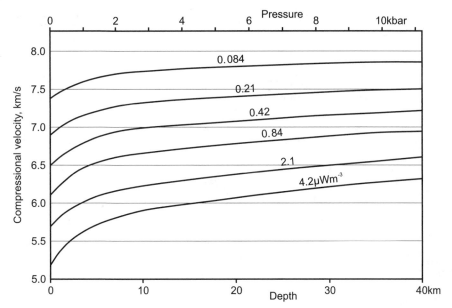

Figure 2.6. Correlation between depth/pressure, seismic velocity and heat production. Modified after Rybach (1979).

Kern and Siegesmund (1989) questioned the validity of such relationships after obtaining and analysing independent data. Cermák et al. (1990) subsequently combined the data sets of Rybach and Buntebath (1984) and Kern and Siegesmund (1989) and upheld the former's original findings, with some corrections to the coefficients of the linear relationship:

$$\ln(A) = 13.92 - 2.38 V_p \tag{2.13}$$

Cermák and Bodri (1986) demonstrated the modifications required for *in situ* estimates of heat production. Seismic velocity is sensitive to variations in temperature and significant corrections to laboratory correlations are required. In view of the uncertainties involved in exact solutions, Cull (1991) suggested an alternative expression giving minimum values for heat production in Australia according to the expression

$$\ln(A) > 8.85 - 1.33 V_p \tag{2.14}$$

Cermák et al. (1991) and Bodri and Cermák (1993) applied expressions of this type to calculate profiles of heat production based on detailed seismic sections from Europe, North America, Australia, Africa, and India (Figure 2.7; note: profile for India is not shown in the figure). No systematic trends were detected within the resulting data, but minimum deviation was obtained using exponential curves as proposed by Singh and Negi (1980). Cermák and Bodri (1993) found that the data fit an exponential curve best when an additional term, *a*, is included. The *a*-term characterises the quantity of heat removed by the pro-

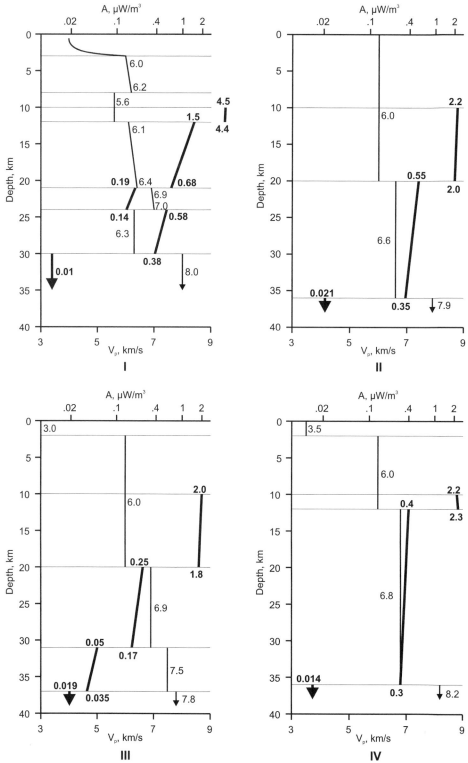

Figure 2.7. Profiles of heat production (bold lines) based on detailed seismic sections (thin lines): I – Europe (Switzerland), II – Australia (Victoria), III – North America (Basin and Range) and IV – Africa (Zaire). Modified after Cermák et al. (1991).

cesses involved in crustal evolution (such as remelting, metamorphosis and leaching by groundwater):

$$A(z) = A_0 \exp(-z/D) - a \tag{2.15}$$

The three parameters (A_0, D and a) are constant within each heat flow province, whereas D varies with depth if the a-term is excluded.

2.2. Frictional Heating along Faults

Radio-isotopic decay is by far the most important mechanism of heat generation within the crust, but other sources may affect surface heat flow in localised regions. The first of these we shall examine is frictional heating along faults.

A fault is a surface along which two bodies of rock move relative to each other. The frictional stress on the fault must be overcome before there can be relative movement. Often, the stress is released in periodic, sudden bursts when much of the energy is dissipated as elastic waves (earthquakes). However, when motion along the fault is by continuous creep (or episodic at such a rate that no appreciable energy is carried as elastic waves), the work done in overcoming the friction is dissipated into the surrounding rock as heat.

The frictional heat ($\mathbf{Q}_f$) generated by a slow-creeping fault is proportional to the distance from the top to the bottom of the fault (d), the rate of slip along the fault (u) and the shear (or frictional) stress acting along the fault (σ_f):

$$\mathbf{Q}_f = d \times u \times \sigma_f \tag{2.16}$$

Generated heat dissipates through the surrounding strata.

EXAMPLE

Brune, Henyey and Roy (1969) modelled the northern end of the San Andreas Fault, California, as a vertical plane 20 km high ($d = 2 \times 10^4$ m) with an average slip rate of 5 cm yr^{-1} ($u = 1.58 \times 10^{-9}$ m s^{-1}). They investigated different stress models for the fault, including depth-constant and depth-varying stress, and calculated the theoretical surface heat flow anomaly for each model (Figure 2.8). No actual heat flow anomaly is observed across the San Andreas Fault (within error margins of about 1.2 mW m^{-2}) so Brune et al. (1969) concluded that the average shear stress along the fault must be less than about 100 bar (10 MPa). Lachenbruch and Sass (1980) re-examined the problem in light of new heat flow data, and arrived at the same conclusion regarding the upper limit of shear stress.

Equation (2.16) suggests that a fault is only likely to exert a significant effect on nearby surface heat flow if either slip rate or shear stress is high

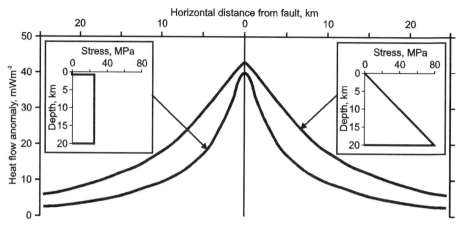

Figure 2.8. Theoretical heat flow anomaly above a vertical plane fault with slip rate 5 cm yr^{-1}. Two possible stress distributions are illustrated: (1) uniform stress (25 MPa) between 1 and 20 km; (2) stress increases linearly (from 0 to 80 MPa) between 0 and 20 km. Modified after Brune et al. (1969).

($u > 10$ cm yr^{-1} or $\sigma_f > 20$ MPa). One tectonic regime where we might expect these conditions to be fulfilled is at a subduction zone. These large, inclined, crustal-scale thrust faults are often associated with rapid convergence rates. For example, the rate of convergence between the Nazca and South American plates, on the western coast of South America, is about 11 cm yr^{-1} (Press and Siever, 1986).

A number of investigators have studied the magnitude of heat generation at convergent plate boundaries, and results are varied. Molnar and England (1990) constructed an idealised analytical model of a subduction setting, taking into account heat carried by downward advection of the lower plate, radiogenic heat produced within the two plates, and background heat flow from the mantle. They concluded that those heat sources and sinks were insufficient to explain observations such as the heat flow between island arcs and subduction trenches, and partial melting in the upper plate of the Himalayan thrust. They proposed that the excess heat must arise from shear heating at the fault surface and that the magnitude of the shear stress at convergent boundaries may be as high as 100 MPa.

Ziagos, Blackwell and Mooser (1985) conversely found that no frictional heat was required to satisfy surface heat flow observations above the subducting Cocos Plate in southern Mexico. Wang et al. (1995) arrived at a similar conclusion for the Cascadia subduction fault lying beneath Vancouver Island. They also observed that compressive stress perpendicular to the strike of the convergent margin is similar both horizontally and vertically, implying a low degree of coupling between the two plates. Both of these observations imply that shear stress on the Cascadia Fault is perhaps as low as 10 MPa. Oleskevich, Hyndman and Wang (1999) also concluded negligible frictional heating beneath Chile, beneath southwest Japan, and on the Cascadia Fault.

The problem of heat generation on fault surfaces has yet to be satisfactorily resolved. It appears likely from the above discussion that different faults may exhibit different behaviour in this respect, perhaps because of different degrees of lubrication related to pore-fluid pressure (e.g. Wang et al., 1995). As numerical modelling techniques improve, and more heat flow data are collected from the vicinity of large faults, the question may be answered. However, for now there is no simple solution as to how much frictional heat is generated by faults.

2.3. Metamorphic Reactions

Another possible source/sink of heat in the crust is exothermic/endothermic metamorphic reactions. The quantity of heat released/absorbed during metamorphic alteration of rocks can be considerable. In general, reactions connected with contact metamorphism (heat, without significant shear stress) are endothermic (absorb heat) while those linked to dynamic metamorphism (involving intense localised stresses, which tend to break up the rock) are exothermic (release heat) (Harker, 1932).

While high-temperature contact metamorphism is invariably endothermic in nature, regional metamorphism generally has both thermal and dynamic components. At low temperature, high shear stresses are possible within a rock and reactions are predominantly exothermic. As temperature increases, however, the reactions become increasingly endothermic as the strength of the rock decreases. In the medium grades, the exchange of heat, positive or negative, is relatively small.

A brief examination of thermodynamic principles is necessary to illustrate the quantitative effects of metamorphism on heat flow. The fundamental equation of the combined first and second laws of thermodynamics defines the total energy within a system:

$$U = TS - PV + \sum_{i=1}^{m} \mu_i n_i \tag{2.17}$$

where U is energy, T is absolute temperature, S is entropy, P is pressure, V is volume, μ is chemical potential of component i of the medium and n is moles of component i.

The last term in Equation (2.17) is known as the Gibbs free energy, G_f, of the system. If we assume temperature, pressure, volume and entropy are only indirectly affected by metamorphic reactions, the change in total energy content of the system is the change in G_f:

$$\Delta U = \Delta G_f \tag{2.18}$$

where

$$G_f = \sum_{i=1}^{m} \mu_i n_i \tag{2.19}$$

It is not necessary to calculate exact values for G_f. For practical applications, we only ever need to know relative values. If ΔG_f is positive for a particular reaction, then the system has absorbed energy, and thus the reaction is endothermic. Conversely, a reaction with a negative ΔG_f is exothermic and liberates heat.

Question: The three minerals andalusite, sillimanite and kyanite share the same chemical formula, Al_2SiO_5, but are stable under different pressure and temperature conditions. Andalusite is stable under normal surface conditions, and Barin (1993) quotes a relative $G_f = -2442.890$ kJ mol^{-1} for andalusite at room temperature and pressure. Sillimanite is the high-temperature polymorph of andalusite, and its relative $G_f = -2441.070$ kJ mol^{-1} under surface conditions. The high-pressure polymorph, kyanite, has $G_f = -2443.881$ kJ mol^{-1}. What is the energy exchange during the metamorphic reactions andalusite $\rightarrow$ sillimanite and andalusite $\rightarrow$ kyanite?

Answer: The energy exchange is found from the difference in relative G_f between the initial and final compound:

ΔG_f for andalusite $\rightarrow$ sillimanite $= (-2441.070 - 2442.890)$ kJ mol^{-1} = 1.820 kJ mol^{-1}.

This is an endothermic reaction, absorbing 1.820 kJ mol^{-1}, or 11.24 kJ kg^{-1} of andalusite.

ΔG_f for andalusite $\rightarrow$ kyanite $= (-2443.881 - 2442.890)$ kJ mol^{-1} = -0.991 kJ mol^{-1}.

This is an exothermic reaction, liberating 0.991 kJ mol^{-1}, or 6.12 kJ kg^{-1} of andalusite.

Devolatilisation reactions absorb more heat than most other types of reaction. Devolatilisation of pelites, for example, requires between 60 and 110 kJ mol^{-1} of volatile released (CO_2 or H_2O). That is an average of about 120 kJ kg^{-1} for a typical pelite, although the exact figure depends upon the particular reaction (Walther and Orville, 1982).

EXAMPLE

Consider the following devolatilisation reaction:

Monticellite + calcite $\Rightarrow$ merwinite + periclase + carbon dioxide

$2CaMgSiO_4 + CaCO_3 \Rightarrow Ca_3MgSi_2O_8 + MgO + CO_2$

$\Delta G_f = (-4340.491 + -568.943 + -394.364) -$
$[(2 \times -2145.731) + -1128.811] = 116.475$ kJ mol^{-1}

The reaction occurs at high temperature and low pressure and is endothermic, absorbing 116.475 kJ mol^{-1}. This is significantly more than the heat of reaction for aluminosilicates.

Most metamorphic reactions are temperature driven and absorb heat. There are only a small number of geological environments where we may expect metamorphic heat generation. The general effect of metamorphism, therefore, is to reduce surface heat flow. The magnitude of the effect depends on the heat of reaction, the reaction rate and the amount of reactant.

Question: A typical basin subsides at 10^{-5} m yr^{-1}. Assuming a constant geothermal gradient, and an ideal mixture of pure monticellite and calcite, at what rate is heat absorbed as the sediment undergoes devolatilisation at depth?

Answer: One metre of sediment passes into unstable temperature and pressure conditions every hundred thousand years. The densities of monticellite and calcite are 3.05 and 2.715 g cm^{-3}, respectively. Their molar masses are 156.46 and 100.09 g mol^{-1}, respectively (Deer, Howie and Zussman, 1990). An ideal mixture contains 7170 mol m^{-3}. During reaction, 116.475 kJ mol^{-1} are absorbed, so heat is absorbed at a rate of 8.352 kJ yr^{-1} m^{-2}, or 0.265 mW m^{-2}.

Surface heat flow is reduced by 0.265 mW m^{-2}, which is insignificant in most situations.

One geological setting where heating rates can reach levels where endothermic reactions may absorb heat at a significant rate is around a hot intrusive body. However, the thermal regime of an igneous intrusion is complicated by many factors that put it beyond the scope of this work. Such factors include fluid flow, release of latent heat during solidification of the magma, the geometry of the body, the extent of the contact aureole with associated heat absorption, and complex basal boundary conditions. These and other concerns connected with the thermal modelling of magmatic intrusions were comprehensively covered by Furlong, Hanson and Bowers (1991) and interested readers should pursue that reference.

2.4. Summary

Crustal sources of heat include radioactive isotopic decay, friction due to intraplate strain and plate motions, and exothermic metamorphic and diagenetic processes. The relative magnitude of each heat source depends on geographic location, but, in general, radiogenic sources are dominant. The amount of radiogenic heat generated by a rock can be estimated from the proportion of uranium, thorium and potassium within it. Acid rocks generate significantly more heat than basic rocks.

An observed linear relationship between heat generation (A_0) and heat flow (**Q**) at the Earth's surface leads to the concept of heat flow provinces. A heat flow province is a region of common tectonothermal history within which consistent values of **q** and D are obtained:

$$\mathbf{Q} = \mathbf{q} + A_0 D$$

The heat flow province concept has been remarkably successful in separating the continental landmasses into regions of shared heat flow characteristics, although simple analytical models for heat production appear to have no justification in reality.

In a petroleum exploration setting, heat generation can be estimated directly from well logs of uranium, thorium and potassium concentration, or via an empirical relationship with total gamma-ray count. For deeper formations, empirical correlations with seismic velocity may be used.

Friction on large faults may be a significant source of heat in some instances, but negligible in others. There is currently no consensus on the issue, and each fault should be examined as an individual case. Metamorphic and diagenetic reactions in general contribute negligibly to surface heat flow.

PART TWO

Measurement Techniques

CHAPTER THREE

Thermal Gradient

Yes, it is well known that the temperature rises about 1°C for every seventy feet downwards, so that if this continues to hold good, the radius of the earth being more than four thousand miles, the temperature at the center would be about two million degrees. Everything there must be in the state of incandescent gas, for gold, platinum, and the hardest rocks cannot resist such a temperature.

A Journey to the Center of the Earth – Jules Verne, 1864.

The parameter people most closely associate with heat flow is temperature, or, more specifically, thermal gradient. By definition, thermal gradient is a vector quantity dependent on the distribution of temperature in three dimensions. The particular gradient measured along a straight line depends on the orientation of that line with respect to the temperature field. The magnitude and orientation of the maximum thermal gradient are found from

$$\nabla T = \frac{\partial T}{\partial x}\mathbf{i} + \frac{\partial T}{\partial y}\mathbf{j} + \frac{\partial T}{\partial z}\mathbf{k} \qquad (3.1)$$

where T is the temperature distribution function in three dimensions, and $\mathbf{i}$, $\mathbf{j}$ and $\mathbf{k}$ are the unit vectors along the x, y and z axes (z is conventionally taken to be vertical).

We need to know the three-dimensional temperature distribution within a body in order to determine the true vector value of the maximum thermal gradient, and hence the magnitude and direction of heat flow. Unfortunately, data of sufficient detail are rarely, if ever, available for geothermal applications, so we are forced to make certain assumptions.

Almost invariably we assume that the direction of maximum gradient within the upper crust is vertical. This is a valid assumption if we accept that the surface of the Earth forms (approximately) a horizontal, constant temperature boundary, which tends to minimise lateral temperature variation at depth. All thermal gradient problems then reduce to one dimension:

$$\nabla T = (\partial T/\partial z) \times \mathbf{k} \qquad (3.2)$$

and the magnitude of the gradient ($\partial T/\partial z$) is the derivative of temperature with respect to depth.

47

To obtain even the most rudimentary estimate of gradient, therefore, we need to know the temperature at two or more depths. Ideally, one of these depths will be at the surface boundary (i.e. the average surface temperature).

Question: In northwest Australia the average surface temperature is $27 \pm 1°C$. A well is drilled in the region to a depth of 2326 m, and allowed to return to thermal equilibrium. The temperature at the bottom of the hole is then measured at $77 \pm 2°C$. What is the magnitude of the average thermal gradient in the hole?

Answer: The gradient is the derivative of temperature with depth, or the change in temperature over the change in depth. If T_0 is the surface temperature and T_1 the temperature at the bottom of the hole:

$$\partial T/\partial z = (T_1 - T_0)/\Delta z = (77 \pm 2 - 27 \pm 1)/2326 = (50 \pm 3)/2326$$
$$= 0.0215 \pm 0.0013°C\,m^{-1} = 21.5 \pm 1.3°C\,km^{-1}$$

The more the temperature field can be constrained, the more accurate and precise will be the resulting gradient values and subsequent heat flow estimates. Therefore, the aim is to obtain as many temperature data as possible in a vertical section. Generally, this requires direct measurement down boreholes drilled into the ground in the region in which we are interested. Three temperature indicators not requiring boreholes are the Curie depth, the stability field of xenoliths, and upper mantle resistivity, but each is fraught with uncertainty.

Figure 3.1 depicts the steps that should be followed in collecting and collating crustal temperature data in order to produce the most accurate possible temperature profile for a particular region. The remainder of the chapter explains each step in detail.

3.1. Direct Measurement Techniques

Direct measurement of underground temperature requires that a temperature-measuring device be lowered down a borehole (or mineshaft, cave or other cavity accessible from the surface). Such devices measure the temperature of the bore fluid, not the surrounding rock, so to obtain meaningful estimates of the ambient temperature of the rock, the bore fluid must be in thermal equilibrium with its surroundings. If the hole has only recently been drilled, the fluid may not have had time to attain thermal equilibrium. Any event that subsequently disturbs the bore fluid also causes a thermal disturbance. The amount of time required for re-equilibration depends on the magnitude of the disturbance.

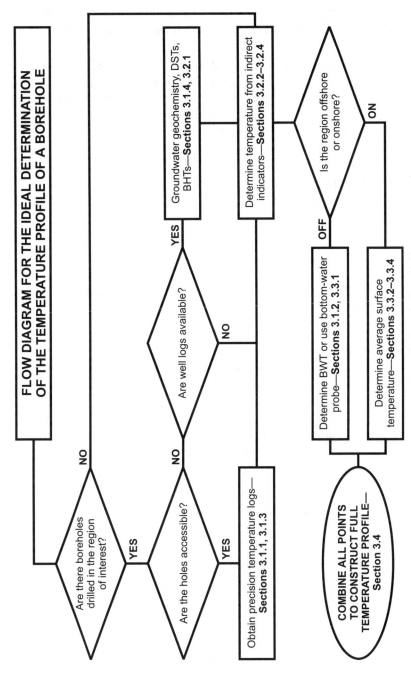

Figure 3.1. Flow chart for the collation of all available temperature data.

The disruption to the thermal equilibrium of the bore-fluid column caused by the circulation of large volumes of fluid during drilling (often for several weeks) can take months to dissipate. Longer recovery times are required for greater drilling times. In general, from 10 to 20 times the drilling time is required before a well is equilibrated to within the accuracy of most equipment (Bullard, 1947).

Production, or removal of fluids from a well, also causes a thermal disturbance, although the magnitude is significantly less than for drilling. The time required to re-equilibrate depends on the construction and production history of the well. A typical production well is cemented, cased, and produces through a 5–7 cm diameter pipe. If production rates are moderate, the thermal disturbance around the well remains small, and equilibrium is regained in the tube within a relatively short time once production is halted. If flow is through a larger tube, or at high rates, the disturbance is proportionately larger and a longer recovery time is necessary. In general, logging can be carried out within a few weeks of production ceasing.

Note: In most practical situations, it is unnecessary to log production wells. Most producing fields have wells that have been shut in for considerable lengths of time and require no further equilibration.

The very act of logging a hole disturbs the bore fluid with the motion of the probe, so while most logging procedures are run upwards from the bottom of the hole, temperature logging should ideally be conducted during descent of the probe. The thermal disturbance due to the passage of the probe is considerably less than that caused by drilling or production, and generally only a day is required for re-equilibration. Logging upwards, or immediately after a previous logging run, will often give satisfactory results if fine detail is not required from the data. Such logs are noisier than undisturbed logs, but medium- and broad-scale temperature trends are retained.

Note: Temperature logging should not be conducted a short time after drilling, and should be conducted downwards. This makes it incompatible with most other logging techniques, which may explain the low percentage of holes that are temperature logged.

3.1.1. Precision Temperature Logs

Temperature can be electronically logged in the same manner as other rock properties such as density, sonic velocity and electrical resistivity. High-precision instruments capable of resolving thermal gradient on a fine scale have been available for many years. Most are electronic in nature, utilising thermistors or platinum resistance sensors as the temperature-sensitive component. A thermistor is simply an electronic component with a temperature-dependent resistance. Once calibrated, a simple resistance measurement is sufficient to determine temperature. Platinum resistance thermometers are more stable than thermistors and have a nearly linear resistance–temperature response over a large temperature range. Unfortunately, their resistance range is quite small (25–50 Ω), so heavy, low-resistance cable is required to maintain accuracy if used in a wire-line probe.

There are two ways of using thermistor sensors for temperature logging. The first is to monitor thermistor resistance directly from the surface using a Wheatstone bridge (Figure 3.2) or other type of ohmmeter. Temperature resolution of 0.01°C or better is possible, but the sensor must be held stationary at each measurement depth until no significant drift in resistance is observed. This may require periods of up to 2 min or more. The second method is to use the thermistor in an oscillatory circuit, so that the frequency of the signal from the probe is proportional to the thermistor resistance. Cable of lower quality can be used and a sampling rate of 2–3 s^{-1} is possible using deconvolution techniques to correct any depth offset due to the thermal time constant of the probe (Beck and Balling, 1988). Logging speeds of 5–10 m min^{-1} are attainable,

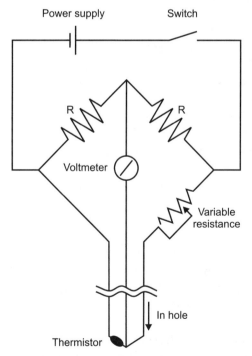

Figure 3.2. Schematic diagram of a Wheatstone bridge circuit for measuring thermistor resistance. Resistance, R, is normally around 10 kΩ.

giving measurements with absolute error of less than 0.5°C and relative error less than 0.05°C (Houseman et al., 1989).

When designing a wire-line temperature probe, the thermistor should be mounted as near as possible to the leading tip to minimise disturbance to the borehole fluids prior to temperature measurement. The probe should be a rugged construction of high-thermal-diffusivity material (e.g. brass) with all electronic components sealed against borehole fluids. The cable connecting the probe to the surface should contain at least four wires to allow circuits that accommodate the effect of cable resistance. A motorised or hand winch completes the basic logging outfit. Productivity can be greatly improved with the integration of a PC for automatic data logging, and various pulleys and frames for smooth field operation. High-precision temperature logs can be easily obtained within a moderate budget.

Self-contained temperature probes are commercially available, with on-board battery, memory and processing chips inside a sealed Dewar flask. These computer tools require only a solid wire slick-line, greatly reducing the bulk and cost of logging cable. Computer tools generally use a platinum temperature sensor, which minimises the internal circuitry required to monitor resistance. Recording is initialised at the surface, and then the tool is simply lowered down the hole. Thermistor resistance is automatically recorded at pre-set time intervals, and when the tool is returned to the surface, the data are downloaded onto an ordinary PC for storage, processing and viewing. Computer tools are more expensive than wire-line tools, but much more versatile. Slick-line is cheaper than wire-line, and can be packed off for logging high-pressure or producing wells. Also, computer tools are generally designed to withstand high temperature and pressure, so are much more suited to logging in deep, hot, pressurised, producing or other bad-environment wells. Precision and accuracy are equivalent to wire-line tools.

A third type of temperature logging tool is based on the Raman effect of temperature-sensitive backscattering of laser light in an optic fibre. It is referred to as a 'Distributed optical fibre Temperature Sensing system' (DTS), and has major advantages over other types of logging. To operate the system, one end of an optic fibre is kept at the surface while the other end is lowered into a borehole and secured into place. The DTS system is then able to produce an instantaneous temperature log along the full length of the fibre without disturbing the surrounding bore fluid. Optic fibre is relatively cheap and can be left secured in the borehole if more than one measurement is to be made. This makes the system ideal for situations requiring regular monitoring or for the study of transient events (e.g. Sakaguchi and Matsushima, 1995; Großwig, Hurtig and Kühn, 1996). DTS systems are currently of lower precision (0.1°C) and depth resolution (0.25–1.0 m) than other logging tools (Wisian et al., 1998), but they provide data unattainable by other means.

An accurate depth log must accompany any temperature log. The DTS system automatically incorporates depth into its output, but for conventional logging systems an accurate record must be kept of the length of line that has been fed from the winch. For some deep or hot wells, corrections may have to be applied to compensate for elastic extension and thermal expansion of the cable. For wire-line logging (where results are fed to the surface in real time), depth and temperature data can be recorded simultaneously and stored on a PC or other memory device. With computer tools, however, time-depth logs must be independently collected and subsequently merged with the time–temperature data to produce the desired depth–temperature pairs.

Note: A high-quality precision temperature log can be used to differentiate between lithology types in a way similar to other logging techniques (e.g. Reiter, Mansure and Peterson, 1980; Blackwell et al., 1999). In a steady-state thermal regime, thermal gradient is inversely proportional to thermal conductivity, so changes in conductivity (i.e. lithology) are reflected in the gradient log (Figure 3.3).

Commercial wire-line logging services sometimes log temperature as part of a conventional logging exercise. Unfortunately, the commercial tool is almost always run concurrently with other wire-line logging tools, usually from 1–3 days after drilling. This is long before the drilling fluids have had time to attain thermal equilibrium, and only with subsequent processing can meaningful temperature data be extracted from the logs.

Von Herzen and Scott (1991) described a method for extrapolating multiple temperature logs collected at different times in the same hole. Their work specifically focussed on data from boreholes drilled in deep water by the Ocean Drilling Program (ODP), but their conclusions are valid for all similar situations. The procedure assumes that the temperature at each depth is relaxing back to equilibrium at a rate inversely proportional to time, similar to the Horner method described later in this chapter. The method essentially extrapolates time to infinity to determine the apparent ambient temperature at all points down the well.

An inherent assumption in Von Herzen and Scott's (1991) work is that each log accurately records the fluid temperature at the time the log was run, but Sawyer, Bangs and Golovchenko (1994) referenced earlier work to point out that this assumption may be flawed. Every temperature logging tool takes a finite time to respond to a change in fluid temperature (proportional to the time constant of the tool), so the output from the tool is not necessarily equal to the temperature at the exact depth of the tool. The discrepancy increases with logging speed and the response time of the tool. Sawyer et al. (1994) went on to consider ways of deconvolving the temperature log to remove the effect of the tool response function and reveal the actual fluid temperature. They concluded that deconvolution is a viable and necessary step in analysing temperature logs obtained soon after drilling. Fluid temperature can be extrapolated to equilibrium only if two or more deconvolved logs exist for the same well.

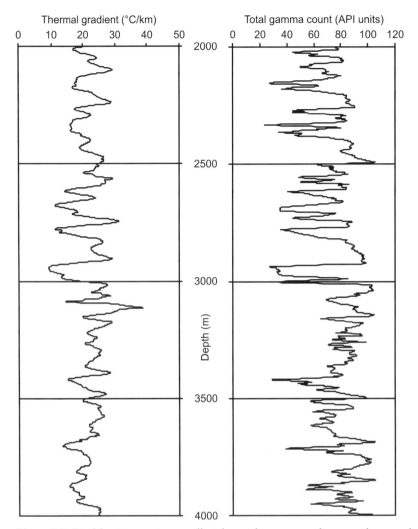

Figure 3.3. Precision temperature gradient log and gamma-ray log over the same interval of well Ferris 1–28 in the Anadarko Basin, Oklahoma. Note the correlation between layers of high GR and gradient (shales) and low GR and gradient (sandstones). Data courtesy of Southern Methodist University Geothermal Laboratory, Dallas, Texas.

Note: Unprocessed temperature logs recorded soon after drilling should not be used in quantitative temperature analyses. Only after considerable processing do they relate to the ambient temperature of the surrounding rocks.

3.1.2. Deep-Water Probes

The relative inaccessibility of deep oceans renders borehole-based geothermal investigations inappropriate. Specialised instruments have therefore evolved to investigate such regions. These instruments generally take the form of thermistor-lined probes that penetrate the soft sediment on the sea floor. Specific designs have evolved in accordance with improvements in technology.

The first deep-ocean probes were developed in the 1950s and can be collectively called 'Bullard-type' probes (Figure 3.4). A typical Bullard probe is a several metre long solid shaft with thermistors regularly spaced along its length (e.g. Bullard, 1954). The usual method of operation requires suspending the probe several tens or hundreds of metres above the ocean floor until it reaches thermal equilibrium, then allowing it to free-fall and penetrate the sediment. The penetration results in frictional heating along the length of the probe, the effect of which can take up to half an hour or longer to dissipate before the ambient temperature becomes apparent. Once equilibrated, the thermal gradient can be calculated across each thermistor interval. A Bullard probe is prone to bending upon impact with the ocean floor and slow to produce results.

The first adaptation of the Bullard probe was the development of 'Ewing-type' probes (Figure 3.5). A Ewing probe operates in much the same way as a Bullard probe, but differs in construction (e.g. Gerard, Langseth and Ewing,

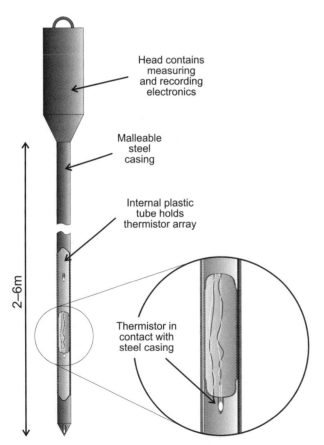

Figure 3.4. A cut-away view of a Bullard-type heat flow probe, showing thermistors positioned in contact with the steel casing. Thermistors are generally spaced about 25 cm apart.

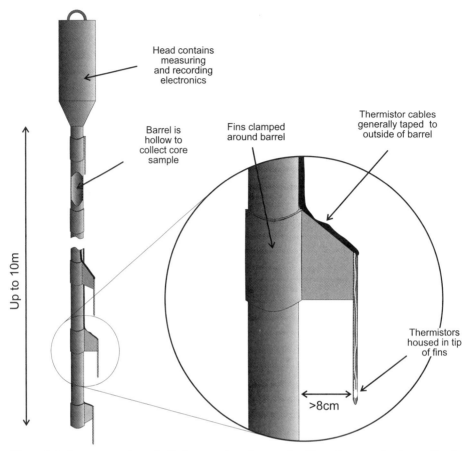

Figure 3.5. A cut-away view of a Ewing-type heat flow probe. Thermistor spacing is usually about 25 cm.

1962). A Ewing probe is built around a hollow shaft designed to collect a core sample for thermal conductivity analysis at the same time as the temperature data. Thermistors are housed within thin vertical straws mounted on fins projecting radially from the probe. This arrangement greatly reduces the time taken for frictional heat to disperse around the thermistors. A thermistor in the tip of a straw will not be affected by the frictional heat of the coring tube if it is at least 8 cm from the coring tube and ten straw diameters from the fin (Jessop, 1990). Equilibrium temperature is reached within a few minutes.

In the 1970s, the deep-sea probe evolved further with the development of 'Lister-type' probes (Figure 3.6). Lister probes are also sometimes referred to as 'violin bow' probes because of their general construction, which consists of a thin (~1 cm diameter) sensor tube supported approximately 5 cm away from a thick (5–10 cm) strength member (e.g. Hyndman et al., 1978). The sensor tube contains individual thermistors for measuring temperature at intervals of 1 m or more. In the spaces between temperature data points there are arrays of thermistors for measuring the harmonic mean value of temperature. The latter

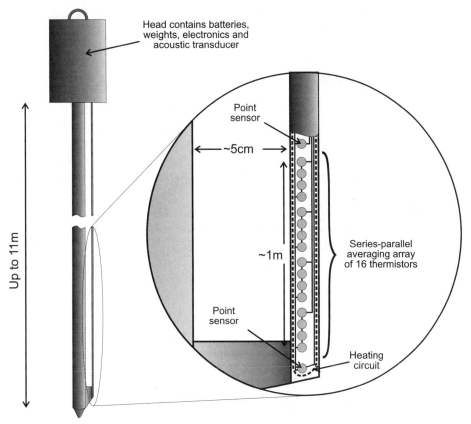

Figure 3.6. A cut-away view of a Lister 'violin bow' heat flow probe (horizontal scale expanded on inset).

are included to estimate the average thermal conductivity between temperature data points (see Section 4.3.4).

Note: The very short depth increment between thermistors on deep-water probes (typically 25 cm) necessitates the use of very precise thermistors in order to calculate an accurate thermal gradient. A precision of 0.001°C or better is required to resolve a gradient of $40°\,\mathrm{C\,km^{-1}}$.

Advances in digital technology and engineering have allowed the evolution of probes that do not need to be brought back to the surface after each recording. Data are either stored in memory on the probe or transmitted in real time via a digital telemetry link to the accompanying ship. Measurements can be made at a rate of up to one per hour by raising the probe only far enough to allow it to attain an equilibrium position beneath the ship prior to the next drop. Earlier probes had to be fully returned to the ship between

measurements, allowing a maximum survey rate of only four locations per day
(Davis, 1988).

3.1.3. Borehole Convection

The great majority of down-hole tools measure the temperature of the drilling
fluid, in spite of continuing efforts to perfect a logging tool that directly measures the temperature of the surrounding formation (e.g. the nuclear logging
tool evaluated by Ross et al., 1982). A column of fluid under the influence of a
thermal gradient may support convection cells, thus causing the temperature
profile within the fluid to differ from that within the surrounding rocks. Hales
(1937) first examined the possibility of convection within a borehole in relation
to geyser eruptions. His results suggested a rule relating fluid properties and the
radius of the borehole to a critical thermal gradient, above which convection
may be expected:

$$\frac{\partial T}{\partial z} = \frac{C \nu \kappa}{g \alpha r^4} \, {}^\circ\text{C} \, \text{m}^{-1} \tag{3.3}$$

where $\partial T / \partial z$ is the critical thermal gradient; g is gravitational acceleration; α, ν
and κ are the thermal expansion coefficient, kinematic viscosity and thermal
diffusivity of the fluid; r is the borehole radius, and C is a constant that equals
2.16×10^{-4} for a long narrow tube (if SI units are used throughout).

For water at 95°C, this reduces to approximately (Jeffreys, 1937)

$$\frac{\partial T}{\partial z} = \frac{1.4 \times 10^{-9}}{r^4} \, {}^\circ\text{C} \, \text{m}^{-1} \tag{3.4}$$

where r and z are measured in metres.

The coefficient on the right of Equation (3.4) alters with temperature
because of its dependence on viscosity, decreasing for higher temperatures
and increasing to about 1.4×10^{-8} for water at 20°C. The theory was subsequently tested and defended by Auld (1948). Interest in the problem reignited
some 20 years later, when Diment (1967) developed virtually the same solution
as Hales (1937), but with an extra term for the adiabatic temperature gradient,
which Hales had discarded as insignificant. According to Diment, convection
cells are established when

$$\frac{\partial T}{\partial z} = \frac{g \alpha T}{c_p} + \frac{C \nu \kappa}{g \alpha r^4} \, {}^\circ\text{C} \, \text{m}^{-1} \tag{3.5}$$

where T and c_p are the absolute temperature and specific heat of the fluid, and
all other terms are as for Equation (3.3).

The adiabatic temperature gradient term is only about 0.15°C km^{-1} at 20°C
and 0.18°C km^{-1} at 95°C. This is less than 1% of typical geothermal gradients,
implying that Hales was probably justified in neglecting it. Gretener (1967)
tested Diment's theory and once again confirmed the findings.

Note: At surface temperatures, normal geothermal gradients ($\sim 25°C\,km^{-1}$) are sufficient to induce convection in boreholes with a diameter greater than about 5 cm.

The general effect of convection within a borehole is to decrease the signal-to-noise ratio on a temperature log without significantly disrupting broader temperature trends. Regions of higher geothermal gradient generally yield noisier logs (Urban, Diment and Nathenson, 1978), as do sections of open holes that have been washed out (thus increasing the effective radius). Several studies (e.g. Diment, 1967; Gretener, 1967; Wisian et al., 1998) have found that convection cells are usually on the order of 1 m in height, and do not significantly affect the ambient temperature gradient. However, some care may be required in wells of abnormally large diameter or in regions of high gradient.

3.1.4. Bottom-Hole Temperatures in Deep Wells

It takes a considerable time for a well to attain thermal equilibrium after drilling. Typically, this is on the order of ten times the drilling time, which can amount to several months for a deep well. Onshore, it is often possible to retain access to a deep borehole long enough to allow precision temperature logging after a sufficient equilibration time. Offshore, however, wells are often plugged and abandoned after drilling, which renders them inaccessible for future temperature logging. Temperature data from these wells must be obtained either during, or very soon after, drilling.

A drill stem test (DST) is considered to give the most accurate measurement of down-hole formation temperature at times shortly after drilling. A tool left stationary for a period of time collects formation fluid flowing from the well wall. The fluid is assumed to come from a distance away from the thermally disturbed zone around the borehole, and thus is assumed to equilibrate quickly to the ambient temperature of the surrounding rock. However, if the flow of fluid is slow, or if there is a component of expanding gas, the DST temperature may be too low. Also, for many wells DST data are unavailable, so other data must be utilised to estimate temperature at depth.

During excavation of a typical oil exploration well, drilling is interrupted at a number of depths to change the drill bit, run wire-line geophysical logs, and case the well. The suite of tools lowered into the well during logging usually includes a thermometer that records the maximum temperature experienced by the instruments down-hole. The maximum temperature is assumed to be at the bottom of the hole, so this gives a direct measurement of the temperature of the drilling fluid at the current well depth, or a 'bottom-hole temperature' (BHT). If more than one logging run is made to the same depth, more than one BHT may be recorded for that depth. The relationship between BHT and the 'virgin rock temperature' (VRT) is complex and there are almost as many techniques for correcting BHT as there have been publications on the subject.

Hermanrud, Cao and Lerche (1990) conducted a study of the inherent accuracy and precision of twenty two different BHT correction methods. They found the most accurate method (on average, best agreement with DST temperatures) to be that of Cooper and Jones (1959), but that method is also one of the least precise (large uncertainty in each individual result). The Horner method (Lachenbruch and Brewer, 1959) is much more precise, but has an inherent bias that reduces its accuracy.

The Cooper and Jones (1959) method models the physical conditions within the drill hole. It assumes a long hole of small diameter has been drilled quickly into thermally equilibrated strata, and that the hole has been filled with a fluid cooler than the strata. The temperature of the fluid approaches that of the strata as heat flows radially inwards from the walls of the borehole. It is possible to show that for any BHT measurement:

$$\text{BHT} = (\text{VRT} - T_f)[1 - F(\alpha, \tau)] + T_f \tag{3.6}$$

where

T_f = temperature of the drilling fluid, and

$$F(\alpha, \tau) = \frac{4\alpha}{\pi^2} \int_0^\infty \frac{\exp(-\tau u^2) du}{u\{[u J_0(u) - \alpha J_1(u)]^2 + [u Y_0(u) - \alpha Y_1(u)]^2\}}$$

where

$\alpha = 2\rho c / \rho_f c_f$
ρc = density × specific heat of strata
$\rho_f c_f$ = density × specific heat of fluid
$\tau = \lambda \Delta t / \rho c r_w^2$
λ = thermal conductivity of strata
Δt = time since circulation ceased
r_w = radius of borehole
$J_N(u)$ and $Y_N(u)$ = Bessel functions of order N of first and second kinds

BHT versus $[1 - F(\alpha, \tau)]$ is linear, with gradient $(\text{VRT} - T_f)$ and intercept T_f. The uncertainty in calculated VRT depends on the accuracy with which the above parameters are estimated or measured. $F(\alpha, \tau)$ has been evaluated and tabulated in such sources as Jaeger (1956) and Davis (1988), and some values are reproduced in Table 3.1. (Also see the spreadsheet, FUNCTIONS.XLS, available for download from the web site mentioned in the Preface.)

Question: A well was drilled in the Browse Basin, Australia, in 1982. The hole was drilled with diameter $2r_w = 15.2$ cm, to a depth $z = 4756$ m below kelly bushing. The maximum temperatures recorded during subsequent logging runs were $T = 125.5°C$ at $\Delta t = 9$ h, $T = 129.6°C$ at $\Delta t = 14$ h and $T = 131.6°C$ at $\Delta t = 18.3$ h. Samples brought to the surface were found to be

Table 3.1. Representative Values of $F(\alpha, \tau)$

τ	α						
	0.5	1.0	1.5	2.0	4.0	6.0	8.0
0.2	0.75706	0.59473	0.48176	0.40030	0.22833	0.15549	0.11682
0.3	0.70753	0.52924	0.41377	0.33506	0.18155	0.12146	0.09057
0.4	0.66750	0.47985	0.36513	0.29021	0.15206	0.10071	0.07481
0.5	0.63365	0.44045	0.32793	0.25696	0.13148	0.08653	0.06414
0.6	0.60424	0.40790	0.29827	0.23109	0.11619	0.07615	0.05637
0.7	0.57820	0.38034	0.27390	0.21027	0.10433	0.06818	0.05042
0.8	0.55483	0.35658	0.25344	0.19309	0.09482	0.06183	0.04571
0.9	0.53364	0.33582	0.23597	0.17863	0.08700	0.05665	0.04186
1.0	0.51427	0.31746	0.22084	0.16628	0.08045	0.05233	0.03866
2.0	0.38062	0.20612	0.13529	0.09912	0.04669	0.03031	0.02240
3.0	0.30245	0.15222	0.09756	0.07088	0.03326	0.02162	0.01601
4.0	0.25004	0.12021	0.07619	0.05521	0.02594	0.01689	0.01252
5.0	0.21228	0.09903	0.06244	0.04523	0.02130	0.01390	0.01031
6.0	0.18379	0.08401	0.05286	0.03830	0.01809	0.01182	0.00878
7.0	0.16156	0.07283	0.04580	0.03322	0.01573	0.01029	0.00765
8.0	0.14377	0.06421	0.04039	0.02933	0.01393	0.00912	0.00678
9.0	0.12925	0.05736	0.03611	0.02625	0.01250	0.00819	0.00609
10.0	0.11718	0.05179	0.03265	0.02376	0.01134	0.00744	0.00553
15.0	0.07877	0.03471	0.02203	0.01611	0.00775	0.00510	0.00380
20.0	0.05857	0.02600	0.01661	0.01219	0.00590	0.00389	0.00290

recrystallised limestone and siltstone with thermal conductivity $\lambda = 2.3\ \mathrm{W\,m^{-1}K^{-1}}$, density $\rho = 2700\ \mathrm{kg\,m^{-3}}$ and heat capacity $c = 700\ \mathrm{J\,kg^{-1}K^{-1}}$. The drilling fluid had density, $\rho_f = 1670\ \mathrm{kg\,m^{-3}}$ and heat capacity, $c_f = 3000\ \mathrm{J\,kg^{-1}K^{-1}}$. Using the Cooper and Jones (1959) method, what is the VRT at the bottom of the hole?

Answer: To answer the question we first need to calculate α and τ for the three different times:

$\alpha = 2\rho c/\rho_f c_f = (2 \times 2700 \times 700)/(1670 \times 3000) = 0.7545$

$\tau = \lambda\,\Delta t/\rho c r_w^2 = (2.3\Delta t)/(2700 \times 700 \times 0.076^2) = 2.107 \times 10^{-4}\Delta t$

At $\Delta t = 9$ h $= 32{,}400$ s, $\tau = 6.827$

At $\Delta t = 14$ h $= 50{,}400$ s, $\tau = 10.62$

At $\Delta t = 18.3$ h $= 65{,}880$ s, $\tau = 13.88$

The next step is to calculate $[1 - F(\alpha, \tau)]$ for each of the three measurements:

$F(0.7545, 6.827) = 0.10334$,   so $[1 - F(\alpha, \tau)] = 0.89666$ when $T = 125.5°C$

$F(0.7545, 10.62) = 0.06791$,   so $[1 - F(\alpha, \tau)] = 0.93209$ when $T = 129.6°C$

$F(0.7545, 13.88) = 0.05204$,   so $[1 - F(\alpha, \tau)] = 0.94796$ when $T = 131.6°C$

Line of best fit through the three T versus $[1 - F(\alpha, \tau)]$ pairs has intercept $T_f = 19.3°C$, and gradient $VRT - T_f = 118.4°C$, so VRT $= 137.7°C$. The uncertainty in T_f is much higher than in VRT.

The Horner plot method is based on an observed similarity in the behaviour of *in situ* temperature and pressure when disturbed by drilling. The method was originally devised to correct pressure build-up data from drill stem tests (Horner, 1951), but was adopted for temperature correction by Lachenbruch and Brewer (1959). The parameters defining a point on a Horner plot are

$T =$ the bottom-hole temperature

$\Delta t =$ the time elapsed between cessation of fluid circulation and measurement of T

$t_c =$ the time elapsed between cessation of drilling and cessation of fluid circulation

Figure 3.7 shows T plotted against $\ln[1 + (t_c/\Delta t)]$. A straight line of best fit through all points from the same depth yields the VRT for that depth at the T-axis intercept (i.e. $\ln[1 + (t_c/\Delta t)] = 0$, or $\Delta t = \infty$). The gradient of the line is a function of the thermal conductivity of the strata, $\lambda(\mathrm{W\,m^{-1}K^{-1}})$, and the rate at which heat is being supplied to the well, $H(\mathrm{W\,m^{-1}})$:

$$T = VRT + (H/4\pi\lambda) \times \ln[1 + (t_c/\Delta t)] \tag{3.7}$$

The accuracy of a Horner plot depends on the reliability and accuracy of T, Δt and t_c. Almost always, T is recorded during a logging exercise, and is commonly found in well-completion documentation or on well log headers.

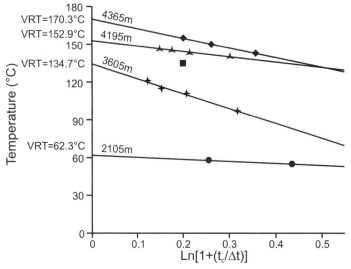

Figure 3.7. Example Horner plot. Bottom-hole temperature data from Echuca Shoals 1 (13°45'01" S, 123°43'25" E) in the Browse Basin, Western Australia, plotted on Horner axes to estimate VRT.

However, the accuracy with which it is recorded is sometimes questionable. In some instances, no change in temperature is recorded between several logging runs many hours apart, or temperatures are rounded off to the nearest 5°C or 10°C. This is unlikely to reflect real conditions, and must be attributed to instrument fault or human error. Unfortunately, the magnitude of these errors is undefinable and data must either be accepted at face value or rejected.

The time since circulation ceased, Δt, is also commonly recorded in well-completion documentation, but is generally rounded off to the nearest half-hour. The accuracy is, therefore, 0.25 h at best. Sometimes Δt is not recorded for a particular temperature measurement, usually when the log has been run for engineering, rather than exploration, purposes (such as to locate the limits of cementing on a well wall). These data cannot be displayed on a Horner plot.

The least reported Horner parameter is the time the drilling fluid was circulated after drilling ceased, t_c. When t_c is not included on the well log header, it can sometimes be retrieved from daily drilling reports, which typically record the respective times at which drilling and circulation ceased. The difference between the two times is t_c. If unavailable, a value for t_c can be estimated using the empirical formula of Hermanrud et al. (1990), although there is a significant uncertainty attached to values of t_c derived in this way:

$$t_c = (1.3 + D)/(1.3 - 0.091 \times D) \tag{3.8}$$

where t_c is in hours and D is depth in kilometres relative to the kelly bushing.

Question: Recalculate VRT from the previous example using the Horner plot method, given that the circulation time was extracted from daily drilling reports, $t_c = 2$ h.

Answer: The required data are summarised below:

$\Delta t = 9$ h, $t_c = 2$ h, $T = 125.5°C$, $\ln[1 + (t_c/\Delta t)] = 0.20067$

$\Delta t = 14$ h, $t_c = 2$ h, $T = 129.6°C$, $\ln[1 + (t_c/\Delta t)] = 0.13353$

$\Delta t = 18.3$ h, $t_c = 2$ h, $T = 131.6°C$, $\ln[1 + (t_c/\Delta t)] = 0.10372$

The line of best fit through the three T versus $\ln[1 + (t_c/\Delta t)]$ pairs has intercept VRT $= 138.0°C$, and gradient -62.6. In this particular case, $t_c/\Delta t < 1/3$ for all points and the result is almost identical to that of the Cooper and Jones (1959) estimate.

Note: The accuracy of any single point on a Horner plot must remain contentious, but if three or more points are available for a single depth a statistical uncertainty can be calculated for VRT based on the spread of points about the line of best fit (e.g. Kreyszig, 1983).

If $(t_c/\Delta t) < 1/3$ for all data points, the Horner technique should give acceptable results (Roux, Sanyal and Brown, 1982). However, a rigorous investigation of the mathematics governing bottom-hole temperature equilibration reveals that, for higher values of $(t_c/\Delta t)$, straight-line extrapolations through Horner data are inherently biased towards underestimating VRT (Dowdle and Cobb, 1975). The magnitude of the bias increases as $(t_c/\Delta t)$ increases.

Roux et al. (1982) developed a procedure to compensate for the theoretical bias in the Horner method on those occasions when $(t_c/\Delta t) > 1/3$. A small number of extra parameters are required for the correction, which is based on linear approximations of theoretical cooling curves. A dimensionless time parameter, t_D, is derived from the thermal diffusivity of the formation, κ $(\mathrm{m^2\,s^{-1}})$, the circulation time, t_c (h), and the borehole radius, r_w (m):

$$t_D = 3600 \times \kappa t_c / r_w^2 \tag{3.9}$$

Where the formation diffusivity or borehole radius are unknown, $\kappa = 10^{-6}\,\mathrm{m^2\,s^{-1}}$ and $r_w = 0.12$ m are suggested as reasonable average values for most situations. A dimensionless temperature factor, T_D, is then derived from

$$T_D = a + (b \times t_D) + (c \times t_D^{1/2}) + (d \times t_D^{1/3}) + (e \times t_D^{1/4}) + (f \times t_D^{1/5}) \tag{3.10}$$

The coefficients, a–f, depend upon the range into which most values of the dimensionless Horner time, $(t_c + \Delta t)/\Delta t$, fall, according to Table 3.2. The bias in the Horner plot is corrected by

$$\text{True VRT} = \text{Horner VRT} - (2.303 \times m \times T_D) \tag{3.11}$$

where m is the gradient of the Horner plot.

> *Note:* The original technique was developed by Roux et al. (1982) assuming English units and a $\log_{10}$ scale on the Horner time axis. The 3600 factor in Equation (3.9) and the 2.303 factor in Equation (3.11) have been added in this text for use with SI units and a natural log scale.

Table 3.2. Coefficients for the Determination of T_D

$(t_c + \Delta t)/\Delta t$	a	b	c	d	e	f
1.25–2	0.4874	0.0027	−0.2862	1.4077	−0.7836	0.7732
2–5	0.2516	−0.0072	0.3650	−0.0001	−3.4989	3.1534
5–10	2.3502	0.0024	−0.0609	4.7833	−5.9058	0.0365

Source: Roux et al. (1982).

This correction is sensitive to data quality and it is doubtful whether most real data are of sufficient accuracy to warrant the correction. If the parameters required for the Roux et al. (1982) correction are known, then the Cooper and Jones (1959) correction is recommended over the Horner plot.

Question: Temperature data were collected from logging runs bottoming in a shale unit at 2700 m, in a hole of 30 cm diameter. Measurements on the shale indicated a thermal conductivity, $\lambda = 1.8 \, \mathrm{W \, m^{-1} \, K^{-1}}$, density $\rho = 2500 \, \mathrm{kg \, m^{-3}}$ and heat capacity $c = 700 \, \mathrm{J \, kg^{-1} \, K^{-1}}$. The drilling fluid had density $\rho_f = 1740 \, \mathrm{kg \, m^{-3}}$ and heat capacity, $c_f = 3100 \, \mathrm{J \, kg^{-1} \, K^{-1}}$. Given the temperature data below, find the VRT using the Horner correction, the Cooper and Jones (1959) method, and the Roux et al. (1982) correction to the Horner plot.

Answer: The required data are summarised below:

$t_c = 18 \, \mathrm{h}$

At $\Delta t = 7 \, \mathrm{h}$, $T = 64°\mathrm{C}$, $(t_c + \Delta t)/\Delta t = 3.5714$, $\ln[1 + (t_c/\Delta t)] = 1.2730$

At $\Delta t = 10 \, \mathrm{h}$, $T = 66°\mathrm{C}$, $(t_c + \Delta t)/\Delta t = 2.8000$, $\ln[1 + (t_c/\Delta t)] = 1.0296$

At $\Delta t = 12 \, \mathrm{h}$, $T = 68°\mathrm{C}$, $(t_c + \Delta t)/\Delta t = 2.5000$, $\ln[1 + (t_c/\Delta t)] = 0.9163$

At $\Delta t = 15 \, h$, $T = 70°\mathrm{C}$, $(t_c + \Delta t)/\Delta t = 2.2000$, $\ln[1 + (t_c/\Delta t)] = 0.7885$

At $\Delta t = 18 \, \mathrm{h}$, $T = 72°\mathrm{C}$, $(t_c + \Delta t)/\Delta t = 2.0000$, $\ln[1 + (t_c/\Delta t)] = 0.6931$

First we will calculate the VRT using the Horner correction [Equation (3.7)]. The necessary data are already laid out above and it is just a matter of calculating the intercept of the line of best fit through the five T versus $\ln[1 + (t_c/\Delta t)]$ pairs. The intercept, and hence VRT, turns out to be 80.95°C, and the gradient of the line is $m = -13.777$.

Now we will calculate VRT using the Cooper and Jones (1959) method [Equation (3.6)]:

$\alpha = 2\rho c/\rho_f c_f = (2 \times 2500 \times 700)/(1740 \times 3100) = 0.6489$

$\tau = \lambda \Delta t/\rho c r_w^2 = (1.8 \times \Delta t)/(2500 \times 700 \times 0.15^2) = 4.5714 \times 10^{-5} \Delta t$

At $\Delta t = 7 \, \mathrm{h} = 25{,}200 \, \mathrm{s}$, $\tau = 1.152$

At $\Delta t = 10 \, \mathrm{h} = 36{,}000 \, \mathrm{s}$, $\tau = 1.646$

At $\Delta t = 12 \, \mathrm{h} = 43{,}200 \, \mathrm{s}$, $\tau = 1.975$

At $\Delta t = 15 \, \mathrm{h} = 54{,}000 \, \mathrm{s}$, $\tau = 2.469$

At $\Delta t = 18 \, \mathrm{h} = 64{,}800 \, \mathrm{s}$, $\tau = 2.962$

$F(0.6489, 1.152) = 0.41178$, so $[1 - F(\alpha, \tau)] = 0.58822$ when $T = 64°\mathrm{C}$

$F(0.6489, 1.646) = 0.34443$, so $[1 - F(\alpha, \tau)] = 0.65557$ when
$\quad T = 66°C$
$F(0.6489, 1.975) = 0.31072$, so $[1 - F(\alpha, \tau)] = 0.68928$ when
$\quad T = 68°C$
$F(0.6489, 2.469) = 0.27076$, so $[1 - F(\alpha, \tau)] = 0.72924$ when
$\quad T = 70°C$
$F(0.6489, 2.962) = 0.23969$, so $[1 - F(\alpha, \tau)] = 0.76031$ when
$\quad T = 72°C$

The line of best fit through the five T versus $[1 - F(\alpha, \tau)]$ pairs has intercept $T_f = 36.003°C$, and gradient $VRT-T_f = 46.743°C$, so VRT $= 82.75°C$. This is $1.8°C$ higher than the Horner estimate.

Now we will calculate the Roux et al. (1982) correction [Equation (3.11)].

We know the Horner VRT $= 80.95°C$, and the gradient $m = -13.777$.

From Equation (3.9):
$$t_D = 3600 \times \kappa t_c / r_w^2$$
$$\kappa = \lambda / \rho c = 1.8/(2500 \times 700) = 1.0286 \times 10^{-6} \text{ m}^2 \text{ s}^{-1}$$
so
$$t_D = 3600 \times 1.0286 \times 10^{-6} \times 18)/0.15^2 = 2.962$$

The five dimensionless Horner times, $[(t_c + \Delta t)/\Delta t]$, fall within the range 2–5, so the coefficients on the second row of Table 3.2 are used to calculate T_D from Equation (3.10):
$$T_D = 0.2516 + (-0.0072 t_D) + (0.3650 \times t_D^{1/2}) + (-0.0001 \times t_D^{1/3}) + (-3.4989 \times t_D^{1/4}) + (3.1534 \times t_D^{1/5}) = 0.1864$$

Then, from Equation (3.11):
$$\text{True VRT} = \text{Horner VRT} - (2.303 \times m \times T_D) = 80.95 - (2.303 \times -13.777 \times 0.1864) = 86.86°C$$

In this particular case, the Roux et al. (1982) correction yields a result several degrees higher than the Cooper and Jones (1959) correction, which is in turn a couple of degrees higher than the Horner correction. The value chosen for 'true' VRT comes down to personal preference.

There is a large body of temperature data for which Δt is unknown. A number of authors have attempted to utilise these data by producing empirical corrections as a function of depth, calibrated against corrected BHTs in the same region. For example, Willett and Chapman (1987) used ninety-five wells with multiple temperature measurements in the Uinta Basin, Utah, to derive the correction

$$\text{VRT} = \text{BHT} + T_C$$

$$T_C = az + bz^2 + cz^3 + dz^4 \tag{3.12}$$

where $a = 6.93$, $b = -1.67$, $c = 0.101$, $d = 0.0026$, T_C is in degrees Celsius and z is depth in kilometres.

Question: A particular well in the Uinta Basin has a measured BHT of 67.3°C at a depth of 1967 m. There are no time data recorded for the measurement. According to Willett and Chapman (1987), what is the best estimate for VRT implied by the temperature measurement?

Answer: According to Equation (3.12), at $z = 1.967$ km,
$$T_C = 6.93 \times (1.967) - 1.67 \times (1.967)^2 + 0.101 \times (1.967)^3 + 0.0026 \times (1.967)^4 = 7.98°C$$
So:
$$VRT = 67.3 + 7.98 = 75.3°C.$$

Such schemes have high uncertainties, and should only be utilised in the regions for which they are calibrated.

3.2. Indirect Temperature Indicators

Several indirect means of estimating temperature at depth have developed as supplements to direct temperature measurements. Such indirect temperature indicators include dissolved solids in groundwater, the Curie depth for magnetic minerals, temperature pressure equilibrium conditions for xenoliths, and the electrical resistivity of the upper mantle.

3.2.1. Groundwater Geochemistry

The water solubility of many compounds increases with temperature. It might be supposed, then, that the ambient temperature of a formation could be estimated from the amount of dissolved material in the pore-water. Such relationships have been sought for a number of common components of pore-water, including silica (SiO_2), for which Swanberg and Morgan (1979) derived the following empirical expression:

$$T = [1315/(5.205 - \log_{10} \mathbf{SiO_2})] \pm 0.5°C \tag{3.13}$$

where T is the calculated water temperature (K) and $\mathbf{SiO_2}$ is the amount of dissolved silica (ppm by weight) in the formation water. The quoted precision is valid in the range 125–250°C.

Question: A water sample from 4644 m depth in the well Buffon 1 in the Browse Basin, Western Australia, was analysed and found to contain 102 ppm SiO_2. What temperature does this imply?

Answer: From Equation (3.13):

$$T = [1315/(5.205 - \log_{10}(102))] = 411.4\,\text{K} = 138.3 \pm 0.5°\text{C}$$

By comparison, a drill stem test at 4246 m yielded a temperature of 129.4°C and a Horner plot for data from 4495 m implied a VRT of 136.7°C.

A number of other groundwater geothermometers have also been proposed (Carmichael, 1989). Alternative thermometers for silica (Figure 3.8) include:

$$T = [1533.5/(5.768 - \log_{10} \mathbf{SiO_2})] \pm 2.0 \qquad (3.14)$$

$$T = [1015.1/(4.655 - \log_{10} \mathbf{SiO_2})] \pm 2.0 \qquad (3.15)$$

Question: What temperatures do the latter two silica geotherm-ometers suggest for the sample in the previous example?

Answer: Using Equation (3.14):

$$T = [1533.5/(5.768 - \log_{10}(102))] = 407.9\ \text{K} = 134.8 \pm 2.0°\text{C}$$

Using Equation (3.15):

$$T = [1015.1/(4.655 - \log_{10}(102))] = 383.6\ \text{K} = 110.4 \pm 2.0°\text{C}$$

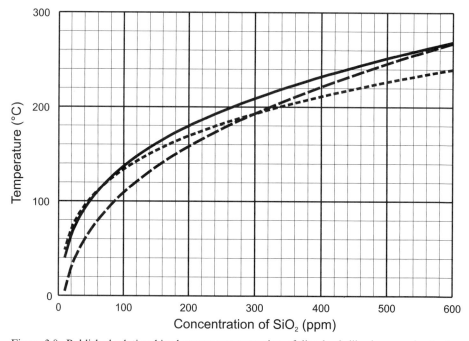

Figure 3.8. Published relationships between concentration of dissolved silica in groundwater (ppm) and temperature of the water (°C). Solid line = Equation (3.13), short-dashed line = Equation (3.14), long-dashed line = Equation (3.15).

Note: Equation (3.13) yields results similar to Equation (3.14) at low temperature and similar to Equation (3.15) at high temperature.

For sodium (Na), potassium (K) and calcium (Ca),

$$T = [855.6/(0.8573 + \log_{10}(\mathbf{Na/K}))] \pm 2.0 \tag{3.16}$$

where **Na** and **K** are concentrations in parts per million by weight, and temperatures are in kelvin.

Fournier and Truesdell (1973) suggested the following:

$$T = [846/(0.5964 + \log_{10}(M_{Na}/M_K))] \tag{3.17}$$

$$T = [1647.3/(2.24 + \log_{10}(M_{Na}/M_K) + \beta \log_{10}(\sqrt{M}_{Ca}/M_{Na}))] \tag{3.18}$$

where

M_X = molar concentration of element X (i.e. moles per litre)
$\beta = 4/3$ for $(\sqrt{M}_{Ca}/M_{Na}) > 1$ and $T < 373.15$ K (100°C)
$\beta = 1/3$ for $(\sqrt{M}_{Ca}/M_{Na}) < 1$ or $T > 373.15$ K (100°C)
T is in kelvin

Question: A water sample from a spring in Yellowstone National Park yielded the following concentrations of ions:

$M_{Na} = 0.0183$ mol L^{-1}
$M_K = 0.00187$ mol L^{-1}
$M_{Ca} = 0.0000774$ mol L^{-1}
$SiO_2 = 692$ ppm

What do each of the above equations suggest is the equilibrium temperature of the water sample?

Answer: For Equations (3.13), (3.14) and (3.15):

$T = [1315/(5.205 - \log_{10}(692))] = 556.05$ K $= 283 \pm 2.0$°C
$T = [1533.5/(5.768 - \log_{10}(692))] = 523.76$ K $= 250.6 \pm 0.5$°C
$T = [1015.1/(4.655 - \log_{10}(692))] = 559.32$ K $= 286.17$°C

For Equation (3.16) we must first convert molar concentration into ppm concentration for Na and K:

0.0183 mol L^{-1} of sodium $= 0.0183 \times 22.98977 \times 1000$ mg L^{-1}
$= 420.7$ ppm

0.00187 mol L^{-1} of potassium $= 0.00187 \times 39.0983 \times 1000$ mg L^{-1} $= 73.11$ ppm

so

Na/K $= 5.754$

Substituting this value into Equation (3.16):

$T = [855.6/0.8573 + \log_{10}(5.754))] = 529.04$ K $= 256 \pm 2.0$°C

For Equation (3.17):

$M_{Na}/M_K = 0.0183/0.00187 = 9.786$

so

$T = [846/(0.5964 + \log_{10}(9.786))] = 533.08K = 259.93°C$

For Equation (3.18), we must first determine the value of β:

$(\sqrt{M_{Ca}})/M_{Na}) = \sqrt{(0.0000774)}/0.0183 = 0.48075$, so $\beta = 1/3$

Also:

$(M_{Na}/M_K) = 0.0183/0.00187 = 9.786$

Substituting these values into Equation (3.18):

$T = [1647.3/(2.24 + \log_{10}(9.786) + (1/3) \times \log_{10}(0.48075))] =$
527.2 K $= 254.1°C$

There is reasonable agreement and the average result is $265 \pm 15°$
C to one standard deviation.

Groundwater geothermometers are most useful in high-temperature environments where direct temperature measurement is difficult. They are primarily developed for use in geothermal reservoir studies, where the energy content of a high-enthalpy reservoir may be estimated by extracting a fluid sample to the surface and analysing its geochemistry.

Note: Corrections may be required if any portion of the original fluid is lost through steam before the sample is analysed (Fournier and Rowe, 1966). This will have the effect of increasing the concentration of solids in the remaining portion of fluid, thus leading to an overestimate of temperature.

In a petroleum setting, geochemical thermometers are only applicable in situations where the pore water is in chemical equilibrium with the surrounding rocks at *in situ* temperature conditions.

3.2.2. Curie Depth

The *Curie temperature* is the point at which a mineral loses its ferromagnetic properties. Pure magnetite has a Curie temperature of 580°C, but titaniferous inclusions may reduce this to as low as 300°C. Ferromagnetic minerals within andesites and alkali-basalts generally have Curie temperatures in the range 100–300°C, and intermediate to mafic compositions are in the range 300–450°C. Curie temperatures in the range 620–1100°C have been recorded for Fe–Co–Ni alloys (Gasparini et al., 1979).

The *Curie depth* is the depth at which crustal rocks reach the Curie temperature. Under certain conditions, the Curie depth can be used as a temperature data point to constrain deep thermal gradients.

Curie depth is generally determined via Fourier analysis of aeromagnetic data. Spector and Grant (1970) published a rigorous statistical analysis of the power spectrum (the square of the Fourier amplitude spectrum) of the total magnetic intensity field produced by a theoretical suite of rectangular blocks buried within the Earth. They concluded the theoretical possibility of determining Curie depth from total magnetic field intensity data, provided the data covered an area of at least 200×200 km, with a maximum grid spacing of 1 km. They were unable to implement the theory at the time due to a lack of computing power, but computers have increased in power and many authors have since published Curie depth estimates based on Spector and Grant's method (e.g. Gasparini et al., 1979; Agrawal, Thakur and Negi, 1992).

Curie depth does not necessarily represent an isotherm. As explained above, different rock types have different Curie temperatures, so the Curie depth may also correspond to a composition boundary. Correlation with deep seismic or gravity models may be necessary to determine which of the two cases applies in a particular situation. Where Curie depth correlates with an inferred velocity or density boundary, it is likely to reflect the change in composition, especially if the data suggest intermediate to mafic rocks underlying acidic rocks. However, where Curie depth does not coincide with a velocity or density boundary, it may be interpreted as the Curie temperature isotherm (580°C in most continental regions). The Curie depth–temperature pair then represents a deep datum for thermal gradient modelling.

3.2.3. Xenoliths

A xenolith is a piece of country rock picked up by magma as it rises through the crust. The pressure–temperature stability conditions indicated by a xenolith's mineral assemblage provide information on the depth and temperature of its origin point. Thus, xenoliths provide an independent estimate of temperature at depths down to hundreds of kilometres (e.g. O'Reilly and Griffin, 1985).

Amundsen, Griffin and O'Reilly (1987) considered the sources of error in such estimates. Many of the critical geobarometer and geothermometer calibrations remain contentious and give rise to significant errors in both depth and temperature estimates. There are further complications related to the nature of the samples. In particular, it is often unclear whether the calculated pressure and temperature relate to the source magma or the surrounding conduit. If it can be established that the host magma rose and was ejected rapidly, xenoliths should be free from the effects of chemical diffusion and represent true *in situ* conditions.

It is also important to note that data derived from a xenolith's mineral assemblage relates to the thermal conditions existing at the time the xenolith was emplaced in the host magma. The thermal conditions at such times are, by implication, transient and do not necessarily represent long-term, steady-state conditions. Nevertheless, deep geotherms have been constructed and interpreted using this method for several places around the world, including Siberia (Boyd, 1984), southern Africa (Pearson, O'Reilly and Griffin, 1995; Figure 3.9) and Finland (Kukkonen and Peltonen, 1999).

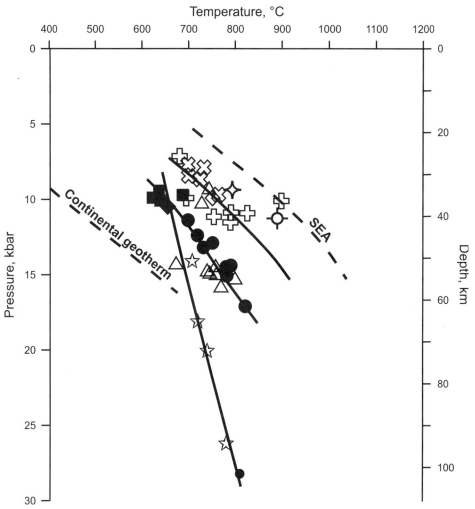

Figure 3.9. Pressure/depth–temperature estimates for South African xenoliths. The xenoliths are divided into those found on a craton (solid symbols) and those found off a craton (open symbols). Symbols represent individual kimberlite pipes. At least three trends are apparent, implying three different sources. Also shown are the extrapolated continental geotherm (assuming $Q = 40$ mW m^{-2} and $\lambda = 3.5$ W m^{-1} K^{-1}) and the geotherm deduced for southeast Australia (SEA) from xenoliths. Modified after Pearson et al. (1995).

3.2.4. Upper Mantle Resistivity

The electrical resistivity of mantle rocks is strongly dependent on temperature (e.g. Majorowicz, Gough and Lewis, 1993). Resistivity thus offers another independent means of estimating temperature in the lower crust and upper mantle. Magnetotelluric methods are required to make the measurements. Table 3.3 lists the electrical resistivity of a 13% Fayalite olivine at different temperatures.

Majorowicz et al. (1993) identified a direct relationship between the 450°C isotherm and the depth to a prominent lower crustal, electrically conductive

Table 3.3. Electrical Resistivity, *R*, of 13% Fayalite Olivine at Different Temperatures, *T*

$T(°C)$	$R(\Omega m)$
660	3000
820	800
950	300
1100	100
1200	50
1400	20
1600	12
1750	10
1850	6
1900	5

Source: Hamilton (1965).

layer in the Canadian Cordillera. In that region, magnetotelluric depth soundings may be used directly to estimate the depth to the 450°C isotherm. Similar relationships might be discovered for other regions of the world, thus enabling independent temperature–depth estimates. In most such cases, however, the depth resolution is poor.

3.3. Surface Temperature

Measurements of temperature at depth are of limited value if the thermal profile cannot be extrapolated to the surface. The thermal gradient in the top portion of the crust remains unconstrained without an estimate of the average surface temperature. Offshore, surface temperature is buffered to a large extent by the thermal bulk of the oceans, and annual variation, due to currents, is typically no more than a few degrees Celsius. Onshore, surface temperatures vary markedly with latitude, altitude, topography and time, further complicated by the cyclic heating patterns of the days and seasons.

3.3.1. Offshore

When we speak of 'surface temperature' for offshore thermal profiles, we refer to the top of the sediment column, or the bottom of the water. This is quite distinct from the average sea surface temperature. We must use the bottom-water temperature as the top boundary of our conductive heat flow models because above that level the assumption of conductive heat flow is no longer valid.

The temperature of the sediment–water interface is only rarely noted during the drilling of an offshore well; this leaves the thermal gradient from the surface down to the first temperature datum unknown. Furthermore, if there is only one down-hole temperature datum, then the thermal gradient cannot be estimated at all. It is therefore of considerable importance that we estimate, with a reasonable degree of accuracy, bottom-water temperature (BWT) at the offshore drilling site.

Although simple in concept, estimating BWT in a specific locality can be difficult. Oceanographic atlases (e.g. Wyrtki, 1971) provide some water temperature data, but these are often for water depths greater than 3 km, or for points within the water column – both useless for petroleum exploration. Oil exploration companies sometimes commission BWT surveys in regions of interest, but these data are rarely published. Usually, the only option is to guesstimate a value for BWT, based on the reasonable assumption that BWT lies somewhere between freezing and the average sea surface temperature.

EXAMPLE

A typical example of usual practice was published by Gallagher (1990), who assumed a BWT of 5°C for the well Volador 1, in Bass Strait, Australia, but gave no indication as to how he arrived at such an estimate.

The mean temperatures of the Earth's three major oceans are all within 1°C of each other for any specific depth along latitudinal lines between 40° S and 40° N (Defant, 1961), and only rare evidence has been found of major fluctuations in sea-floor temperature over short time frames (e.g. Fisher et al., 1999). This implies a first-order relationship between BWT, latitude and water depth.

To investigate the relationship, we collated all BWT data published between 1984 and 1989 in the *Journal of Deep-Sea Research*. The data revealed that for any latitude, $L(°)$, sea-floor temperature, T_{sf}, is related to water depth, z(m), by an equation of the following form (Figure 3.10):

$$\ln(T_{sf}) = A + B \times \ln(z) \tag{3.19}$$

where

$$A = 4.63 + 8.84 \times 10^{-4}L - 7.24 \times 10^{-4}L^2$$

$$B = -0.32 + 1.04 \times 10^{-4}L + 7.08 \times 10^{-5}L^2$$

The relationship works best when T_{sf} is expressed as degrees above the freezing point, T_f:

$$T_{sf} = \text{BWT} - T_f \tag{3.20}$$

where (Foldvik and Kvinge, 1974; Toole, 1981)

$$T_f \approx -1.90 - 7.64 \times 10^{-4}z \,°\text{C} \tag{3.21}$$

Equations (3.19)–(3.21) define a set of curves (Figure 3.11) relating BWT to latitude and water depth. Error margins of ±2°C cover most of the spread (due to seasonal variations and local effects of currents) in the original temperature

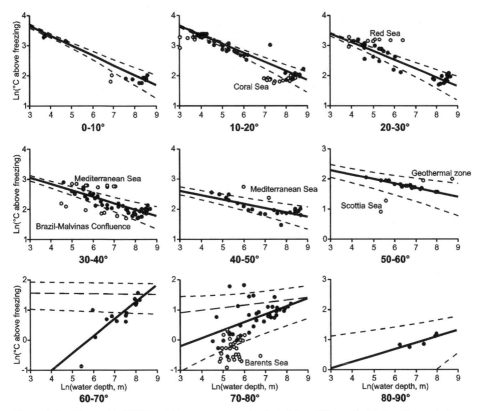

Figure 3.10. Plots of ln(BWT $- T_f$) versus water depth (m) for different latitudes. Open circles are from enclosed seas, shallow or other anomalous zones and are not included in the model calculations. Solid lines are lines of best fit through the data. Dashed lines are model values and ±2°C error margins.

data (Figure 3.10). The model is supported by a published BWT measurement of 26°C at a depth of 64 m in the Arafura Sea at latitude 10° S (Moore, Bradshaw and Edwards, 1996). The model predicts 25 ± 2°C.

In very shallow water, Figure 3.11 predicts high BWTs – in excess of expected surface temperature at all latitudes. The surface layer of ocean water, generally 20–100 m thick, has close to homogeneous physical properties such as temperature and salinity (e.g. Wyrtki, 1971; Hamilton, 1986; Weller and Price, 1988), so in our model the homogeneous zone is assumed to extend to the depth at which BWT equals mean surface temperature. That is, if Figure 3.11 predicts BWT in excess of surface temperature, BWT is assumed to equal surface temperature.

Note: Water in enclosed seas, such as the Mediterranean, Red and Barents seas, is more homogeneous than open ocean water. In enclosed seas, BWT can be assumed to equal surface temperature, regardless of water depth.

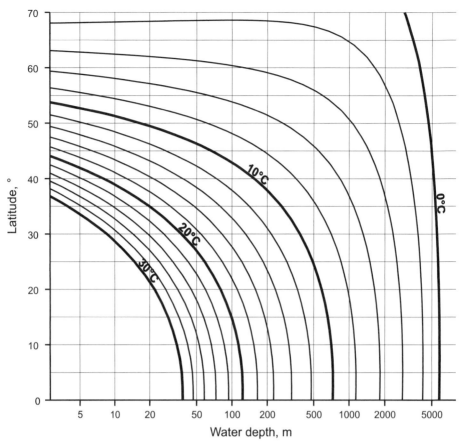

Figure 3.11. BWT (°C) as a function of water depth (m) and latitude (°), as predicted by Equations (3.19)–(3.21). Contour interval = 2°C.

Sea surface temperature has been routinely studied since early in the twentieth century. The development of satellite imagery and digital computers has allowed sophisticated models of global water surface temperature to be built (Boville and Gent, 1998; Figure 3.12). Note that the temperature of the Pacific Ocean is lower by several degrees than that of the Atlantic at high northern latitudes and that the global average is not a simple function of latitude. For these reasons, the global map (Figure 3.12) should be used to estimate average sea surface temperature at specific locations, and not a global average function.

Question: Volador 1 was drilled in 260 m of water in Bass Strait, Australia. Given that the latitude of the well is 38° S, what is the approximate bottom-water temperature? How does this compare with Gallagher's (1990) estimate of 5°C?

Answer: From Figure 3.11, at a water depth of 260 m and a latitude of 38°C, BWT = 9 ± 2°C. From Figure 3.12, surface temperature

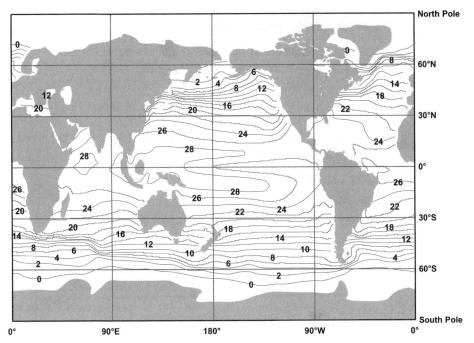

Figure 3.12. Average sea surface temperature model (C). Modified after Boville and Gent (1998).

in Bass Strait is approximately 15°C. Predicted BWT does not exceed observed surface temperature, so stands at $9 \pm 2°C$.

This value for BWT is significantly higher than Gallagher's estimate, and decreases his thermal gradient in the top kilometre of Volador 1 by approximately 10%, bringing it in line with gradient estimates from greater depth.

This method for estimating BWT removes some of the guesswork and arbitrariness from the procedure, although a margin of uncertainty remains.

3.3.2. Onshore

When we speak of surface temperature onshore, we refer to the temperature in the top layer of soil or rock – not the air. The temperature of the rock generally exceeds average air temperature by a few degrees on land due to surface albedo. For example, Howard and Sass (1964) found that for eleven extrapolations of near-surface onshore thermal gradient, average surface temperature exceeded average air temperature by a mean amount of 3.0°C.

Global averages are not applicable for onshore regions because of the great variation observed in climate along latitudinal lines. Consider, for example, the average minimum and maximum temperatures for Beijing, China (8–16°C), Boulder, Colorado (3–18°C) and Athens, Greece (14–23°C) which are all at about the same latitude.

(Data from http://www.onlineweather.com)

Local meteorological records are the best source for local climate data. The average surface temperature, T_0, can be estimated from

$$T_0 = 3 + (T_{av.min} + T_{av.max})/2 \qquad (3.22)$$

where

$T_{av.min}$ = average annual minimum temperature
$T_{av.max}$ = average annual maximum temperature

Question: A well is drilled in the onshore Canning Basin, Western Australia at 20° S, 125° E. What is a good estimate for average surface temperature?

Answer: The Australian Bureau of Meteorology maintains a web site at

http://www.bom.gov.au

that provides excellent records of climate data across the whole of Australia. With a minimum of effort, we can find the average annual minimum and maximum temperatures for the region around our drill site:

$T_{av.min} = 20°C$
$T_{av.max} = 34°C$
From Equation (3.22):
$T_0 = 3 + (20 + 34)/2 = 30°C$

3.3.3. Daily and Seasonal Cycles

Note: The following two sections contain large mathematical terms with Greek characters, exponential and trigonometric functions. Those afraid of maths may wish to skip straight to Section 3.4.

The daily passage of the sun across the sky has an obvious effect on the air temperature at the Earth's surface. The temperature generally rises during the day and falls at night in a more or less regular cycle. Likewise, the change in temperature between summer and winter is regular and periodic. Not surprisingly, the temperature in the top layer of the Earth is affected by these diurnal and seasonal perturbations. Carslaw and Jaeger (1959, p. 64) considered the effect of periodic surface heating on a half-space and derived the following solution:

$$T_\theta = T_0 \times \exp(-\varepsilon z) \sin(\omega t - \varepsilon z) \tag{3.23}$$

This equation describes the departure, T_θ, from a mean value of temperature at a particular depth, z, and time, t, resulting from a surface heating cycle with amplitude T_0 (half of the peak–trough amplitude) and frequency ω ($\omega = 2\pi/P$; P = period). The thermal properties of the medium are included in the ε term, where $\varepsilon = (\pi/P\kappa)^{1/2}$ (κ = thermal diffusivity of the medium).

Equation (3.23) looks complicated, but can be looked at as the product of two relatively simple terms and a constant. The constant, T_0, is the amplitude of the surface temperature cycle. The exponential term, $\exp(-\varepsilon z)$, describes a simple decay in the amplitude of the temperature perturbation with depth. The sine term, $\sin(\omega t - \varepsilon z)$, describes the time lag between the temperature perturbation at the surface and at depth. In simple terms, the equation states that the natural temperature fluctuations between day and night, summer and winter, propagate into the crust, but the effect decays exponentially with depth. Also, there is a time delay between the temperature fluctuation at the surface and at depth because heat takes time to propagate into the Earth.

We can easily find the depth at which the temperature fluctuation is in phase with the surface cycle. The sine term tells us that this is when $\varepsilon z = 2\pi$. So

$$z = 2\pi/\varepsilon = 2\pi/(\pi/P\kappa)^{1/2} = (4\pi P\kappa)^{1/2} \tag{3.24}$$

This defines an *effective wavelength*, z_{wl}, for a temperature cycle near the surface of the Earth. The magnitude of the temperature perturbation (relative to the surface perturbation) at a depth of one effective wavelength is given by the exponential term in Equation (3.23):

$$exp(-\varepsilon z_{wl}) = \exp(-2\pi) = 0.0019 \tag{3.25}$$

At deeper than one effective wavelength, the temperature perturbation caused by a cyclic surface temperature can be ignored.

Question: Basalt plains cover much of central Victoria, Australia. The basalt has thermal diffusivity $\kappa = 7.9 \times 10^{-7}\,\mathrm{m^2\,s^{-1}}$. What is the effective wavelength of the temperature perturbations caused by the daily and seasonal cycles?

Answer: For the diurnal cycle, the period of the temperature perturbation is 1 day, or $P = 8.64 \times 10^4$ s. A complete-cycle phase shift is attained at the depth given by Equation (3.24):
$$z = (4\pi P\kappa)^{1/2} = (4\pi \times 8.64 \times 10^4 \times 7.9 \times 10^{-7})^{1/2} = 0.93 \text{ m}$$
The daily temperature fluctuation only affects the top metre of basalt. The seasonal cycle is 1 year, or $P = 3.1557 \times 10^7$ s, and:
$$z = (4\pi \times 3.1557 \times 10^7 \times 7.9 \times 10^{-7})^{1/2} = 17.7 \text{ m}$$
The annual fluctuation affects temperature down to almost 20 m.

Obviously, where temperature is disturbed, temperature gradient is also likely to be affected. Thermal gradient is of great importance in determining heat flow, so we should also investigate to what depth the disturbance to gradient is significant. The apparent thermal gradient, $(\partial T/\partial z)_a$, is

$$(\partial T/\partial z)_a = (\partial T/\partial z) + (\partial T_\theta/\partial z) \tag{3.26}$$

The magnitude of the perturbation can be calculated by differentiating Equation (3.23) with respect to depth:

$$(\partial T_\theta/\partial z) = T_0 \times (-\varepsilon)\exp(-\varepsilon z) \times [\sin(\omega t - \varepsilon z) + \cos(\omega t - \varepsilon z)] \tag{3.27}$$

As for Equation (3.23), we have a constant, an exponential term and a time-dependent phase term. The disturbed gradient is equal to the undisturbed gradient at depth z when $(\partial T_\theta/\partial z) = 0$:

$$\sin(\omega t - \varepsilon z) = -\cos(\omega t - \varepsilon z) = -\sin(\pi/2 - \omega t + \varepsilon z) \tag{3.28}$$

This relationship is only true for z at certain times, $t = t_{min}$:

$$t_{min} = (2\pi n + 3\pi/4 + \varepsilon z)/\omega \text{ and } (2\pi n + 7\pi/4 + \varepsilon z)/\omega \tag{3.29}$$

where n is an integer. For each value of n there are two values of t_{min}.

Question: At what times is the thermal gradient equal to the undisturbed gradient when measured at 15 m depth in the area of central Victoria described in the previous example?

Answer: This is too deep for a significant diurnal disturbance, so we need only consider the annual cycle:

$$\omega = 2\pi/P = 2\pi/(3.1557 \times 10^7) = 1.991 \times 10^{-7}\,\text{s}^{-1}$$
$$\varepsilon = (\pi/P\kappa)^{1/2} = (\pi/(3.1557 \times 10^7 \times 7.9 \times 10^{-7}))^{1/2} = 0.355\,\text{m}^{-1}$$

So:

$$\varepsilon z = 0.355 \times 15 = 5.325$$

From Equation (3.29):

$$t_{min} = (2\pi n + 3\pi/4 + \varepsilon z)/\omega \text{ and } (2\pi n + 7\pi/4 + \varepsilon z)/\omega$$
$$= 3.1557 \times 10^7 \times n + 3.858 \times 10^7 \text{ and } 3.1557 \times 10^7 \times n$$
$$+ 5.436 \times 10^7$$

where both times are in seconds. This means that the thermal gradient at 15 m depth is equal to the undisturbed gradient after approximately 81 days and 264 days, and at subsequent 1-year intervals.

Maximum disturbance to thermal gradient can also be calculated from Equation (3.27). Maximum disturbance is attained when $\sin(\omega t - \varepsilon z) = \cos(\omega t - \varepsilon z)$. This is true at times, $t = t_{max}$:

$$t_{max} = (\pi n + \pi/4 + \varepsilon z)/\omega \tag{3.30}$$

where n is an integer. The amplitudes of the maxima are found by substituting $t = t_{max}$ back into Equation (3.27):

$$|(\partial T_\theta/\partial z)|_{max} = T_0 \times (-\varepsilon) \times \exp(-\varepsilon z) \times [\sin(\pi n + \pi/4) + \cos(\pi n + \pi/4)]$$
$$= T_0(\varepsilon)\exp(-\varepsilon z) \times [\sqrt{2}]$$

$$(3.31)$$

The negative signs disappear from the equation because we are dealing with magnitudes only. Rearranging Equation (3.31), we can find the threshold depth, z_{min}, at which the maximum departure from mean gradient is no longer significant. That is, the depth at which

$$|(\partial T_\theta/\partial z)|_{max} = 0.01 \times |(\partial T/\partial z)| = T_0(\varepsilon)\exp(-\varepsilon z_{min}) \times [\sqrt{2}]$$
$$z_{min} = -\frac{1}{\varepsilon}\ln\left|\frac{0.01}{T_0\varepsilon\sqrt{2}}\frac{\partial T}{\partial z}\right|$$

$$(3.32)$$

Question: The average surface temperature in winter in central Victoria is 8.7°C. The average summer temperature is 22.1°C. The average thermal gradient in the region is 25°C km^{-1}. How deep must our temperature observations be made if gradients are required with an accuracy better than 1%, independent of the time of measurement?

Answer: We must solve Equation (3.32) to find z_{min}. To do so we have to calculate the relevant parameters:

$T_0 = $ half the peak to trough amplitude of the temperature perturbation $= (22.1 - 8.7)/2 = 6.7$ K

$|(\partial T/\partial z)| = $ average thermal gradient $= 0.025$ K m^{-1}

$\varepsilon = 0.355$ m^{-1} from the previous example

We can now calculate z_{min}:

$z_{min} = -(1/0.355) \times \ln[(0.01/(6.7 \times 0.355 \times 1.4142) \times 0.025] = 26.8$ m

Thermal gradient observations should be made deeper than 26.8 m. Observations made at depths less than 18 m may yield gradient estimates with errors greater than 25%.

Although the theory of periodicity is well understood, in practice the thermal diffusivity of near-surface layers is usually poorly known. Precise values of t_{min} and t_{max} cannot be calculated. Also, t_{min} and t_{max} are only relevant for single observation points yet gradients are measured in terms of $\partial T/\partial z$, requiring temperatures from at least two depths. For these reasons, plus the difficulty in defining the time origin, $t = 0$, it is best to make measurements at depths where the maximum possible annual perturbation is negligible.

Note: It is perilous to rely on temperature or gradient data recovered from very shallow holes. These data are likely affected by diurnal and seasonal temperature cycles, and are unlikely to reflect accurately the deeper gradient. As a rule of thumb, only temperatures from deeper than about 30 m can be considered free from seasonal fluctuations.

3.3.4. Climate Changes

Any permanent change in surface temperature will have an effect on the near-surface thermal gradient. Significant changes in surface temperature may result from variations in insolation caused by recent cultural changes to the land surface (e.g. Hyndman and Everett, 1968; Figure 3.13), long-term climate changes (e.g. Gosnold, Todhunter and Schmidt, 1997), or the advance and retreat of sheet ice associated with periods of glaciation (e.g. Beck, 1977).

One-off surface temperature changes propagate into the ground in a way similar to diurnal and seasonal cycles. The depth to which the effect is notice-

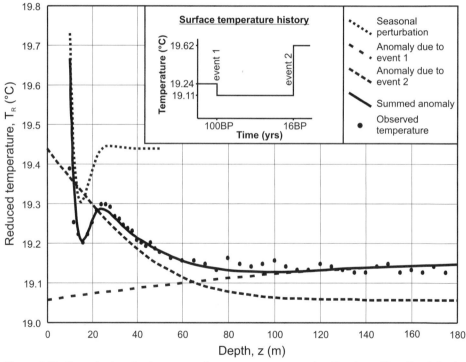

Figure 3.13. Perturbations in the near-surface temperature record at Berrigan, New South Wales. Reduced temperature (observed temperature −0.019 × depth) shows the summed effects of seasonal periodicity, a sharp rise in surface temperature of 0.51°C at 16 BP (effect of short-term quarrying), and an abrupt drop in surface temperature of 0.13°C at 100 BP (effect of land clearing). Data from Cull (1980).

able depends on the magnitude of the temperature step, the time since the event and the thermal diffusivity of the ground. Insolation or climatic changes can be modelled as discrete events, each with an associated step function in surface temperature. Carslaw and Jaeger (1959, p. 58) gave the following solution:

$$T_\theta = T_0 \times \mathrm{erfc}[z/(2\sqrt{\kappa t}))] \tag{3.33}$$

where T_θ is the departure from original equilibrium temperature at depth z and time t after an instantaneous change in surface temperature of T_0; κ is thermal diffusivity of the medium; and erfc(x) is the complimentary error function, defined as

$$\mathrm{erfc}(x) = 1 - \mathrm{erf}(x) = 1 - \frac{2}{\sqrt{\pi}} \int_0^x \exp(-\xi^2)d\xi \tag{3.34}$$

The effect of more than one temperature step is found by simple addition:

$$T_\theta = \Sigma T_{\theta i} \tag{3.35}$$

where $T_{\theta i}$ is the temperature deviation due to the ith event.

Question: Some 15,000 years ago, a sequence of basalt layers accommodated a stable thermal gradient of 30°C km^{-1}. At that time, an ice age began, dropping average surface temperature from 5°C to 0°C. The ice age ended 10,000 years ago and average surface temperature rose to 10°C. What is the present-day temperature at 500 m depth, given the thermal diffusivity of basalt, $\kappa = 7.9 \times 10^{-7}$ m^2 s^{-1}?

Answer: We want to find the present temperature at $z = 500$ m. Original surface temperature was 5°C and original thermal gradient was 30°C km^{-1} = 0.03 K m^{-1}.

Original temperature at 500 m was
$T = 5 + (500 \times 0.03) = 20$°C

The temperature deviation due to the first event, a step of −5°C in average surface temperature 15,000 years ago, is found from Equation (3.33):
$T_0 = -5$°C
$t = 15{,}000 \times 3.1557 \times 10^7 = 4.7336 \times 10^{11}$ s
$T_{\theta 1} = -5 \times \mathrm{erfc}[500/(2\sqrt{(7.9 \times 10^{-7} \times 4.7336 \times 10^{11}))}] =$
$-5 \times \mathrm{erfc}[0.40882] = -5 \times 0.56316 = -2.8158$ K

The parameters for the second event, a step of 10°C in average surface temperature 10,000 years ago, are
$T_0 = 10$°C
$t = 10{,}000 \times 3.1557 \times 10^7 = 3.1557 \times 10^{11}$ s
$T_{\theta 2} = 10 \times \mathrm{erfc}[500/(2\sqrt{(7.9 \times 10^{-7} \times 3.1557 \times 10^{11}))}] =$
$10 \times \mathrm{erfc}[0.50070] = 10 \times 0.47889 = 4.7889$ K

So present-day temperature at $500\,\mathrm{m} = 20 - 2.8158 + 4.7889 = 21.97°C$

Whenever temperature varies, thermal gradient is affected. If we call the thermal gradient β, then the change in gradient, $\Delta\beta$, due to a change in surface temperature, T_0, is described by

$$\Delta\beta = -T_0 \times [(\pi\kappa t)^{-1/2} \times \exp(-z^2/(4\kappa t))] \tag{3.36}$$

and, again, the effect of more than one event can be found by simple addition:

$$\Delta\beta = \Sigma\Delta\beta_i \tag{3.37}$$

where $\Delta\beta_i$ is the deviation in gradient due to the ith event.

Question: Using the previous example, what is the present day thermal gradient at 500 m depth?

Answer: The relevant parameters are

    Initial gradient $= \beta_0 = 0.030\,\mathrm{K\,m^{-1}}$
    $\kappa = 7.9 \times 10^{-7}\,\mathrm{m^2\,s^{-1}}$
    $z = 500\,\mathrm{m}$

For the first event:

    $T_0 = -5\,\mathrm{K}$
    $t = 15{,}000 \times 3.1557 \times 10^7 = 4.7336 \times 10^{11}\,\mathrm{s}$

From Equation (3.36):

$$\Delta\beta_1 = -(-5) \times [(3.14159 \times 7.9 \times 10^{-7} \times 4.7336 \times 10^{11})^{-1/2}$$
$$\times \exp(-500^2/(4 \times 7.9 \times 10^{-7} \times 4.7336 \times 10^{11}))] = 0.00390\,\mathrm{K\,m^{-1}}$$

And for the second event:

    $T_0 = 10\,\mathrm{K}$
    $t = 10{,}000 \times 3.1557 \times 10^7 = 3.1557 \times 10^{11}\,\mathrm{s}$

$$\Delta\beta_2 = -(10) \times [(3.14159 \times 7.9 \times 10^{-7} \times 3.1557 \times 10^{11})^{-1/2}$$
$$\times \exp(-500^2/(4 \times 7.9 \times 10^{-7} \times 3.1557 \times 10^{11}))] = -0.00879\,\mathrm{K\,m^{-1}}$$

Present gradient at 500 m:

    $\beta = \beta_0 + \beta_1 + \beta_2 = 0.030 + 0.00390 - 0.00879 = 0.0251 = 25.1°C\,\mathrm{km^{-1}}$

This represents greater than 15% departure from the original steady-state gradient.

There is a growing body of literature concerned with using near-surface temperature profiles as indicators and records of global warming. Near-surface ground temperature profiles have an advantage over long-term air temperature records in that short-period fluctuations are effectively filtered from the record and only true long-term effects remain. Temperature profiles in the top 200 m can be inverted to reconstruct average surface temperature trends for the past 100–150 years (e.g. Gosnold et al., 1997). In general, the deeper section of a profile defines a constant thermal gradient that, by extrapolation, suggests a surface temperature lower than the present mean. The magnitude and timing of the implied climate change can be derived from a model of a step- or ramp-increase in surface temperature.

EXAMPLE
Cermák et al. (1996) reported on the results of more than 300 medium-depth (~500 m) temperature profiles from around Europe. The data point to a rise in surface temperature of 1–2°C over the past 100 years, as well as a few other climatic events. The latter include a 'little ice age' in the seventeenth and early eighteenth centuries, a 'little climatic optimum' in the thirteenth-century, a possible warm period around 400 AD and a possible cold interval between 700 and 1000 AD.

Milankovic (1969) suggested that ice ages are periodic and relate to perturbations in the Earth's rotation and orbit. This was supported by Hays, Imbrie and Shackleton (1976), who studied the frequency spectrum of the variation in oceanic ^{18}O isotope through time. They identified a number of periodic oscillations in global temperature on long time scales. Small-amplitude temperature fluctuations with periods of 19,000, 23,000 and 40,000 years exist on top of a dominant ice-age cycle with a period of 100,000 years.

If glaciations are periodic they can be treated in the same way as diurnal or seasonal heating cycles. Equation (3.32) tells us the depth in the crust (diffusivity, $\kappa = 10^{-6}$ m^2 s^{-1}) to which a typical thermal gradient ($\partial T/\partial z = 25°$C km^{-1}) could be affected by a glaciation cycle with a 10^5-year periodicity ($P = 3.1557 \times 10^{12}$ s) and 10°C peak-to-trough amplitude ($T_0 = 5°$C):

$$\varepsilon = (\pi/P\kappa)^{1/2} = 9.9776 \times 10^{-4} \, \text{m}^{-1}$$

$$z_{\min} = -\frac{1}{\varepsilon}\ln\left|\frac{0.01}{T_0\varepsilon\sqrt{2}} \frac{\partial T}{\partial \mathbf{z}}\right| = 3350 \, \text{m} \tag{3.38}$$

Geothermal gradients measured at depths down to several kilometres could theoretically contain significant errors in regions that have experienced recent glaciation (Crain, 1968). The magnitude of the error might be expected to vary geographically, decreasing towards more temperate zones (Cull, 1979). However, no systematic trends have been detected in heat flow data obtained from depths down to 3000 m in Canada (Sass, Lachenbruch and Jessop, 1971a). It appears that successive climatic events in Canada during the Pleistocene produced compensating perturbations, resulting in only small residual changes. Alternatively, continuous ice cover may have insulated the ground during the Pleistocene glacial periods so that present-day temperatures are not significantly warmer (Gates, 1976). Beck (1977) reviewed the principal facts of climatic history elsewhere in the Northern hemisphere and concluded that climatic perturbations in the geothermal gradient probably remain significant to depths of about 2000 m, causing errors up to 20% in the top 400 m.

Note: In spite of theoretical predictions of widespread errors due to climatic changes, the magnitude of perturbations in near-surface temperature gradients appears to vary geographically.

3.4. Data Integration

Temperature data from many different sources must eventually be integrated into a single thermal profile. A well-studied hole might have a surface temperature estimate, a number of corrected BHTs, one or two DSTs, several groundwater geochemistry estimates, a regional Curie depth determination and some regional xenolith stability field data. It is quite likely that these data will show considerable scatter due to the different errors associated with each. It then becomes necessary to decide which data to accept in preference to which other data.

In general, data should be ranked in order of decreasing accuracy, with the least accurate data being given lowest priority. Although it may vary somewhat for individual cases, the order of decreasing priority will generally be the following:

1. Precision temperature logs
2. Drill stem tests
3. Corrected BHTs
    a. Cooper and Jones (1959) correction with reliable data
    b. Horner plot correction with reliable data
    c. Horner plot correction with less reliable data
    d. Other corrections
4. Groundwater geochemistry
5. Curie point
6. Mantle resistivity temperature correlation
7. Xenolith equilibrium point
8. Uncorrected BHTs or non-equilibrium temperature logs

The most precise thermal profile is constructed using the maximum amount of complementary data available. Less accurate data that do not conform to the trend of better-constrained data should be rejected (Figure 3.14).

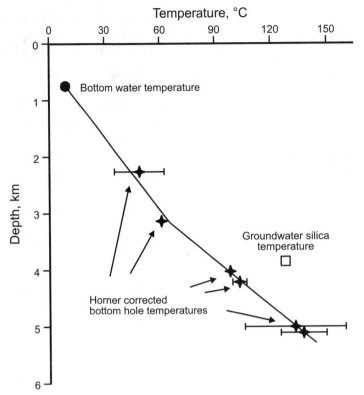

Figure 3.14. Temperature profile of Barcoo 1, Browse Basin, Western Australia, rejecting the SiO$_2$ datum, which is apparently anomalous.

3.5. Summary

Thermal gradient is a vector quantity dependent on the distribution of temperature in three dimensions. Generally, we assume that the direction of maximum gradient within the upper crust is vertical, so the gradient is the derivative of temperature with respect to depth ($\partial T / \partial z$). The aim is to obtain as many temperature data as possible in a vertical section.

Direct measurement of underground temperature generally requires that a temperature-measuring device be lowered down a borehole. Logging should not be attempted until at least 10–20 times the drilling time has elapsed to allow the thermal disruption of the bore fluid circulation during drilling to equilibrate. High-precision, electronic instruments capable of resolving thermal gradient on a fine scale yield the most accurate and precise temperature data possible. These instruments fall into one of three classes: wire-line tools in constant electrical contact with the surface; self-contained computer tools with on-board memory for recording a time–temperature log; distributed optical fibre temperature-sensing systems.

Specialised instruments have evolved to investigate heat flow in the deep oceans. These instruments generally take the form of thermistor-lined probes that penetrate the soft sediment on the sea floor. The specific design can usually be categorised as one of three types: Bullard probes (solid shafts with regularly spaced thermistors); Ewing probes (incorporating a hollow shaft to collect a core sample for thermal conductivity analysis); Lister probes (consisting of a thin sensor tube supported away from a thick strength member).

At surface temperatures, normal geothermal gradients are sufficient to induce convection in boreholes with diameter greater than about 5 cm. However, the general effect of convection within a borehole is simply to decrease the signal-to-noise ratio on a temperature log without significantly disrupting broader temperature trends.

The bulk of temperature data in a petroleum setting comes from bottom-hole temperatures (BHTs). These are usually recorded soon after drilling and are affected by the thermal effects of mud circulation. The data must be corrected to get an estimate of equilibrium formation temperature. The most common means of correction is the Horner plot, which displays temperature versus a dimensionless time parameter on semi-log axes. An extrapolated straight line through the data estimates the temperature after infinite recovery time. For temperature data collected less than three times the mud circulation time after drilling ceases, the Horner plot may underestimate true temperature. In this situation another correction should be used. The Cooper and Jones (1959) and Roux et al. (1982) corrections are two possibilities.

Groundwater geothermometers are primarily intended for use in geothermal reservoir studies. The amount of dissolved solids in groundwater is related to the temperature at which the solution equilibrated. The geothermometers work best under conditions of high temperature ($> 200°C$), when conventional temperature-measuring systems may not operate.

A small number of methods are available for estimating temperature in the deep crust without requiring direct access. The Curie depth is the depth at which minerals lose their magnetic properties; this can be estimated through spectral analysis of regional aeromagnetic data. Where Curie depth does not coincide with a velocity or density boundary, it may be interpreted as the Curie temperature isotherm (580°C in most continental regions).

A xenolith is a piece of country rock picked up by magma as it rises through the crust. The pressure–temperature stability conditions indicated by a xenolith's mineral assemblage provide information on the depth and temperature of its origin point.

Magnetotelluric methods can yield a value for the electrical resistivity of the upper mantle; this is strongly dependent on temperature. However, the depth resolution of such temperature estimates is poor.

Offshore, surface temperature is defined at the sediment–water interface. It can be estimated, within error margins of 2°C, as a function of latitude and water depth. In shallow water and enclosed seas it is equal to average sea surface temperature. Onshore, local meteorological records are the best resource for estimating surface temperature. Generally, it varies with latitude, altitude, topography and time, further complicated by long-term climatic trends and the cyclic heating patterns of the days and seasons. Near-surface temperature profiles record changes in surface temperature as deviations from equilibrium gradients. These deviations can be inverted to reconstruct the recent climatic history of the region.

CHAPTER FOUR

Thermal Conductivity

If experimenters will find ... the variations of conductivity of the earth's crust up to its melting point, it will be easy to modify the solution given above, so as to make it applicable to the case of a liquid globe gradually solidifying from without inwards, in consequence of heat conducted through the solid crust to a cold external medium.

On the Secular Cooling of the Earth – Prof. William Thomson, 1862.

We have looked at how heat generation and thermal gradient are measured or otherwise approximated within rocks. The last remaining parameter required to define steady-state heat flow is thermal conductivity. Simply put, thermal conductivity, λ, is a measure of how easily heat is transmitted through a material. It is a tensor operator that relates the heat flow vector to the thermal gradient vector within a body, and it is an inherent physical property. Many rocks are anisotropic, with conductivity dependent upon the direction of heat flow, but geothermal problems usually involve only the vertical component.

Thermal conductivity must be estimated over the same section that thermal gradient and heat generation are known. Where temperature is defined at discrete depths (e.g. Horner-corrected bottom-hole temperatures), an average thermal conductivity is required between each of those depths. Where a continuous temperature log is available, we require a continuous conductivity log. Heat flow remains undefined in any section where there is a gap in the thermal conductivity record.

As an aid to understanding the concept of thermal conductivity, a review of heat transfer theory is included in the first section of this chapter. The review covers phonon conduction theory and radiation, and explains why the former is the dominant mechanism for heat conduction within the crust, and why conductivity is temperature dependent.

Note: Section 4.1 and its sub-sections are included for those readers interested in gaining more of an intuitive feel for the subject. It may be skimmed over with no loss by those interested more in the practical aspects of measurement and application.

4.1. Heat Transfer Theory

Any satisfactory theory relating thermal conductivity to other physical proper-ties of a body must include consideration of all processes capable of transport-ing heat. It is usually assumed that crustal materials are rigid and that only phonons and photons may propagate heat. In contrast, magnetic data suggest that the lower mantle is electrically conductive, implying large numbers of free electrons, which can also transfer heat. Other processes operating at depth include diffusion of electron-hole pairs (Price, 1955) and migration of low-level excitons (Jamieson and Lawson, 1958; Whitmore, 1960). These latter processes are significant only at very high temperatures (thousands of degrees Celsius) and are not relevant to crustal research.

Each heat transfer process is independent, so for crustal rocks λ can be expressed as:

$$\lambda = \lambda_{Ph} + \lambda_r \tag{4.1}$$

where λ_{Ph} relates to phonon conduction and λ_r to radiation of photons.

In addition, in some areas significant amounts of heat can be transferred by non-conductive means. Mass transfer of heat via fluid convection within the crust will be examined in Section 6.3.5.

4.1.1. Phonon Conduction

At a molecular level, phonon conduction, λ_{Ph}, may be visualised as vibrations propagating along interatomic bonds. The vibrational states of the atoms in a crystalline solid are many and complex, depending on the distribution of neigh-bouring atoms and the number and types of external forces. It is often con-venient to consider these vibrational states as particles (phonons) that form a gas. Elementary kinematic theory (e.g. Brown, 1968) can then be used to express phonon conduction:

$$\lambda_{Ph} = (v \times c \times l)/3 \tag{4.2}$$

where v is the mean phonon velocity, c is the specific heat and l is the mean free path in the lattice.

Note: Specific heat can be obtained from the Debye theory, which is beyond the scope of this text. Interested readers are referred to Ziman (1960, p. 44).

In a perfect infinite lattice, the mean free path is limited only by collisions between two or more phonons. It is important to note that phonons exhibit wave-like behaviour in their relationship to the lattice. If phonons were limited to particle behaviour then momentum would always be conserved, the energy transported in a thermal gradient would never be dispersed and the conductivity would be infinite. However, when two phonons interact, the resulting angular frequency is the sum of the two:

$$w_1 + w_2 = w_3$$

but the resulting wave vector $\mathbf{q}$ ($= 2\pi/\text{wavelength}$) is not necessarily a simple sum:

$$\mathbf{q}_1 + \mathbf{q}_2 = \mathbf{q}_3 + (2\pi/a)n_{\text{wavelength}} \tag{4.3}$$

where a is the lattice spacing and $n_{\text{wavelength}}$ is an integer.

The last term in Equation (4.3) is simply an expression of how often the molecular sequence is repeated in the lattice. A phonon can be shifted by any integer number of wavelengths without disturbing the relative positions of the atoms that form the lattice. When $n_{\text{wavelength}} = 0$, collisions between phonons are called Normal (N) processes and momentum is conserved as if they were particles. However, for $n_{\text{wavelength}} \neq 0$, the direction of flow changes due to the collision and some energy is dispersed within the lattice. These 'Umklapp' (U) processes are the major source of thermal resistance. The N-processes serve to redistribute the phonon frequencies and, in that sense, contribute to U-processes.

Klemens (1958) reviewed much of the early work of Debye and Peierls and confirmed that U-processes cause the thermal conductivity to vary inversely with temperature. Leibfried and Schlömann (1954) determined the constants of proportionality and the dependence on other parameters. Their results can be expressed as follows (Abeles, 1963):

$$\lambda_{Ph} = (D \times a \times \theta^3)/(T \times \gamma^2) \tag{4.4}$$

where

    D = a numerical constant
    a = the lattice spacing (m)
    θ = the Debye temperature (K)
    T = temperature (K)
    $\gamma = \alpha/(\rho c \beta)$ = the Gruneisen parameter (α = thermal expansivity, ρ = density, c = specific heat, β = compressibility)

It follows that λ_{Ph} is strongly dependent on θ (Webb, Wilkinson and Wilks, 1952; Klemens, 1958) and is almost inversely proportional to temperature; $\lambda \propto T^{-1}$ (Figure 4.1). Lawson (1957) and Williams and Anderson (1990) respectively derived two alternative relationships for λ_{Ph}:

$$\lambda_{Ph} = [a \times (\rho v)^3][3 \times \gamma^2 \times T \times \rho^{1/2}] \tag{4.5}$$

$$\lambda_{Ph} = [a \times v \times \mu^2]/[3 \times B \times T] \tag{4.6}$$

where most terms are as for Equation (4.4), and

    v = the speed of sound
    $\mu = \rho v_s^2$ = the shear modulus (v_s = shear velocity)
    $B = \rho(v_p^2 - (4/3) \times v_s^2)$ = the bulk modulus (v_p = compressional velocity)

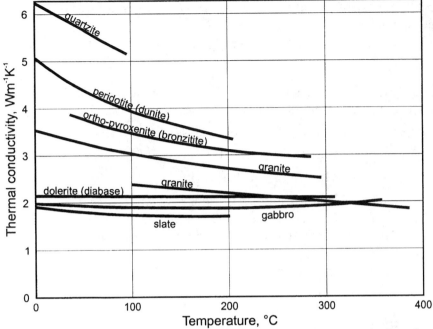

Figure 4.1. The temperature dependence of thermal conductivity in some rocks. Data from Birch and Clark (1940).

Note: All terms in Equation (4.6) can be estimated from electric well logs, which are discussed in Section 4.4.

Equations for λ_{Ph} can also be used to investigate the effect of pressure on thermal conductivity. Mooney and Steg (1969) derived the pressure derivative

$$\frac{\partial \lambda}{\partial P} = \lambda_0 \left(3\gamma_0 + \frac{2}{3} \right) \beta \qquad (4.7)$$

where

γ = the Gruneisen parameter as defined for Equation (4.4)
$\beta = 1/B$, where B is the bulk modulus as defined for Equation (4.6)
P = pressure

Question: Granite has the following physical properties: $\lambda_0 = 3.5 \ \mathrm{W\,m^{-1}\,K^{-1}}$ (Beardsmore, 1996), $B = 45 \times 10^9 \ \mathrm{N\,m^{-2}}$ (Giancoli, 1984), $\gamma_0 \approx 0.2$, $\rho = 2.7 \times 10^3 \ \mathrm{kg\,m^{-3}}$ (Giancoli, 1984), under surface conditions. What is the pressure derivative of thermal conductivity for granite? Given an average granitic composition for the crust, what is the depth derivative of thermal conductivity in the crust?

Answer: The pressure derivative of thermal conductivity is derived from Equation (4.7):

$$\partial\lambda/dP = 3.5 \times (3 \times 0.2 + 2/3) \times (45 \times 10^9)^{-1} = 9.852 \times 10^{-11}$$
$$\text{W}\,\text{m}^{-1}\,\text{K}^{-1} \text{ per Nm}^{-2}$$

The depth derivative of pressure in granitic crust is found from the product of the density and the gravitational acceleration, $\rho \times g$:

$$\partial P/\partial z = (2.7 \times 10^3) \times (9.8) = 2.646 \times 10^4 \text{ N}\,\text{m}^{-2} \text{ per metre}$$
$$\partial\lambda/\partial z = \partial\lambda/\partial P \times \partial P/\partial z = 9.852 \times 10^{-11} \times 2.646 \times 10^4$$
$$= 2.607 \times 10^{-6} \text{ W}\,\text{m}^{-1}\,\text{K}^{-1} \text{ per metre}$$

For granitic rock, λ increases with depth at a rate of about 2.61×10^{-3} W m^{-1}K^{-1} per kilometre. This is only about 0.1% of λ_0 per kilometre depth, and can be effectively ignored in the top 50 km of crust.

4.1.2. Radiation

The importance of thermal radiation as a means of transferring heat in Earth materials was not fully appreciated prior to Clark's (1957) expression of theory that was first suggested in the nineteenth century. The governing functions are derived from Stefan's law, which states that in a black body the total energy radiated per unit time, E_R, can be expressed as

$$E_R = ST^4 \tag{4.8}$$

where $S = \sigma A$ (σ = the Stefan–Boltzmann constant = 5.670400×10^{-8} W m^{-2} K^{-4}, A = the surface area).

If two black bodies at temperatures T_1 and T_2 are separated by a distance d, the net transfer of energy from the hotter to the cooler can be written as

$$fE_{R1} - fE_{R2} = ST_1^4 - ST_2^4 \tag{4.9}$$

where f is the fraction of energy radiated in the required direction and is assumed to be the same for both bodies. Equation (4.9) can be simplified in terms of energy flow, $\mathbf{Q}_s (= E_{R1} - E_{R2})$:

$$Q_s = Sf^{-1}(T_1^4 - T_2^4) = Sf^{-1}(T_1^3 + T_1^2 T_2 + T_1 T_2^2 + T_2^3)(T_1 - T_2) \tag{4.10}$$

Radiative conductivity, λ_r, is defined by

$$\mathbf{Q}_s = \lambda_r(\partial T/\partial z) = \lambda_r \times (T_1 - T_2)/d \tag{4.11}$$

If d is sufficiently small, $T_1 \approx T_2$ and, combining Equations (4.10) and (4.11), radiative conductivity can be approximated:

$$\lambda_r = (4 \times d \times ST^3)/f \tag{4.12}$$

Clark's (1957) derivation is more detailed and can be approximated as

$$\lambda_r = (16 \times n^2 \times \sigma \times T^3)/(3\varepsilon) \tag{4.13}$$

where n is the refractive index and ε is the opacity or absorption coefficient ($\varepsilon^{-1} = d$, mean free path).

The opacity, ε, is sensitive to wavelength, so to negate the wavelength dependence we can substitute a mean extinction coefficient (ε_{av}) integrated over all wavelengths.

It seems intuitive that the T^3 term in Equation (4.13) would cause λ_r to quickly dominate over λ_{Ph} with increasing temperature. This conviction led Lubimova (1958) and MacDonald (1959), among others, to propose non-convective thermal models for the mantle, assuming thermal conductivities greatly in excess of surface values. Clark (1957), however, proposed temperature dependence for ε_{av} and this was later experimentally verified by Fukao, Mizutani and Uyeda (1968), who examined the absorption spectra of olivine at high temperatures (Figure 4.2). Clark assumed that electrons are mainly responsible for energy absorption and, since increasing temperature causes the conduction bands to become more densely populated, ε_{av} may be related to electrical conductivity, σ_0, according to the following expression:

$$\varepsilon_{av} = \varepsilon_{av0} + [120\pi\sigma_0 \times \exp(-E/2\kappa T)]/n \tag{4.14}$$

where κ is Boltzmann's constant ($1.3806503 \times 10^{-23}$ J K^{-1}), and E is the energy gap.

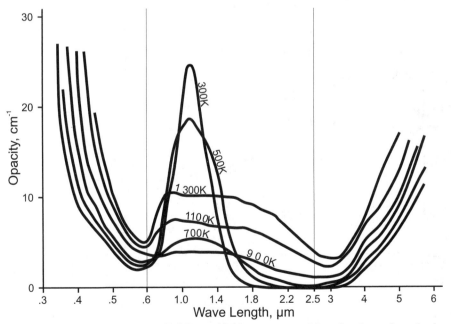

Figure 4.2. Absorption spectra of olivine at high temperatures. Note the change in scale along the x-axis. Modified after Fukao et al. (1968).

It is not necessary to understand the meaning of all the parameters to see that a continual increase in ε_{av} with temperature is suggested, at a rate depending on the values of E and ρ_0. The absorption spectra of olivine (Figure 4.2) indicate that the increase in ε_{av} with T is substantial, and sufficient to depress λ_r so that the phonon component of conductivity remains dominant to at least 800 K.

Subsequent measurements and estimates of λ_r have shown that its contribution to total thermal conductivity remains insignificant below about 600 K (Figure 4.3), but contributes approximately one quarter of the total thermal conductivity at 800 K. This temperature range covers most of the Earth's crust, and in all but the deepest portions λ_r can be effectively ignored. However, in the deepest crust, or around hot intrusive bodies, λ_r may constitute a significant portion of total conductivity.

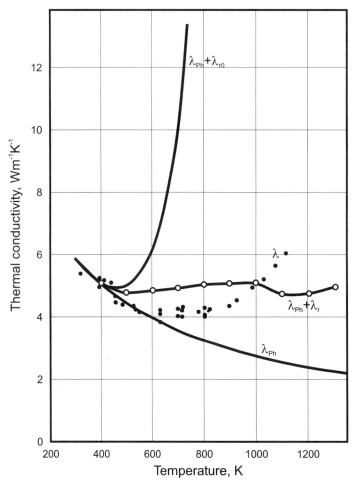

Figure 4.3. Olivine thermal conductivity at high temperature. λ = Thermal conductivity estimated from diffusivity measurements; λ_{Ph} = phonon component of conductivity, λ_r = radiative component of conductivity calculated from high-temperature spectral data, λ_{r0} = radiative component of conductivity calculated from room temperature spectral data. Modified after Fukao et al. (1968).

Question: Given that the mean refractive index of olivine is 1.67, and the mean opacity at 800 K is 2.02 cm^{-1} (Fukao et al., 1968), what is the radiative thermal conductivity of olivine at 800 K?

Answer: The relevant parameters are:

$n = 1.67$

$\varepsilon_{av} = 202$ m^{-1}

$T = 800$ K

so from Equation (4.13):

$\lambda_r = 16 \times (1.67)^2 \times 5.67032 \times 10^{-8} \times (800)^3 / (3 \times 202) = 2.14$ W m^{-1} K^{-1}

Given that the phonon conductivity of olivine at 800 K is approximately 3.0 W m^{-1} K^{-1} (Figure 4.3), the total calculated conductivity is approximately 5.14 W m^{-1} K^{-1}, with about 40% arising from the radiative component.

4.2. Mixing Laws

An understanding of the molecular theory of heat transfer helps us, in a qualitative way, to visualise the process of thermal conduction within individual crystals. However, it does not help us towards our goal of quantifying thermal conductivity of *in situ* rock formations. The thermal conductivity of a rock is prohibitively difficult to model using phonon conduction theory because it strongly depends on the geometrical relationships between the different mineral components. For example, an interbedded sequence of sandstone and shale has a different conductivity to a shaly sand or sandy shale with the same bulk proportions of minerals.

Estimating the mean conductivity of a rock or formation requires the choice of a mixing model that best describes its geometry. There are three important models to consider (Figure 4.4).

4.2.1. Harmonic Mean

The first model applies to beds layered perpendicular to the direction of heat flow. This model best describes a vertical well drilled through a sequence of sub-horizontal strata, where each bed has a different thermal conductivity. The mean conductivity, λ_B, of the sequence is the *harmonic mean* of the individual beds:

$$\frac{1}{\lambda_B} = \sum_{i=1}^{n} \frac{\phi_i}{\lambda_i} \quad (= \text{thermal resistance; see Section 6.2}) \qquad (4.15)$$

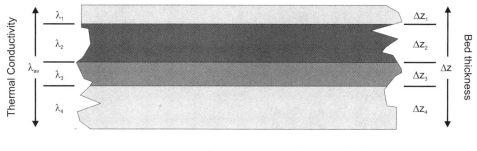

a) $\lambda_{av} = \Delta z / [\Delta z_1/\lambda_1 + \Delta z_2/\lambda_2 + \Delta z_3/\lambda_3 + \Delta z_4/\lambda_4]$

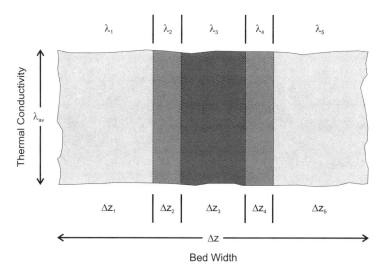

b) $\lambda_{av} = [\Delta z_1 \times \lambda_1 + \Delta z_2 \times \lambda_2 + \Delta z_3 \times \lambda_3 + \Delta z_4 \times \lambda_4 + \Delta z_4 \times \lambda_4] / \Delta z$

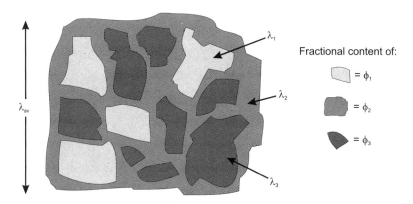

c) $\lambda_{av} = \lambda_1^{\phi 1} \times \lambda_2^{\phi 2} \times \lambda_3^{\phi 3}$ or $\lambda_{av} = [\phi_1 \times \sqrt{\lambda_1} + \phi_2 \times \sqrt{\lambda_2} + \phi_3 \times \sqrt{\lambda_3}]^2$

Figure 4.4. The three key mixing models: (a) harmonic mean, for example, horizontally layered beds; (b) arithmetic mean, for example, vertical dyke and contact metamorphic zone; (c) geometric/square-root mean, for example, mixture of lithologies.

where λ_B is the mean thermal conductivity, λ_i is the thermal conductivity of the *i*th bed and ϕ_i is the thickness of the *i*th bed/total thickness of the sequence $(0 \le \phi_i \le 1; \Sigma\phi_i = 1)$.

Equation (4.15) can also be written as

$$\frac{1}{\lambda_B} = \frac{1}{Z}\sum_{i=1}^{n}\frac{z_i}{\lambda_i} \tag{4.16}$$

where z_i is the thickness of the *i*th bed and Z is the total thickness of the sequence $(Z = \Sigma z_i)$.

This model also describes interbedded units (e.g. sandstone–shale, limestone–shale) when the proportion of each component is known.

Question: A particular sequence of rocks, 1 km thick, is composed of horizontally layered, interbedded sandstone $(\lambda = 2.7 \text{ W m}^{-1} \text{ K}^{-1})$ and shale $(\lambda = 1.7 \text{ W m}^{-1} \text{ K}^{-1})$ in proportion 3:1. The top 200 m of the sequence is limestone $(\lambda = 3.0 \text{ W m}^{-1} \text{ K}^{-1})$. What is the best estimate of the mean thermal conductivity across the 1 km sequence?

Answer: The thicknesses of the components are:
    sandstone $= 600$ m
    shale $= 200$ m
    limestone $= 200$ m
So, from Equation (4.16):
    $(1/\lambda_B) = (1/1000)[(600/2.7) + (200/1.7) + (200/3.0)] = 0.4065$
hence $\lambda_B = 2.46 \text{ W m}^{-1} \text{ K}^{-1}$

4.2.2. Arithmetic Mean

The second model applies when beds are arranged parallel to the direction of heat flow. In most geological situations, this requires vertical contacts between formations. Such a situation may arise through faulting, igneous intrusion, tight folding, salt pluming, or other circumstances. The mean thermal conductivity for this model is best approximated using an *arithmetic mean*:

$$\lambda_B = \sum_{i=1}^{n} \phi_i \lambda_i \tag{4.17}$$

where all symbols are as in Equation (4.15).

Question: A vertical fault juxtaposes a limestone unit $(\lambda = 3.0 \text{ W m}^{-1} \text{ K}^{-1})$ against a shale $(\lambda = 1.8 \text{ W m}^{-1} \text{ K}^{-1})$. What is the mean vertical thermal conductivity in the region of the fault?

Answer: We can assume that the two units are evenly distributed on either side of the fault. That is, $\phi_{shale} = \phi_{limestone} = 0.5$. Then, from Equation (4.17):

$$\lambda_B = 0.5 \times 3.0 + 0.5 \times 1.8 = 2.4 \; \mathrm{W\,m^{-1}\,K^{-1}}$$

4.2.3. Geometric or Square-Root Mean

The third model describes situations where several components of known conductivity are randomly orientated and distributed within a mixture. This model generally applies to a rock composed of a mixture of different minerals, and is the most difficult to approximate mathematically. The most popular way to estimate mean conductivity is by a *geometric mean*. The geometric mean of an n-component system is the product of the thermal conductivity of each component raised to the power of its fractional proportion:

$$\lambda_B = \prod_{i=1}^{n} \lambda_i^{\phi_i} \tag{4.18}$$

where all symbols are as in Equation (4.15).

Roy, Beck and Touloukian (1981) preferred a *square-root mean*, which has a greater physical basis than the geometric mean:

$$\sqrt{\lambda_B} = \sum_{i=1}^{n} \phi_i \sqrt{\lambda_i} \tag{4.19}$$

Note: The square-root-mean is *not* equivalent to the commonly used root-mean-square.

Either of these equations can be used to estimate the bulk thermal conductivity of a rock from its mineralogy and pure mineral thermal conductivities. Pure mineral thermal conductivities can either be measured directly or derived from tables (e.g. Horai and Simmons, 1969; Touloukian et al., 1970b; Table 4.1). Pore fluid must be included as a component of the bulk rock.

Question: The matrix of a dolomitic limestone is estimated to contain 25% dolomite ($\lambda = 5.51 \; \mathrm{W\,m^{-1}\,K^{-1}}$) and 75% calcite ($\lambda = 3.59 \; \mathrm{W\,m^{-1}\,K^{-1}}$). The porosity of the rock at 1000 m is 10%, and the pores are filled with water ($\lambda = 0.65 \; \mathrm{W\,m^{-1}\,K^{-1}}$). The temperature at 1000 m is 40°C. What is the best estimate of mean thermal conductivity for the rock at 1000 m using first the geometric mean and then the square-root mean?

Answer: Porosity must be included in the calculation, so the proportions of the components are

    water $= 10\%$

    dolomite $= 22.5\%$

    calcite $= 67.5\%$

Using Horai and Simmons' (1969) mineral conductivity data (refer to Table 4.1), Equation (4.18) gives:

$$\lambda_B = (3.59)^{0.675} \times (5.51)^{0.225} \times (0.65)^{0.1} = 3.33 \text{ W m}^{-1} \text{ K}^{-1}$$

The square-root mean, Equation (4.19), gives:

$$\lambda_B = [(0.675 \times 3.59^{1/2}) + (0.225 \times 5.51^{1/2}) + (0.1 \times 0.65^{1/2})]^2 = 3.56 \text{ W m}^{-1} \text{ K}^{-1}$$

This example indicates a moderate (6–7%) difference in the results of the two mixing models, illustrating the importance of choosing the correct model. Unless there are reasons to do otherwise, it is recommended to use the square-root mean, based on physical principles.

Table 4.1. Thermal Conductivity, λ, of Some Common Minerals

Mineral	Temp. range (K)	λ (W m^{-1} K^{-1})	Comments[Source]
Quartz	273.2–623.2	$4134 \times T^{-1.053}$	Parallel to c-axis[1]
	273.2–623.2	$820 \times T^{-0.861}$	Perpendicular to c-axis[1]
		7.69	Average[2]
Calcite	273.2–374.2	$264.5 \times T^{-0.727}$	Perpendicular to c-axis[1]
	273.2	5.51	Parallel to c-axis[1]
	332	5.16	Average[1]
		3.59	Average[2]
Dolomite		5.51	Average[2]
Orthoclase		2.32	Average[2]
Albite		2.37	Average[2]
Anorthite		1.68	Average[2]
Plagioclase		1.91	Average[2]
Sillimanite	333.2	2.60	Average[1]
Cordierite	320.8–398.1	$116.3 \times T^{-0.635}$	Average[1]
Salt	273–460	$4610 \times T^{-1.146}$	Average[1]
Topaz	314.2–419.8	$9946 \times T^{-1.094}$	Average[1]
Forsterite		5.12	Average[1]
Wollastonite	317.2–397.2	2.65	Average[1]
Zircon	318.8–411.6	4.03	Parallel to c-axis[1]
	318.7–414.2	4.14	Perpendicular to c-axis[1]
Tourmaline	398.2–723.2	$0.492 \times T^{0.297}$	Parallel to c-axis[1]
	393.2–729.2	$0.108 \times T^{0.556}$	Perpendicular to c-axis[1]

Sources: [1] Touloukian et al. (1970b), [2] Horai and Simmons (1969).

Note: The particular mixing model to use depends on the geometry of the problem. The model should be chosen to best approximate the actual situation.

4.3. Measurements of Rock Conductivity

Our ultimate aim is to determine the average thermal conductivity between temperature data points. The best way to achieve this is to construct an accurate conductivity profile of the entire section under investigation, and subsequently to examine those parts of the section between temperature data. The investigation should follow a general course summarised by the flow diagram in Figure 4.5. The specific steps taken through the diagram will depend on the data available. Each step is explained in detail in the following sections.

Note: Almost all geothermal data (temperature, palaeotemperature, heat generation and thermal conductivity) are obtained from boreholes and drill cores. The bulk of the following discussion refers mainly to such data. Most of the descriptions and explanations, however, are easily adapted to surface samples if necessary.

4.3.1. The Lithological Column and Porosity Data

The first step towards interpreting the thermal conductivity of a sequence penetrated by a borehole is to understand clearly the lithological distribution. Thermal conductivity is closely related to lithology and, in some cases, the lithological column may be the only available data set. As a first approximation, lithology alone can be used to estimate thermal conductivity (Table 4.2).

Question: The rock sequence between two temperature data points in a well is composed entirely of interbedded shale and limestone. The mud log indicates that shale accounts for approximately 20% of the sequence. As a first approximation, what is the average conductivity of the sequence between the temperature data points?

Answer: As a first approximation, we can estimate the conductivity of the shale to be $1.8 \pm 0.2 \ \mathrm{W\,m^{-1}\,K^{-1}}$, and the limestone, $3.0 \pm 0.3 \ \mathrm{W\,m^{-1}\,K^{-1}}$. It is also safe to assume that the lithological proportions are only accurate to $\pm 5\%$ or worse. The harmonic mean gives the best estimate of conductivity across horizontally layered lithologies:

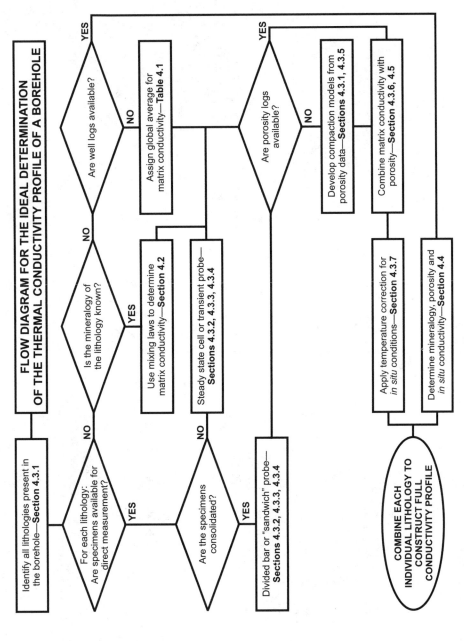

Figure 4.5. Flow chart for determining a vertical thermal conductivity profile.

FLOW DIAGRAM FOR THE IDEAL DETERMINATION OF THE THERMAL CONDUCTIVITY PROFILE OF A BOREHOLE

Identify all lithologies present in the borehole—**Section 4.3.1**

For each lithology: Are specimens available for direct measurement?

Is the mineralogy of the lithology known?

Are well logs available?

NO → Assign global average for matrix conductivity—**Table 4.1**

YES → Use mixing laws to determine matrix conductivity—**Section 4.2**

Are the specimens consolidated?

NO → Steady state cell or transient probe—**Sections 4.3.2, 4.3.3, 4.3.4**

YES → Divided bar or "sandwich" probe—**Sections 4.3.2, 4.3.3, 4.3.4**

Are porosity logs available?

NO → Develop compaction models from porosity data—**Sections 4.3.1, 4.3.5**

Combine matrix conductivity with porosity—**Section 4.3.6, 4.5**

Apply temperature correction for *in situ* conditions—**Section 4.3.7**

Determine mineralogy, porosity and *in situ* conductivity—**Section 4.4**

COMBINE EACH INDIVIDUAL LITHOLOGY TO CONSTRUCT FULL CONDUCTIVITY PROFILE

103

Table 4.2. Comparison of Published Compilations of Thermal Conductivities ($W\,m^{-1}\,K^{-1}$)

Lithology						Sources					
	1[a]	2	3	4	5	6	7	8	9	10	11
Sandstone	7.1	4.2 ± 1.4	3.1 ± 1.3		3.7 ± 1.2	2.8		3.7 ± 1.2			4.7 ± 2.8
Claystone	2.9				2.0						1.8
Mudstone	2.9										1.9 ± 0.4
Shale	2.9	1.5 ± 0.5	1.4 ± 0.4		2.1 ± 0.4	1.4		2.1 ± 0.4			1.8 ± 1.2
Kaolinite									1.8 ± 0.3		
Glauconite									0.5 ± 0.2		
Siltstone	2.9	2.7 ± 0.9	3.2 ± 1.3		2.7 ± 0.2	2.7 ± 0.9		2.7 ± 0.2			
Limestone	3.1	2.9 ± 0.9	2.4 ± 0.9	2.21	2.8 ± 0.4		3.4 ± 3.0	2.8 ± 0.3			2.5 ± 0.6
Marl	3.2	2.1 ± 0.7	3.0 ± 1.1		2.7 ± 0.5						2.4 ± 0.5
Dolomite		5.0 ± 0.6	3.1 ± 1.4		4.7 ± 0.8		4.8 ± 1.5	4.7 ± 1.1			3.7 ± 1.8
Halite		5.5 ± 1.8	5.7 ± 1.0		5.4 ± 1.0			5.4 ± 0.3			5.9
Chert		4.2 ± 1.5	1.4 ± 0.5		1.4 ± 0.5						
Quartzite				6.0			5.0 ± 2.4	5.9 ± 0.8		3.5 ± 0.4	5.6 ± 1.9
Granite							3.4 ± 1.2			3.5 ± 0.4	2.8 ± 0.6
Basalt	1.8			1.7			1.7 ± 0.6			2.0 ± 0.2	1.5
Tuff				1.7 ± 0.3							
Conglomerate		2.4 ± 0.8	3.2 ± 1.8		2.1 ± 1.0						
Coal		0.3 ± 0.1	0.2 ± 0.2	0.2 ± 0.04	0.2 ± 0.1	0.3 ± 0.1					
Loose sand				2.44 ± 0.8							
Typical sediment			2.3 ± 2.0								

[a]Matrix conductivity values, only representing bulk conductivity when $\phi = 0$.

Sources: 1 = Beardsmore (1996), 2 = Majorowicz and Jessop (1981), 3 = Beach, Jones and Majorowicz (1987), 4 = Raznjevic (1976), 5 = Reiter and Jessop (1985), 6 = Taylor, Judge and Allen (1986), 7 = Roy et al. (1981), 8 = Reiter and Tovar (1982), 9 = Touloukian et al. (1970b), 10 = Drury (1986), 11 = Barker (1996).

$$1/\lambda_B = (0.2/1.8) + (0.8/3.0) = 0.378,$$
$$\text{so } \lambda_B = 2.6 \pm 0.6 \text{ W m}^{-1}\text{K}^{-1}$$

This figure should only be used as a first approximation, as evidenced by the high uncertainty ($\sim$23%). Sandstones have even higher errors associated with them due to the greater range of possible mineralogy.

A lithological column is generally not difficult to locate for a commercially drilled well. The well completion report will usually carry a diagram of the formations penetrated, the lithology of the formations and the depths of the contacts. A more accurate record of lithologies can often be found on the mud log, and lithological information is sometimes interpreted from electrical well logs. The precision and accuracy of lithological records vary greatly between wells – possible errors and uncertainties in lithological proportions and contact depths should be closely noted.

Ideally, all lithologies under investigation should be physically sampled for thermal conductivity and porosity analyses. The greater the number of samples collected from each formation, the more statistically reliable will be the results. Drill cuttings may be used if they are the only samples available, but cores are preferable. In reality, there are many cases when neither is available and alternative means must be employed to estimate conductivity. These situations are discussed in later sections.

Porosity data are used to constrain compaction models for different lithologies, as described in Section 4.3.5, and are very relevant to thermal conductivity studies. The porosity of core samples should be measured prior to thermal conductivity testing, and existing data should also be extracted from well completion reports. Porosity data relating to the same formation and lithology should be gathered together and collated on a regional scale.

4.3.2. Sample Preparation

Similar lithologies from different regions can have significantly different conductivities, owing to minor variations in mineralogy and porosity. Whenever possible, therefore, the thermal conductivity of the actual formations penetrated should be measured. Even then, the conductivity of a particular formation can vary dramatically with porosity. The conductivity of the rock matrix, however, is assumed to remain relatively constant within a formation. It is the matrix conductivity, therefore, that laboratory tests aim to measure. The matrix conductivity is subsequently combined with compaction models to construct *in situ* conductivity profiles.

Methods for measuring the thermal conductivity of rocks can be broadly divided into two types. Steady-state methods generally compare an unknown conductivity to a known one, while transient methods usually involve examining the temperature response of a sample to a heat impulse. Steady-state techniques are generally more accurate and yield a value of thermal conductivity in the desired direction. That is, samples can be prepared to measure the conductivity parallel to the maximum thermal gradient. Most transient techniques average conductivity over two or three dimensions, but can be applied to a much broader group of samples, including those that are unconsolidated.

Conventional steady-state thermal conductivity measurement is possible only on consolidated rocks. Suitable samples should have cylindrical plugs, of a diameter equal to that of the available measuring device, extracted along their vertical axes (Section 4.3.3). The ends of each plug should be ground perpendicular to the cylinder axis and polished to ensure good thermal contact.

Note: Soft or friable rocks are often under-represented in thermal conductivity studies. Careful sampling should ensure that each lithology is adequately represented, including samples of unconsolidated rocks collected for transient or cell measurements (explained in the following sections). Ideally, enough samples should be measured to determine reliably the average matrix conductivity of each lithology. That is, the standard deviation of measurements on the same lithology should be within 10–15% of the mean.

A true estimate of *in situ* conductivity requires that the samples be prepared with the same pore-fluid volume and composition as the rocks *in situ*. However, pore-fluid characteristics are not always known or reproducible, so the usual aim is to measure the conductivity of the granular matrix of the rock. This is achieved by measuring the conductivity of water-saturated samples and subsequently removing the effect of the water using an appropriate mixing law.

Prior to saturation, samples should first be dehydrated. A fan-forced drying oven at 102°C is sufficient for this purpose. Each sample can be assumed dry when its mass ceases to diminish with additional drying time. The time taken to dry a sample depends on its porosity, permeability and initial water saturation, and can vary from hours to weeks (Beardsmore, 1996).

Note: The thermal conductivity of all samples should be measured prior to dehydration, then again prior to saturation. Although matrix conductivity cannot be accurately deduced, some specimens may not survive the saturation procedure, and all available conductivity data become important.

All samples should be water-saturated before final thermal conductivity measurements are made. Saturation can be achieved by submerging the samples in distilled water inside a sealed vacuum chamber and evacuating the air.

When air is no longer observed escaping from the samples they may be assumed fully evacuated. A maximum of a couple of hours is sufficient for most rocks. The samples should then be released from vacuum but left submerged to absorb water at atmospheric pressure for a time at least equal to that taken to evacuate. Saturation time may be hastened by applying external pressure, but this may also damage the samples.

The mass change (in grams) between the dehydrated and saturated states of the sample can be attributed entirely to the influx of water into the pore spaces, and therefore it equals the pore volume of the sample in cubic centimetres ($\rho_{H_2O} = 1.0\,g\,cm^{-3}$). The act of saturation therefore yields a direct measurement of the porosity, ϕ, of the sample (ratio of pore volume to total sample volume).

Note: Although natural pore fluids may be saline, distilled water is recommended for its precisely known density and thermal conductivity. This technique gives an accurate porosity estimate only if the rock is permeable.

Question: A cylindrical plug of limestone has a diameter of 2.5 cm and is 1.6 cm high. Its dry mass is 18.831 g. After saturation, the mass of the plug is 19.711 g. What is the porosity of the sample, ϕ?

Answer: The porosity of the sample is the mass change in grams between dry and saturated, divided by the volume of the sample (cm^3). That is:

$\phi = (19.711 - 18.831)/(\pi \times 1.25^2 \times 1.6) = 0.88/7.854 = 0.112$

Porosity $= 11.2\%$

The dehydration/saturation procedure is best suited to consolidated rocks. When the rocks contain a significant proportion of very fine or poorly cemented material, they may experience a degree of swelling or loss of competence. If a sample swells during saturation, it absorbs more water and yields an overestimate of porosity. During the subsequent conductivity measurement (with a small retaining pressure), the sample may be recompressed and lose water. Matrix conductivity is then calculated higher than true matrix conductivity. The effect can be minimised by sealing the sides of the sample against fluid loss using thin tape, grease or another sealing agent, but measured matrix conductivities of clay-rich lithologies should be viewed as maximum estimates.

Another hazard when dehydrating clay-rich lithologies is that the clay minerals may begin to lose water molecules bound within the crystal lattice, effectively altering the mineralogy. Drying should be conducted with care, and a careful record kept of mass versus time. If it becomes evident that the clay is losing bound water (e.g. a sudden increase in the rate of mass loss), the sample should be removed from the oven immediately.

Some samples may lose competence when water-saturated and thus be unsuitable for conventional steady-state methods. Also, in many instances, cores are not recovered from wells and drill cuttings may be the only materials available for analyses. Many authors (e.g. Sass, Lachenbruch and Munroe, 1971b; Middleton, 1993) advocate the use of a hollow cell for steady-state thermal conductivity measurement of unconsolidated material, requiring alternative methods of sample preparation. The unconsolidated sediment or drill cuttings should be clean of drilling mud but retain as much as possible of the initial lithological mix. This is not always a simple requirement to fulfil, especially if the sediment contains a high percentage of fine-grained material. The sediment must be packed tightly into the cell, and the remaining pore space filled with water under vacuum conditions. The change in mass reveals the exact volume of water added.

For transient methods, the sample preparation depends on the type of apparatus used. Often a needle probe is employed and little or no preparation of the sample is required. Soft sediments can be retained in the core pipe and the probe can be inserted through a small hole drilled into the pipe. Drill cuttings need to be washed clean of drilling fluid and packed into a suitable container into which a needle can be inserted. Other types of transient probes (discussed in Section 4.3.4) have their own preparation requirements.

Note: The important thing to remember in any thermal conductivity measurement is that the sample should be representative of a particular lithology, and should be free from contaminating drilling fluid or other foreign matter. Any special preparation made to the sample, especially the nature of the pore fluid, should be carefully noted and taken into account when calculating matrix or total thermal conductivity.

4.3.3. Steady-State Method

Steady-state thermal conductivity measurements are usually made using a divided-bar apparatus – a tool designed to measure the thermal conductivity of discs or cylindrical plugs of material. The device, first described by Benfield (1939) is easy to construct and operate, and results are usually accurate to within 5%.

A typical divided bar (Figure 4.6) exerts a small retaining pressure (enough to ensure good thermal contacts) along a cylindrical assembly 2–4 cm in diameter. The top and bottom sections of the bar are constructed such that they can be maintained at constant but different temperatures. One way of achieving this is to construct them out of hollow brass cylinders, through which water of constant temperature may be pumped. Alternatively, a thermostatically controlled electrical heating circuit might be used. The central section of the

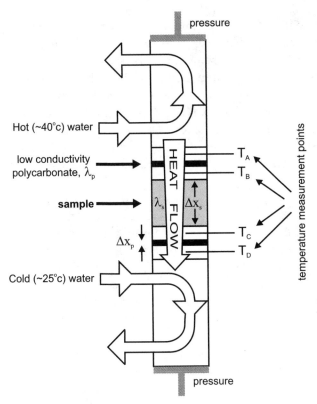

Figure 4.6. A typical divided-bar apparatus.

instrument comprises two thin discs of a low-thermal conductivity material and the sample to be measured, separated from each other by two brass (or similar high-conductivity material) discs of the same diameter. Temperature is measured at four points, one on each side of the two standard discs, by insertion of a needle-probe thermistor or thermocouple junction into narrow holes drilled into the brass. Typically, the low-conductivity material will be some variety of polycarbonate.

The top and bottom of the assembly are usually maintained at constant temperatures of around 40°C and 25°C, respectively. The thermistor used to measure temperature should be calibrated to a precision better than ±0.01°C over the range 20–45°C.

Note: It is important to design the divided bar with the warm end at the top, so as not to induce convection of fluids within the sample being measured. Convection can significantly increase the measured conductivity.

The premise governing the construction of the divided bar is that heat flow is constant across a thermal conductivity boundary. In other words, if a constant thermal gradient is applied across a sample of unknown conductivity in series with a material of known conductivity, heat flow will equilibrate to be constant through the two materials. The heat flow can be determined in the known sample and used to calculate the conductivity of the unknown sample.

By calculating thermal conductivity in this way, we are making three assumptions. Firstly, heat conduction along the bar is assumed to be 100% efficient, with no loss of heat through the sides of the bar or sample. It is thermodynamically impossible to achieve zero side loss, but loss can be minimised with insulation and the effect is partially neutralised by the instrument design (explained below).

Our second assumption is that the total temperature drop across the brass sections of the instrument is negligible compared with the temperature drop across the polycarbonate and core sample. This assumption is acceptable due to the low thermal resistance of the brass sections (typically 10^{-4} m^2 K W^{-1}; e.g. Sears, Zemansky and Young, 1978) compared with the sample (typically 10^{-2} m^2 K W^{-1}).

Thirdly, the two standard conductivity discs must be identical in thickness and thermal conductivity. This is reasonable if the discs are cut from adjacent positions on the same sheet of polycarbonate material.

If these three assumptions are accepted, we may proceed to determine the unknown thermal conductivity. Referring to Figure 4.6, let the following terms be defined:

$$\Delta T_s = T_B - T_C = \text{temperature drop across the sample}$$

$$\Delta T_1 = T_A - T_B = \text{temperature drop across the top polycarbonate}$$

$$\Delta T_2 = T_C - T_D = \text{temperature drop across the bottom polycarbonate}$$

$$\Delta x_s, \lambda_s = \text{thickness (m), thermal conductivity (W m}^{-1}\text{ K}^{-1}\text{) of sample}$$

$$\Delta x_p, \lambda_p = \text{thickness (m), thermal conductivity (W m}^{-1}\text{ K}^{-1}\text{) of}$$
$$\text{polycarbonate}$$

$$Q = \text{heat flow along bar}$$

From our first assumption, Q is constant along the length of the assembly. Also applying the third assumption:

$$Q = Q_{\text{top polycarbonate}} = Q_{\text{sample}} = Q_{\text{bottom polycarbonate}}$$
$$Q = \lambda_p \times \Delta T_1/\Delta x_p = \lambda_s \times \Delta T_s/\Delta x_s = \lambda_p \times \Delta T_2/\Delta x_p \tag{4.20}$$

To help compensate for side loss of heat, heat flow across the top and bottom polycarbonate discs is averaged, and assumed to be equal to the heat flow across the sample:

$$0.5 \times \lambda_p \times (\Delta T_1 + \Delta T_2)/\Delta x_p = \lambda_s \times \Delta T_s/\Delta x_s \tag{4.21}$$

A simple rearrangement gives:

$$\lambda_s = [(\Delta T_1 + \Delta T_2)/\Delta T_s] \times \Delta x_s \times [\lambda_p/(2\Delta x_p)] \tag{4.22}$$

The only term on the right-hand side of Equation (4.22) that cannot be directly measured is $[\lambda_p/(2\Delta x_p)]$, which relates to the physical properties of the polycarbonate discs. The term, henceforth referred to as C, should remain constant for the life of the equipment. We describe below how C can be determined.

Also of importance is the thermal resistance at the contacts between the bar and the sample. The contact resistance, R, increases the apparent resistance of the sample, $\Delta x_s/\lambda_{s(measured)}$, and should be removed when calculating true sample conductivity, $\lambda_{s(corrected)}$:

$$\lambda_{s(corrected)} = \Delta x_s/[(\Delta x_s/\lambda_{s(measured)}) - R] \tag{4.23}$$

C and R can be determined at the same time by measuring several standard samples of known thickness and thermal conductivity (e.g. fused silica glass – $\lambda_{s(corrected)} = 1.36$ W m^{-1} K^{-1}; Touloukian et al., 1970b).

Rewriting Equation (4.22):

$$\lambda_{s(measured)} = [(\Delta T_1 + \Delta T_2)/\Delta T_s]\Delta x_s C \tag{4.24}$$

Equation (4.23) can now be written:

$$\frac{\Delta x_s}{\lambda_{s(corrected)}} = \frac{\Delta T_s}{(\Delta T_1 + \Delta T_2)} \times \frac{1}{C} - R \tag{4.25}$$

Equation (4.25) is linear. Several measurements of standards of different thickness should be sufficient to define a straight line of gradient $1/C$ and intercept R. Equation (4.25) can then be used to determine the conductivity of future samples.

Question: We built a new divided-bar apparatus. The first step was to calibrate it to find the constant, C, and contact resistance, R, for the device. To this end, we used four discs of fused silica glass, $\lambda_s = 1.36$ W m^{-1} K^{-1}. We also measured a granite sample of unknown conductivity. The raw data we collected from each of these experiments is tabulated below.

	Δx_s (m)	$T_A(°C)$	$T_B(°C)$	$T_C(°C)$	$T_D(°C)$
Silica disc 1	0.00920	42.921	38.320	31.458	26.900
Silica disc 2	0.01215	42.932	38.660	30.802	26.961
Silica disc 3	0.01613	42.943	39.061	30.445	27.329
Silica disc 4	0.02380	42.900	39.350	29.135	26.839
Granite	0.02468	42.868	37.706	30.947	26.839

Find C and R for our new instrument, and determine the conductivity of the granite.

Answer: Remembering that $\Delta T_1 = T_A - T_B$, $\Delta T_2 = T_C - T_D$ and $\Delta T_s = T_B - T_C$, solution for C and R involves substituting the values for the silica standards into Equation (4.25). As an intermediate step, we can tabulate $\Delta x_s/\lambda_s$ and $\Delta T_s/(\Delta T_1 + \Delta T_2)$:

	$\Delta T_s/(\Delta T_1 + \Delta T_2)$	$\Delta x_s/\lambda_s$
Silica disc 1	0.74921	0.006765
Silica disc 2	0.96857	0.008934
Silica disc 3	1.23121	0.011860
Silica disc 4	1.74734	0.017500
Granite	0.72913	

The four points for the silica discs define a straight line ($r = 0.9998$) and suggest $C = 92.45$ and $R = 0.001435$.

The conductivity of the granite can be found by substituting these values back into Equation (4.25):

$$0.02468/\lambda_s = (0.72913/92.45) - 0.001435$$
$$\lambda_s = 3.83 \ \mathrm{W\,m^{-1}\,K^{-1}}$$

If λ_s denotes the measured conductivity of a water-saturated sample, then the conductivity of the rock matrix, λ_m, is found from the square-root mean law:

$$\lambda_m = [(\sqrt{\lambda_s} - \phi\sqrt{\lambda_w})/(1 - \phi)]^2 \tag{4.26}$$

where λ_w is the conductivity of water (Touloukian et al., 1970a):

$$\lambda_w = -7.42 \times 10^{-6}T^2 + 5.99 \times 10^{-3} \times T - 0.522 \tag{4.27}$$

where T is temperature (K).

Question: The bulk thermal conductivity of a saturated limestone ($\phi = 11.2\%$) is measured using the divided bar from the above example: $\lambda_{s(corrected)} = 2.64 \ \mathrm{W\,m^{-1}\,K^{-1}}$ at 32°C. Given that the sample is 1.57 cm long, what is the conductivity of the limestone matrix, λ_m?

Answer: λ_w is found from Equation (4.27) after converting temperature to kelvin:

$$\lambda_w = -7.42 \times 10^{-6} \times (305.15)^2 + 5.99 \times 10^{-3} \times (305.15) - $$
$$0.522 = 0.615 \ \mathrm{W\,m^{-1}\,K^{-1}}$$

Matrix conductivity is then found using Equation (4.26):

$$\lambda_m = [(\sqrt{\lambda_s} - \phi\sqrt{\lambda_w})/(1 - \phi)]^2 = [(\sqrt{2.64} - $$
$$0.112 \times \sqrt{0.615})/(1 - 0.112)]^2 = (1.537/0.888)^2$$
$$\lambda_m = 3.00 \ \mathrm{W\,m^{-1}\,K^{-1}}$$

For unconsolidated material, a hollow cell is often recommended as a substitute for a consolidated sample. The cell is generally constructed from a hollow cylinder of Perspex or clear plastic, sealed at both ends with brass (or other highly conductive material) discs. The cell is constructed to have the same diameter as the divided bar, so that when packed full with saturated sediment or rock chips it mimics a consolidated sample of the same lithology. The thermal conductivity of the total cell is found by normal operation of the divided bar. The conductivity of the sediment aggregate, λ_{ag}, is subsequently derived by removing the effects of the cell wall (conductivity $= \lambda_{\text{cell}}$) and pore water (Figure 4.7).

The conductivity of the total cell is the arithmetic mean of the aggregate and the cell (refer to Figure 4.7 for variable names):

$$\lambda_{\text{total}} = \lambda_{ag} \times \frac{\pi r_i^2}{\pi r_o^2} + \lambda_{\text{cell}} \times \frac{\pi r_o^2 - \pi r_i^2}{\pi r_o^2} \tag{4.28}$$

So

$$\lambda_{ag} = (\lambda_{\text{total}} - \lambda_{\text{cell}})\left(\frac{r_o}{r_i}\right) + \lambda_{\text{cell}} \tag{4.29}$$

Then the conductivity of the matrix is found using the square-root mean law:

$$\lambda_m = [(\sqrt{\lambda_{ag}} - \phi\sqrt{\lambda_w})/(1 - \phi)]^2 \tag{4.30}$$

or the geometric mean law:

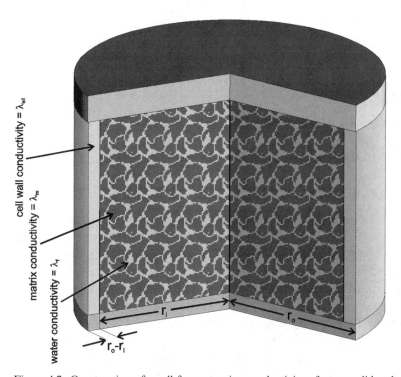

Figure 4.7. Construction of a cell for measuring conductivity of unconsolidated material.

$$\ln(\lambda_m) = [\ln(\lambda_{ag}) - \phi \times \ln(\lambda_w)]/(1 - \phi) \tag{4.31}$$

where λ_{total} = conductivity of the full cell and ϕ is proportion of water within the cell.

The uncertainties associated with the cell method are inherently greater than those associated with core measurements. The extra uncertainty due to the conductivity of the cell and volume of intergranular water are added to the systematic error of the divided-bar technique.

Question: Drill cuttings collected over a depth range of 10 m are packed into a hollow cell designed to fit into a divided bar of 4 cm diameter. The walls of the cell are 2.5 mm thick and composed of a polycarbonate material with thermal conductivity $\lambda_{cell} = 0.42$ W m^{-1} K^{-1}. Careful measurement shows that the proportion of water in the cell is 13.6%. The conductivity of the entire cell at 32°C is determined to be 2.63 W m^{-1} K^{-1}. What is the mean conductivity of the sediment matrix, λ_m?

Answer: The relevant variables are

$\lambda_{total} = 2.63$ W m^{-1} K^{-1}

$\lambda_{cell} = 0.42$ W m^{-1} K^{-1}

$r_o/r_i = 20/17.5 = 1.143$

$\phi = 0.136$

$\lambda_w = 0.615$ W m^{-1} K^{-1} from previous example

So from Equation (4.29):

$\lambda_{ag} = (2.63 - 0.42) \times (1.143)^2 + 0.42 = 3.307$ W m^{-1} K^{-1}

λ_m is then found using Equation (4.30):

$\lambda_m = [(\sqrt{\lambda_{ag}} - \phi\sqrt{\lambda_w})/(1 - \phi)]^2 = [(1.8185 - (0.136 \times 0.7842))/0.864]^2$

$\lambda_m = 3.93$ W m^{-1} K^{-1}

or from Equation (4.31):

$\ln(\lambda_m) = [\ln(\lambda_{ag}) - \phi \times \ln(\lambda_w)]/(1 - \phi) = [1.196 - (0.136 \times -0.48613)]/0.864 = 1.461$

$\lambda_m = 4.31$ W m^{-1} K^{-1}

In this case we would suggest $\lambda_m = 4.1 \pm 0.2$ W m^{-1} K^{-1}

The cell method makes the implicit assumption that the strata over the sampled depth range are more or less homogeneous, and yields the approximate geometric or square-root mean conductivity of the sampled sediments. However, many strata are layered and are thus more accurately modelled using the harmonic mean law, which is less than or equal to the geometric mean at all times. The greater the inhomogeneity of the formation, the greater the discrepancy between the geometric and harmonic means.

Another concern with the cell method is its inability to allow for anisotropy within formations. The conductivity of anisotropic rocks and minerals randomly arranged in an aggregate will not equal the vertical conductivity of the *in situ* formation. Sass et al. (1992) proposed the following relationship:

$$\lambda_{ag} = \lambda_x^{2/3} \times \lambda_z^{1/3} \tag{4.32}$$

where λ_{ag} is the aggregate conductivity, λ_x is the conductivity parallel to bedding and λ_z is conductivity perpendicular to bedding. Deming (1994) combined this relationship with an observed decrease in anisotropy with increasing conductivity to arrive at the following empirical formula:

$$\lambda_z = \exp[(\ln(\lambda_{ag}) - 0.6267)/0.5480] \tag{4.33}$$

This equation gives an estimate of vertical matrix conductivity in formations for which only an aggregate of drill cuttings is available. The formula is intended for use on aggregates with conductivity in the range 1.0–4.0 $\mathrm{W\,m^{-1}\,K^{-1}}$, and care should be taken 'in the case of unusual sedimentary lithologies or igneous rocks' (Deming, 1994). Equation (4.33) is not recommended for very low-conductivity rocks (e.g. coal).

Question: An aggregate of drill cuttings is taken from a shale-rich sequence and packed randomly into a cell for thermal conductivity measurement. The matrix conductivity of the aggregate is 2.32 $\mathrm{W\,m^{-1}\,K^{-1}}$. What is an estimate of the *in situ* vertical matrix conductivity of the formation?

Answer: From Equation (4.33):
$$\lambda_z = \exp[(\ln(2.32) - 0.6267)/0.5480] = 1.48 \ \mathrm{W\,m^{-1}\,K^{-1}}$$

Unfortunately, Equation (4.33) makes no allowance for isotropic units within the aggregate. This means that, while the relationship may work well for highly anisotropic lithologies like shale, it makes an unnecessary correction to isotropic sediments. It is suggested that the correction only be applied to that proportion of the sediment that is anisotropic. While not mathematically precise, it should still improve the result.

Question: The aggregate in the above example is found to contain 20% massive sandstone. What might be a better estimate of *in situ* vertical matrix conductivity given this new evidence?

Answer: Given that 20% of the matrix is isotropic, the anisotropy correction should only be applied to 80% of the formation. That is, 20% of the formation has vertical conductivity 2.32 $\mathrm{W\,m^{-1}\,K^{-1}}$ and 80% has 1.48 $\mathrm{W\,m^{-1}\,K^{-1}}$. These should be combined using a harmonic mean:
$$\lambda_z = 1/((0.2/2.32) + (0.8/1.48)) = 1.60 \ \mathrm{W\,m^{-1}\,K^{-1}}$$

4.3.4. Transient Methods

Steady-state techniques are unsuitable for poorly consolidated sediment or *in situ* measurements. Yet in many cases, especially sea-floor measurements, such situations are encountered where a thermal conductivity estimate is required to convert temperature data into a heat flow measurement. A conductivity measurement system known as the *transient* technique has evolved to handle these cases.

The idea behind transient measurement techniques is that the thermal conductivity of a body can be deduced from the rate at which its temperature changes in response to an applied heat source. Devices utilising transient techniques vary only in shape and position of the heat source in relation to the temperature measurement point. The most commonly used transient tool is a line-source needle probe (Figure 4.8), first described by DeVries and Peck (1958), and then by Von Herzen and Maxwell (1959). It can be shown (e.g. Carslaw and Jaeger, 1959) that if a line source of heat and a temperature sensor are packaged closely together in a hypodermic needle and inserted into a material with thermal conductivity λ_s, then

$$\lambda_s = (Q/4\pi)(\partial \ln(t)/\partial T) \tag{4.34}$$

where Q is applied heat (W m^{-1}), t is time (s) and T is temperature (K).

If Q is known accurately, λ_s can be determined directly once $\ln(t)$ versus T attains linearity. This is usually within 30 s, although a minimum of 200 s is generally required to establish unambiguous gradients.

Question: A 10 cm needle probe with a heating coil powered by a 5 V battery is inserted into a sandstone block. A current, $I = 250$ mA, is switched on at time $t = 0$ s, and the temperature, T is recorded over a 200 s period.

t (s)	T (°C)	t (s)	T (°C)
0	20.967	80	22.203
10	21.379	100	22.268
20	21.746	120	22.321
30	21.911	140	22.366
40	22.003	160	22.405
50	22.068	180	22.439
60	22.121	200	22.470

What is the thermal conductivity of the sandstone?

Answer: The power of the heating coil, $P = V \times I = 5 \times 0.25 = 1.25$ W. The heat source is 10 cm long, so the heating rate, $Q = P/0.1 = 12.5$ W m^{-1}. The $\ln(t)$ versus T curve from 60 to 200 s is very close to linear with a gradient of 3.445. λ_s is found directly from Equation (4.34):

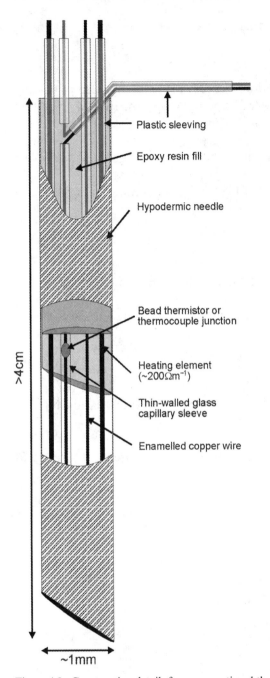

Plastic sleeving

Epoxy resin fill

Hypodermic needle

Bead thermistor or
thermocouple junction

Heating element
($\sim 200 \Omega m^{-1}$)

Thin-walled glass
capillary sleeve

Enamelled copper wire

>4cm

~1mm

Figure 4.8. Construction details for a conventional thermal conductivity needle probe. Not to scale.

$$\lambda_s = (Q/4\pi) \times (\partial \ln(t)/\partial T) = (12.5/4\pi) \times (3.445) =$$
$$3.43 \, \text{W} \, \text{m}^{-1} \, \text{K}^{-1}$$

The disadvantage of the conventional needle probe is that steady heating currents must be maintained over extended periods, with the risk that resulting temperature gradients may induce convection within samples of high permeability (resulting in overestimates of conductivity). To reduce measurement duration and the possibility of convection, the maximum gradient technique was developed using a dual-needle configuration (Cull, 1975).

At a distance r from a line source of heat, the temperature is (Carslaw and Jaeger, 1959)

$$T = \frac{Q}{4\pi(\rho C \kappa)} \int_0^t \exp(-r^2/4\kappa t)\partial t \tag{4.35}$$

where ρ, C and κ are the density, heat capacity and thermal diffusivity of the host material, respectively.

Differentiating Equation (4.35), the rate of temperature increase can be expressed as:

$$(\partial T/\partial t) = (Q/4\pi\lambda t) \times \exp(-r^2/4\kappa t) \tag{4.36}$$

This equation has a maximum value, M, at time t_m (Figure 4.9). By further differentiating and equating to zero, we obtain the following relationship:

$$\kappa = r^2/4t_m \tag{4.37}$$

This expression for κ can now be substituted back into Equation (4.36), and conductivity can be expressed as

$$\lambda = Q \times \exp(-1)/(4\pi M t_m) \tag{4.38}$$

which reduces to:

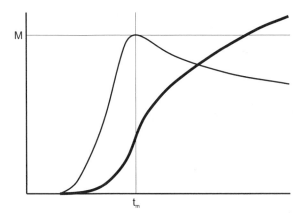

Figure 4.9. Temperature (thick line) and heating rate (thin line) versus time at a position offset from a line heat source.

$$\lambda = 0.0293 \times \mathbf{Q}/M t_m \tag{4.39}$$

If M and t_m can be determined and $\mathbf{Q}$ is known, then thermal conductivity is easily calculated.

The above theory can be put into practice using a dual-needle probe, with one needle housing the heating wire and the other the thermocouple or thermistor (Figure 4.10). Distance between the two needles, r, should be set such that t_m is significant compared with the accuracy of the measuring device. For an average rock of $\kappa = 10^{-6}$ m^2 s^{-1}, Equation (4.37) indicates that $t_m > 1$ s when $r > 2$ mm. A small safety margin is beneficial, so a needle spacing of around 3 mm is recommended.

Cull (1975) observed a systematic error in t_m, attributed to contact resistance in the probe. A correction factor can be calculated by evaluating Equation (4.37) for two measurements of the same sample but for different values of r:

$$\kappa_{r1} = \kappa_{r2}$$

so

$$\frac{r_1^2}{4(t_{m1(\text{observed})} - t_D)} = \frac{r_2^2}{4(t_{m2(\text{observed})} - t_D)} \tag{4.40}$$

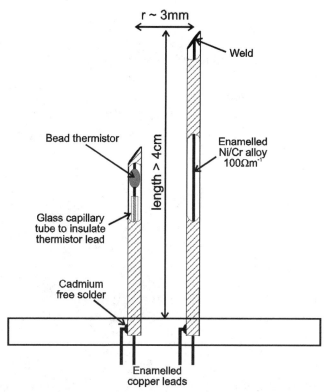

Figure 4.10. Construction details for a dual-needle probe.

Once the constant time delay, t_D, is established, future measurements of t_m can be corrected:

$$t_m = t_{m(\text{observed})} - t_D \qquad (4.41)$$

Question: A series of determinations of M and t_m were made using a sample of granite with different spacing between the heat source and measurement point. The heating circuit carried a current $I = 0.86$ A along a wire with resistivity $R = 60 \ \Omega \, m^{-1}$. Calculate the conductivity of the sample given the following results:

Offset (mm)	$M \ (°C \, s^{-1})$	$t_m(\text{observed})$ (s)	t_m (s)
1.5	0.467	2.3	1.08
2.0	0.331	2.7	1.48
2.5	0.246	3.3	2.08
2.0	0.195	3.9	2.68
3.5	0.170	4.2	2.98
4.0	0.136	4.9	3.68

Answer: The first step is to calculate the time-delay, t_D, as defined by Equation (4.41). Equation (4.39) indicates that the product of M and t_m should be constant, regardless of the offset distance. That is:

$$M \times (t_{m(\text{observed})} - t_D) = k$$

where k is a constant.

This rearranges to a simple linear relationship:

$$t_{m(\text{observed})} = (k/M) + t_D$$

Using the M and $t_{m(\text{observed})}$ values from the table of results above, the relationship is best fit using $t_D = 1.22$ s. This value of t_D was used to calculate t_m in the fourth column above.

The heating power $Q = I^2 R = 0.86^2 \times 60 = 44.38 \ W \, m^{-1}$. Equation (4.39) is used to calculate λ using Q, M and t_m for each offset distance:

offset (mm)	$\lambda (W \, m^{-1} \, K^{-1})$
1.5	2.58
2.0	2.65
2.5	2.54
3.0	2.49
3.5	2.57
4.0	2.60

The average is $\lambda = 2.57 \ W \, m^{-1} \, K^{-1}$. If we assume an uncertainty of ± 0.1 s in t_m, $\pm 0.005 °C \, s^{-1}$ in M, and ± 0.02 A in I, then the uncertainty in λ comes to around $\pm 10\%$, or $\lambda = 2.6 \pm 0.3 \ W \, m^{-1} \, K^{-1}$. This was a real example (Cull, 1975). A divided bar measurement on the same sample indicated a conductivity $\lambda = 2.74 \ W \, m^{-1} \, K^{-1}$, well within the limits of the needle-probe result.

The needle probes described above can be difficult to use in consolidated rock because they cannot be easily embedded. It is possible to dispense with the needle package, however, and cement the same configurations onto polythene or Teflon strips. Very thin 'probes' can be sandwiched between two halves of a rock cut smoothly with a diamond saw (Cull, 1974). Plastic strips are sufficiently soft and flexible that, with a few bars of pressure applied between the two halves of rock, they fill any small gaps or irregularities left after slicing and occupy only a small space. With wires cemented in the conventional needle-probe configuration (temperature sensor very close to the heat source), the effect of the plastic is the same as the wall resistance usually found at the needle surface. It does not affect the long-time thermal response from which conductivity is obtained. For the dual needle or maximum-gradient configuration, the thickness of plastic ($\sim 10^{-4}$ m) is a small fraction of the distance between the heater and sensor ($\sim 3 \times 10^{-3}$ m), and does not result in significant delays in sensor response.

The dual-needle or maximum-gradient configuration may be unsuitable for measurements in coarse-grained material. A large offset distance between heat source and sensor is required so that true bulk properties are measured, rather than the properties of a single grain. However, the offset distance cannot be increased indefinitely without violating the assumption of an infinite line source of heat. For such samples, measurements are best made using a circular heat source with a temperature sensor at the centre (Figure 4.11). No line-source end-effects are encountered while large circle radii are equivalent to large offset distances.

With circle diameters less than 1 cm, adequate signal-to-noise ratios can be sustained using a heater power of approximately 0.25 W (Cull, 1975). The temperature record is then differentiated in the same way as for the line-source configuration.

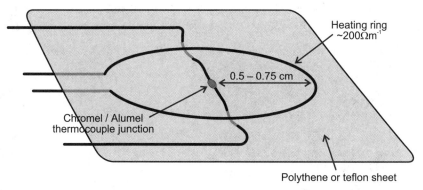

Figure 4.11. Ring source sandwich probe for embedding into small samples of rock.

Note: If plastics are used in the construction of sandwich probes, there is a limit to the heating power that can be sustained before local melting becomes probable. It is suggested that 0.25 W should not be exceeded.

Carslaw and Jaeger (1959) and Somerton and Mossahebi (1967) considered the temperature rise at the centre of a circular source of heat. The solution can be expressed as follows:

$$T = (\mathbf{Q}_T/4\pi r\lambda)\,\text{erfc}(r^2/4\kappa t)^{1/2} \tag{4.42}$$

where $\mathbf{Q}_T$ (W) is the total amount of heat produced per unit time by a ring of radius r. Following the same procedures as for a line source, the rate of temperature change can be expressed as follows:

$$(\partial T/\partial t) = [\mathbf{Q}_T/(8(\pi^3\kappa t^3)^{1/2}\lambda)] \times \exp(-r^2/4\kappa t) \tag{4.43}$$

Equation (4.43) has a maximum value, M, when $(\partial^2 T/\partial t^2) = 0$, at time

$$t_m = r^2/6\kappa \tag{4.44}$$

Using this relationship, κ can be eliminated from Equation (4.43) and conductivity can be expressed as follows:

$$\lambda = (\mathbf{Q}_T/2\pi r)[4(\exp(3) \times \pi/6)^{1/2} \times t_m M]^{-1} \tag{4.45}$$

Evaluating the constants, this reduces to

$$\lambda = 0.0771 \times \mathbf{Q}/Mt_m \tag{4.46}$$

where $\mathbf{Q}$ (W m^{-1}) = heat liberated per unit length of wire per unit time = $\mathbf{Q}_T/2\pi r$. This corresponds to Equation (4.39) for the line-source solution.

Question: A circular heating coil of 1 cm diameter is constructed with a temperature sensor in the centre. The loop is glued onto a Teflon strip and designed to supply heat at a rate of 0.25 W. The assembly is sandwiched between two halves of a limestone block and sealed with a moderate retaining pressure. Temperature, T, measured at time, t, after heating commenced is tabulated below.

t (s)	T (°C)	t (s)	T (°C)
0	20.967	35	21.044
5	20.967	40	21.068
10	20.968	45	21.092
15	20.972	50	21.116
20	20.983	55	21.139
25	21.000	60	21.161
30	21.021	65	21.183

What is the approximate thermal conductivity of the limestone?

Answer: First we must determine the heating rate of the loop:

$$\mathbf{Q} = \mathbf{Q}_T/2\pi r = 0.25/(2\pi \times 0.005) = 7.9577 \text{ W m}^{-1}$$

The maximum value of $(\partial T/\partial t)$ is at time $t_m = 42.5 \pm 1.5$ s, when the gradient $M = 0.0241 \pm 0.0001$ °C per 5 s $= 0.00482 \pm 0.00002$ K s^{-1}. Substituting into Equation (4.46):

$\lambda_s = 0.0771\mathbf{Q}/Mt_m = (0.0771 \times 7.9577)/(0.00482 \times 42.5) =$
$\quad 3.00 \pm 0.12 \text{ W m}^{-1}\text{K}^{-1}$

Direct observation of t_m results in underestimates of thermal conductivity using Equation (4.46) (Cull, 1974). Corrections must be applied in a manner similar to those described in Equations (4.40) and (4.41) for the line-source configuration.

Question: The circuit in the above example is found to have an inherent time lag, $t_D = 1.5$ s. What is the corrected value of λ_s?

Answer: From Equation (4.41):

$t_m = t_{m(\text{observed})} - t_D = 42.5 - 1.5 = 41 \pm 1.5$ s.

So:

$\lambda_s = (0.0771 \times 7.9577)/(0.00482 \times 41.0) = 3.10 \pm$
$\quad 0.13\text{W m}^{-1}\text{K}^{-1}$

Thermal conductivity measured on several small samples from a single rock often differs significantly because of local variations in composition, especially in coarse-grained rocks. Additionally, the parent strata may contain large fracture systems, or may be composed of aggregates, in which case the thermal behaviour of small samples bears little relevance to *in situ* conditions.

Note: These sampling limitations also apply to thermal conductivity measured using the steady-state method.

In such cases, direct *in situ* measurement of bulk conductivity is preferable. In response to this need, line-source probes have been developed for use in boreholes (Beck, Anglin and Sass, 1971), but they require a considerable amount of power and long observation times. Point-source probes reduce the duration of measurement and have the additional advantage that measurements can be made at the deepest point in narrow holes. This is not possible with a line source.

The basic assembly for a point-source probe generally consists of a silicon rubber rod, so that no thermal short circuit can exist between the heater and sensor (Figure 4.12). The heater consists of several windings of Nichrome resistance wire (total less than 2 Ω) and the temperature sensor is a thermistor.

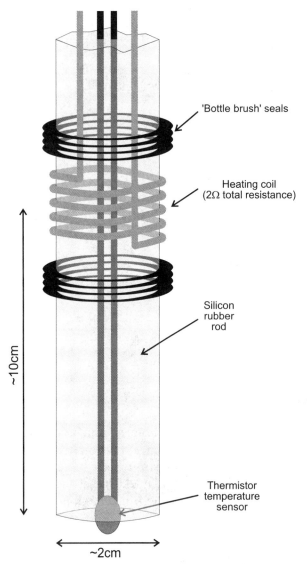

Figure 4.12. Point-source *in situ* thermal conductivity probe.

A 12 V car battery supplies sufficient power for the heater. Bottlebrush seals along the edge of the rod limit convection in water-filled holes (Beck et al., 1971). For best results, the probe diameter should be close to that of the hole in which it is used. This further limits convection and ensures rapid transfer of heat between the probe and the surrounding rock.

The point-source solution as given by Carslaw and Jaeger (1959) is

$$T = [\mathbf{Q}_T/(8C\rho(\pi\kappa)^{3/2})] \times \int_0^t t^{-3/2} \exp(-r^2/(4\kappa t))\partial t \qquad (4.47)$$

where $\mathbf{Q}_T$ is heat output from the source (W).

Following the maximum-gradient techniques described above, this expression can be differentiated to give

$$(\partial T/\partial t) = [\mathbf{Q}_T/(8C\rho(\pi\kappa)^{3/2})]t^{-3/2} \times \exp[-r^2/(4\kappa t)] \tag{4.48}$$

This is simply a transformation of Equation (4.43), because the point-source problem is simply the inverse of the ring-source problem. It follows that Equation (4.46) can be used to express conductivity:

$$\lambda_s = 0.0123 \times \mathbf{Q}_T/rMt_m \tag{4.49}$$

where r is the distance to the temperature sensor.

For a sensor distance of 10 cm in material of diffusivity near 1×10^{-6} m^2 s^{-1}, it is apparent from Equation (4.44) that λ_s cannot be determined in less than about 25 min. For sensor separations of 20 cm, the necessary time increases to around 110 min.

It may be possible to reduce measurement time by adding a second differentiating phase in the procedure. The final output is then the second derivative of temperature, and the reduction procedures are identical to those used above for the first derivative. The curve $\partial^2 T/\partial t^2$ has a maximum value when $\partial^3 T/\partial t^3 = 0$, which can be shown to be at a time t'_m given by

$$t'_m = r^2/16.33\kappa \tag{4.50}$$

Eliminating κ, it can be further shown that

$$\lambda = 0.004\mathbf{Q}_T/rM'(t'_m)^2 \tag{4.51}$$

With r equal to 10 cm, it is clear from Equation (4.50) that measurement duration is reduced to approximately 10 min. However, if power is dissipated at 200 W in rock of conductivity 2.5 W m^{-1} K^{-1}, the sensor (at a distance of 10 cm) must be capable of resolving temperature gradients of approximately 0.001°C min^{-1}.

In Equations (4.49) and (4.51), corrections are required to t_m and t'_m similar to those described by Equation (4.41) for the line source. Correction factors, t_D and t_D', can be found by calibrating the probe in a material of known conductivity.

Another situation where *in situ* measurement is preferable is on the ocean floor. A Lister-type deep-ocean heat flow probe (see Section 3.1.2) contains circuitry (see Figure 3.6) that enables it to make an *in situ* estimate of average thermal conductivity over a fixed depth interval. It works on a principle similar to a line-source needle probe, except that instead of recording the temperature increase due to an applied heat source, it records the temperature decay after a heat pulse. An array of thermistors is used to average the temperature response over a depth interval (usually ~1 m), then thermal conductivity, λ, is found from

$$\lambda = 2\alpha\tau \times F(\alpha, \tau) \times \mathbf{H}/(4\pi\Delta Tt) \tag{4.52}$$

where

$\alpha = 2 \times$ ratio of specific heat capacity of the sediment to that of the probe
$= 2$ (usually)
$\tau = \kappa t/r^2 =$ thermal time constant ($\kappa =$ thermal diffusivity of sediments
($m^2\,s^{-1}$), $r =$ radius of probe (m))
$F(\alpha, \tau)$ values were given in Table 3.1
$\mathbf{H} =$ magnitude of heat pulse ($J\,m^{-1}$)
$\Delta T =$ temperature above equilibrium (K)
$t =$ effective time since heat pulse (the origin, $t = 0$, must be empirically established)

An iterative approach is usually employed to find λ, beginning with an initial guess at κ, then using λ calculated for a number of different times to refine the initial κ estimate before a final calculation of λ (Davis, 1988).

Question: In a particular heat pulse experiment, a thermistor array returned the following time–temperature data:

t (s)	ΔT (K)
120	0.795
280	0.380
370	0.293
420	0.260
560	0.198

Given that the probe had a diameter of $2r = 1$ cm and the heat pulse was $\mathbf{H} = 1080\,J\,m^{-1}$, what was the conductivity of the sediments?

Answer: To proceed we must make an initial guess at the diffusivity of the sediments. We will use $\kappa = 3 \times 10^{-7}\,m^2\,s^{-1}$.

Given that $\mathbf{H} = 1080$, $\alpha = 2$ and $\kappa/r^2 = 0.012$, Equation (4.52) reduces to

$$\lambda = 12.96 \times F(2, 0.012t)/(\pi \times \Delta T)$$

Using the above data, we can calculate λ for each time t:

t (s)	$F(2, 0.012t)$	λ ($W\,m^{-1}\,K^{-1}$)
120	0.12788	0.66358
280	0.06431	0.69815
370	0.05032	0.70848
420	0.04490	0.71241
560	0.03450	0.71880

The calculated conductivity is increasing with time, which indicates that our initial guess at κ was too low. We recalculate using $\kappa = 5 \times 10^{-7}\,m^2\,s^{-1}$. Now $\kappa/r^2 = 0.02$ and Equation (4.52) is

$$\lambda = 21.6 \times F(2, 0.02t)/(\pi \times \Delta T)$$

Referring again to Table 3.1 and the above data:

t (s)	$F(2, 0.02t)$	λ (W m^{-1} K^{-1})
120	0.08547	0.73918
280	0.04080	0.73821
370	0.03155	0.74035
420	0.02801	0.74070
560	0.02133	0.74068

There is now excellent agreement in calculated conductivity at all times.

$$\lambda = 0.74 \text{ W m}^{-1} \text{ K}^{-1}$$

4.3.5. Compaction Models

To relate the matrix conductivity to the *in situ* conductivity of the formation from whence the sample came, we must investigate the porosity of the formation. Water has a low conductivity compared with most minerals, so the conductivity of a matrix/water mix is very sensitive to the relative proportions of the two components. Porosity is therefore the major variable controlling the thermal conductivity of sedimentary rocks (Figure 4.13). It follows that we need to assess carefully the porosity of a formation before we can accurately model its conductivity.

While the porosity of basalts, granites and other hard rocks generally varies little with depth (except perhaps for a highly vesicular basalt), most sedimentary rocks compact with burial, and porosity generally decreases systematically with depth. Water is forced from the rock, increasing the mean thermal conductivity. If the section under investigation has been surveyed with wire-line tools, then porosity may be derived by electrical well log analysis (see Section 4.4). Otherwise, we must construct compaction models to relate porosity to depth of burial.

There are at least three popular models describing the compaction of sediment with increasing burial. That of Sclater and Christie (1980) states that the porosity, ϕ, decays exponentially with depth of burial, z:

$$\phi = \phi_0 \times \exp(-Az) \tag{4.53a}$$

or the equivalent:

$$\ln(\phi) = \ln(\phi_0) - Az \tag{4.53b}$$

where ϕ_0 = porosity of sediments at time of deposition and A is constant compaction coefficient.

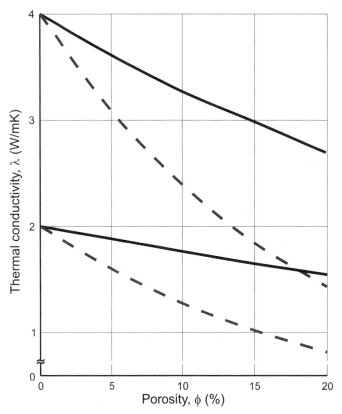

Figure 4.13. Conductivity, λ, versus porosity, ϕ, for two rocks with $\lambda_m = 2$ and $\lambda_m = 4$. Solid line assumes water-filled pores, dashed line air-filled pores. Modified after Jessop (1990).

Question: For North Sea sandstones, Sclater and Christie (1980) suggest $\phi_0 = 0.49$ and $A = 2.7 \times 10^{-4}$ when z is expressed in metres. What is the expected porosity of sandstone buried to 2500 m under the North Sea?

Answer: From Equation (4.53):

$$\ln(\phi) = \ln(0.49) - 2.7 \times 10^{-4} \times 2500 = -0.71335 - 0.67500 = -1.38835$$

So

$$\phi = \exp(-1.38835) = 0.2495.$$

From an initial porosity of 49%, the sandstone has almost halved its pore space to 25%.

Falvey and Middleton (1981) alternatively proposed that the reciprocal of porosity, $1/\phi$, increases linearly with depth:

$$1/\phi = 1/\phi_0 + Bz \tag{4.54}$$

where B is a constant compaction coefficient.

Question: What value of B will give the same porosity for sandstone at 2500 m as the previous example? What, then, is the discrepancy between the two models for porosity at 1500 m?

Answer: As for the previous example, $\phi = 0.2495$, $\phi_0 = 0.49$ and $z = 2500$.

So from Equation (4.54):

$(1/0.2495) = (1/0.49) + B \times 2500$

$B = [(1/0.2495) - (1/0.49)]/2500 = 7.8688 \times 10^{-4}$

At 1500 m, then, Equation (4.54) predicts:

$1/\phi = 1/0.49 + 7.8688 \times 10^{-4} \times 1500 = 3.221$

$\phi = 1/3.221 = 0.31$, or 31% porosity

In contrast, Equation (4.53) predicts:

$\ln(\phi) = \ln(0.49) - 2.7 \times 10^{-4} \times 1500 = -1.1183$

$\phi = \exp(-1.1183) = 0.33$, or 33% porosity

The discrepancy between the compaction models is small.

Baldwin and Butler (1985) suggested that while the Sclater and Christie (1980) model is best for sandstone, the compaction of shale and limestone is best explained using a power law model:

$$z = z_{max}(1 - \phi)^C \tag{4.55}$$

where z_{max} is the depth at which all fluid is expelled and C is a compaction constant.

Note: At $z = 0$, the Baldwin and Butler (1985) relationship, Equation (4.55), predicts 100% porosity.

Transposed into a form similar to Equations (4.51) and (4.52), Equation (4.55) becomes

$$\ln(1 - \phi) = -\ln(z_{max})/C + \ln(z)/C \tag{4.56}$$

Question: Baldwin and Butler (1985) suggested values of $C = 6.35$, $z_{max} = 6020$ m for shale formations less than 200 m thick, and $C = 8$, $z_{max} = 15,000$ m for shales greater than 200 m thick. The different compaction curves reflect the fact that it is difficult for thick shale sequences to lose water. What is the predicted porosity at 3000 m burial for a shale unit (a) 100 m thick, and (b) 500 m thick?

Answer: For part (a), $z = 3000$, $z_{max} = 6020$, $C = 6.35$, so from Equation (4.56):

$\ln(1 - \phi) = (1/6.35) \times \ln(3000/6020) = -0.10968$

$(1 - \phi) = \exp(-0.1097) = 0.896$

$\phi = 0.104$, or 10.4%

For part (b), $z = 3000$, $z_{max} = 15,000$, $C = 8$, so from Equation (4.56):

$\ln(1 - \phi) = (1/8) \times \ln(3000/15000) = -0.20118$

$(1 - \phi) = \exp(-0.20118) = 0.818$

$\phi = 0.182$, or 18.2%

To determine which of these models is most appropriate for a region, real porosity data are required. Data are often included in well completion reports and can be extracted at the same time as lithology information. Alternatively, new data can be measured on core samples prior to steady-state thermal conductivity measurements, as described in Section 4.3.2. Different lithologies compact at different rates, so those data relating to the same formation and lithology should be collated – on a regional scale if enough data are not available from a single well. The data should be plotted successively on $\ln(\phi)$ versus z, $(1/\phi)$ versus z and $\ln(1 - \phi)$ versus $\ln(z)$, axes. The graph producing the most linear plot (determined by linear interpolation) is the most realistic, and values for ϕ_0, z_{max}, A, B or C can be resolved as required for each lithology.

Question: Eleven porosity data for sandstone units in the well Rob Roy 1, Browse Basin, Australia, were included in the completion report for the well, and are listed below:

Depth (m)	ϕ (%)	Depth (m)	ϕ (%)	Depth (m)	ϕ (%)
366.5	42	1364.5	30	1966.5	9
576.0	51	1438.0	19	2048.5	6
1036.5	40	1664.5	11	2112.0	11
1288.0	28	1896.0	6		

Assuming these are the only data available for the region, which compaction model gives the best linear fit, and what is the expected porosity of a sandstone layer at 1500 m?

Answer: See Figure 4.14:

Plotting the data on ϕ versus z axes: $\phi = 0.6 - 2.6 \times 10^{-4} z$, with a correlation $r^2 = 0.89$.

The same data on $\ln(\phi)$ versus z axes: $\phi_0 = 102\%$, $A = 1.2 \times 10^{-3}$, with a correlation $r^2 = 0.82$.

On $(1/\phi)$ versus z axes: $\phi_0 = -28.5\%$, $B = 7.6 \times 10^{-3}$, with a correlation $r^2 = 0.66$.

On $\ln(1 - \phi)$ versus $\ln(z)$ axes: $z_{max} = 2830$ m, $C = 2.803$, with a correlation $r^2 = 0.79$.

Thus, the data are best represented in this case by a simple linear decrease in porosity with depth. At 1500 m, $\phi = 0.6 - (2.6 \times 10^{-4} \times 1500) = 0.210 = 21.0\%$.

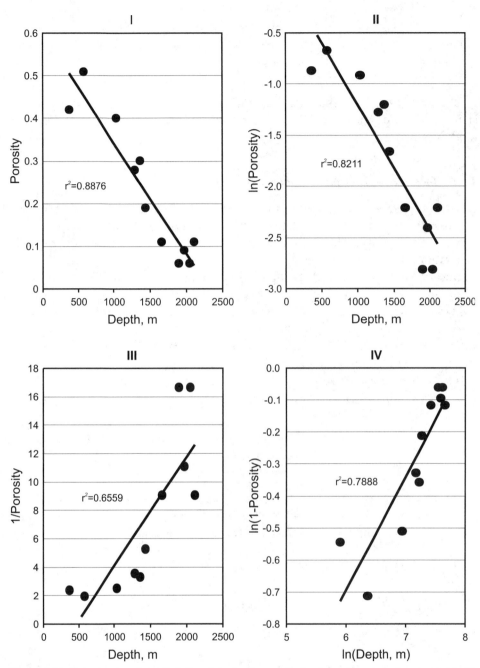

Figure 4.14. Porosity data from Rob Roy 1 in the Browse Basin, Australia, plotted versus depth (I) and on axes corresponding to the compaction models of (II) Sclater and Christie (1980), (III) Falvey and Middleton (1981), (IV) Baldwin and Butler (1985).

[The Sclater and Christie (1980) model predicts $\phi = 16\%$, the Falvey and Middleton (1981) model predicts $\phi = 13\%$, and the Baldwin and Butler (1985) model predicts $\phi = 20\%$.]

As a rule of thumb, the Falvey and Middleton (1981) model works well for shallow burial depths (< 1000 m), while at greater depths one of the other two models is better.

4.3.6. Bulk Rock Thermal Conductivity

Unfortunately, we do not live in an ideal world. In many cases, it is not possible to obtain a sample of a particular lithology on which to measure matrix conductivity. There are also many holes for which there are no well logs from which to derive estimates. Under these circumstances, there are a couple of ways to proceed. Samples or logs can be sought from the same formations in nearby wells and extrapolated into the region of study. Otherwise, we may choose a value of matrix conductivity equivalent to a similar lithology elsewhere in the world. Many compilations of conductivity values exist, but while different compilations agree closely with each other for some lithologies, they vary markedly for others (see Table 4.2). Choosing the correct value for a particular locality is fraught with uncertainty and the error margins on subsequent heat flow estimates should reflect this.

Note: Some compilations are of matrix conductivity while others are of bulk conductivity.

By this stage, we now know the matrix conductivity of all penetrated formations, along with compaction models to calculate the porosity at any given depth. The lithological column is known with some accuracy, and all that remains is to integrate the various data into thermal conductivity versus depth models. For each lithology, the conductivity–depth model depends on which mixing law (geometric or square-root) is preferred:

$$\lambda(z) = \lambda_m^{1-\phi(z)} \lambda_f^{\phi(z)} \tag{4.57}$$

or

$$\lambda(z) = [(1 - \phi(z)) \times \lambda_m^{1/2} + \phi(z) \times \lambda_f^{1/2}]^2 \tag{4.58}$$

where $\phi(z)$ is the compaction function, and λ_m, λ_f are the matrix and pore fluid conductivities, respectively.

The pore fluid is whatever happens to be filling the pore spaces of the rock. Saline water is usually assumed, but often a well will penetrate a column of rock where the pore spaces are partly or fully filled with oil or gas. Near the surface, rocks might be above the water table and pores may be entirely air-filled. Table 4.3 shows the thermal conductivity of several common pore fluids.

Question: In an earlier example, we found that the best compaction model for sandstone in Rob Roy 1 is

$$\phi = 0.6 - 2.6 \times 10^{-4}z$$

where z is in metres.

Several sandstone core samples were taken from the well, and the average matrix conductivity was found to be $\lambda_m = 4.7 \pm 0.3\,\mathrm{W\,m^{-1}\,K^{-1}}$. What is the expected bulk conductivity (before temperature correction) of sandstone saturated with saline water at 1500 m depth?

Answer: The expected porosity at $z = 1500$ was found in the earlier example, $\phi = 0.210$. From Table 4.3, the conductivity of saline water at room temperature ($T = 295\mathrm{K}$) is $\lambda_f = 0.621\,\mathrm{W\,m^{-1}\,K^{-1}}$. We are told that $\lambda_m = 4.7 \pm 0.3\,\mathrm{W\,m^{-1}\,K^{-1}}$.

From Equation (4.57):
$$\lambda_{1500} = 4.7^{0.790} \times 0.621^{0.210} = 3.07 \pm 0.18\,\mathrm{W\,m^{-1}\,K^{-1}}$$
or, from Equation (4.58):
$$\lambda_{1500} = [0.790 \times 4.7^{1/2} + 0.210 \times 0.621^{1/2}]^2 = 3.53 \pm 0.22\,\mathrm{W\,m^{-1}\,K^{-1}}$$
Averaging the two estimates, $\lambda_{1500} = 3.3 \pm 0.3\,\mathrm{W\,m^{-1}\,K^{-1}}$.

We have now determined a value of thermal conductivity for every point at which lithology is known. The accuracy of the estimate at each point depends on the local quality of the data and may vary greatly from one formation to another. It is important to remain aware of the relative precision and accuracy of the conductivity estimate in each formation.

Table 4.3. Thermal Conductivity of Common Pore Fluids

Fluid	Conductivity ($\mathrm{W\,m^{-1}\,K^{-1}}$)	Source
Water		
Fresh	$-7.42 \times 10^{-6} \times T^2 + 5.99 \times 10^{-3} \times T - 0.522$	1
Saline	$-7.42 \times 10^{-6} \times T^2 + 5.99 \times 10^{-3} \times T - 0.5$	
Hydrocarbon		
Oil	0.1	1
Gas	$0.000143 \times T - 0.0159$	1
Air	0.023	2

Note: T = absolute temperature in the range 295–450 K.

Source: 1 = Touloukian et al. (1970a), 2 = Giancoli (1984).

4.3.7. Temperature Correction

Conductivity measured in a laboratory or obtained from global compilations must be corrected for *in situ* temperature conditions. Section 4.1 explained that the rate of phonon conduction through a crystal lattice is inversely proportional to temperature. This translates directly into a temperature dependence of thermal conductivity in rocks, which are mixtures of crystalline grains. The temperature dependence of any particular formation is difficult to predict, however, because it is a complex function of the different responses of its constituent minerals.

Sekiguchi (1984) suggested an empirical correction to be applied to any arbitrary rock over the temperature range 0–300°C. The corrected thermal conductivity, λ (W m^{-1} K^{-1}), at absolute temperature, T (K), is given by

$$\lambda = (T_0 T_m/(T_m - T_0))(\lambda_0 - \lambda_m)((1/T) - (1/T_m)) + \lambda_m \qquad (4.59)$$

where
$\lambda_m = 1.05$ W m^{-1} K^{-1}, a calibration coefficient
$\lambda_0 = $ thermal conductivity at laboratory temperature, T_0
$T_0 = $ temperature (K) at which λ_0 was measured
$T_m = 1473$ K, a calibration coefficient
This equation is relatively insensitive to small uncertainties in temperature.

Question: A sample from a particular depth in a well is found to have a thermal conductivity of 2.3 W m^{-1} K^{-1} in the laboratory ($T_0 = 298$ K). The *in situ* temperature is $T = 50 \pm 5$°C. What is the corrected *in situ* conductivity, and the uncertainty in the temperature correction?

Answer: The simplest way to answer the question is to calculate the corrected conductivity for $T = 45$°C, 50°C and 55°C (318 K, 323 K and 328 K). Given that $\lambda_0 = 2.3$ W m^{-1} K^{-1} at $T_0 = 298$ K, Equation (4.59) gives the following:

$\lambda = (373.578) \times (1.25) \times (0.003143 - 0.000679) + 1.05 =$
    2.20 W m^{-1} K^{-1} for $T = 45$°C
$\lambda = (373.578) \times (1.25) \times (0.003095 - 0.000679) + 1.05 =$
    2.18 W m^{-1} K^{-1} for $T = 50$°C
$\lambda = (373.578) \times (1.25) \times (0.003047 - 0.000679) + 1.05 =$
    2.16 W m^{-1} K^{-1} for $T = 55$°C

So the *in situ* conductivity is 2.18 ± 0.02 W m^{-1} K^{-1}. The sensitivity of the correction decreases with increasing temperature, so the uncertainty in the correction is better than 0.02 W m^{-1} K^{-1} for $T > 50$°C.

The Sekiguchi (1984) correction is intended for any rock, regardless of mineralogy, porosity or pore fluid. It is probably sufficient in most situations because the error introduced due to the correction is minimal compared with other sources of error. However, if a more precise temperature correction is required, each component of the rock can be corrected independently. The temperature dependence of several mineral and fluid conductivities is known with reasonable accuracy (Tables 4.2 and 4.3), and the *in situ* conductivity of these can be independently calculated prior to calculating the bulk conductivity.

The conductivity of individual lithologies tends to vary with temperature according to

$$\ln(\lambda) = \ln(\lambda_0) + \ln(T/T_0)M \tag{4.60}$$

where λ_0, T and T_0 are as for Equation (4.59) and M is a constant dependent upon lithology.

Values of M for several common lithologies are given in Table 4.4. It is interesting to note that the thermal conductivity of some lithologies (basalt, coal and tuff) actually increases with temperature, instead of the expected decrease. For lithologies not in Table 4.4, Equation (4.59) is sufficient.

Question: The conductivity of a sample of coal measured under surface conditions is $0.22 \ \mathrm{W \, m^{-1} \, K^{-1}}$. The coal was sampled from a formation with *in situ* temperature $T = 76°C$. What is the *in situ* conductivity of the coal, λ?

Answer: Referring to Equation (4.60) and Table 4.4:
$\lambda_0 = 0.22 \ \mathrm{W \, m^{-1} \, K^{-1}}$, $T = 76°C = 349.15K$, $T_0 = 295K$ and $M = 0.714$

So
$\ln(\lambda) = \ln(0.22) + \ln(349.15/295) \times 0.714 = -1.3938$
$\lambda = \exp(-1.3938) = 0.25 \ \mathrm{W \, m^{-1} \, K^{-1}}$

The valid temperature range for the temperature corrections given here (~300–500 K) is sufficient for depths down to about 6–10 km, depending on the prevailing thermal gradient. This is sufficient for virtually all boreholes. For sections of the crust we cannot directly sample, thermal conductivity is calcu-

Table 4.4. Coefficient, M, for Equation (4.60) for Different Lithologies

Lithology	M	Lithology	M
Sandstone	Variable	Coal	0.714
Limestone	−0.185	Dolomite	0.0
Clay	0.0	Halite	−1.37
Granite	0.0	Loose sand	0.0
Basalt	0.5	Tuff	0.25

Source: Touloukian et al. (1970b).

lated from empirical relationships that automatically incorporate a temperature correction. For measurements made on lower crustal rocks that have been exposed at the surface, a temperature correction can only be derived from high-temperature–high-pressure laboratory experiments.

4.4. Well Log Analysis

We have shown how to measure the thermal conductivity of both consolidated and unconsolidated rock specimens, and how to estimate conductivity when no specimens are available. A rich source of information we have not yet discussed is electric well logs. Electric well logs are available for many wells from which no physical rock samples have been retrieved, and much effort has been put into developing ways of using these logs to estimate conductivity.

Published methods can be broadly grouped into two classes. The first class relates one or more electric logs directly to thermal conductivity. Most of these are based on empirical observations of relationships between thermal conductivity and other physical properties of rocks. The second class takes the intermediate step of deriving mineralogy from log data, then uses mixing laws to estimate the bulk thermal conductivity. The derived mineralogy can be independently corroborated via the lithological log, increasing confidence in the results.

The best example from the first class of methods was published by Williams and Anderson (1990), who investigated in detail the relationship between thermal conductivity and well logs based on phonon conduction theory (see Section 4.1). They concluded that thermal conductivity could be accurately derived for low-porosity, crystalline rocks using the density (RHOB, $g\,cm^{-3}$), photoelectric capture cross section (PEF, barns/electron, b/e), full-waveform sonic (compression and shear velocity, v_p and v_s, $m\,s^{-1}$) and temperature (T, K) logs. Specifically:

$$\lambda = 0.7531 + 0.1005 \times (av_m\mu^2/3K_ST) \qquad (4.61)$$

where

a = mean interatomic distance $\approx 10^{-9} \times ((5.32 \times PEF^{0.2778} + 13.8)/(1000 \times RHOB))^{1/3} m$

v_m = mean phonon velocity = $3^{1/3}(1/v_p^3 + 2/v_s^3)^{-1/3}$ $m\,s^{-1}$

μ = shear modulus = $1000 \times RHOB \times v_s^2$

K_S = bulk modulus = $1000 \times RHOB(v_p^2 - \frac{4}{3}v_s^2)$

Question: The well Kalyptea Sidetrack 1, in the Browse Basin, Australia, was not cored in any section. It was, however, extensively logged, and the log data from three depths are given below:
At 3469 m:

RHOB = 2.653 $g\,cm^{-3}$,   PEF = 5.12 b/e,   DT = 68.6 μs
    ft^{-1}, $T = 107.6°C$
At 4126 m:
    RHOB = 2.533 g cm^{-3},   PEF = 2.97 b/e,   DT = 81.6 μs
    ft^{-1}, $T = 135.2°C$
At 4266 m:
    RHOB = 2.360 g cm^{-3},   PEF = 5.54 b/e,   DT = 96.1 μs
    ft^{-1}, $T = 143.0°C$

Assuming $v_s = 0.55v_p$, what is the thermal conductivity suggested at these depths?

Answer: First convert DT (sonic travel time, the inverse of compression velocity) into v_m:

$v_{p(3469)} = 68.6\,\mu s\,ft^{-1} = 4443$ m s^{-1}, $v_{s(3469)} = 2444$ m s^{-1},
    $v_{m(3469)} = 2724$ m s^{-1}
$v_{p(4126)} = 81.6\,\mu s\,ft^{-1} = 3735$ m s^{-1}, $v_{s(4126)} = 2054$ m s^{-1},
    $v_{m(4126)} = 2289$ m s^{-1}
$v_{p(4266)} = 96.1\,\mu s\,ft^{-1} = 3172$ m s^{-1}, $v_{s(4266)} = 1744$ m s^{-1},
    $v_{m(4266)} = 1944$ m s^{-1}

Then determine a, μ and K_S:

$a_{(3469)} = 2.029 \times 10^{-10}$ m, $\mu_{(3469)} = 15.85 \times 10^9$ kg m^{-1} s^{-2},
    $K_{S(3469)} = 31.24 \times 10^9$ kg m^{-1}s^{-2}
$a_{(4126)} = 2.024 \times 10^{-10}$ m, $\mu_{(4126)} = 10.69 \times 10^9$ kg m^{-1} s^{-2},
    $K_{S(4126)} = 21.09 \times 10^9$ kg m^{-1}s^{-2}
$a_{(4266)} = 2.116 \times 10^{-10}$ m, $\mu_{(4266)} = 7.178 \times 10^9$ kg m^{-1} s^{-2},
    $K_{S(4266)} = 14.18 \times 10^9$ kg m^{-1}s^{-2}

Now solve for λ using Equation (4.61):

$\lambda_{(3469)} = 0.7531 + 0.1005 \times [2.029 \times 10^{-10} \times 2724 \times (15.85 \times 10^9)^2 / (3 \times 31.24 \times 10^9 \times (273.15 + 107.6))] =$ 1.14 W m^{-1} K^{-1}

$\lambda_{(4126)} = 0.7531 + 0.1005 \times [2.024 \times 10^{-10} \times 2289 \times (10.69 \times 10^9)^2 / (3 \times 21.09 \times 10^9 \times (273.15 + 135.2))] = 0.96$ W m^{-1} K^{-1}

$\lambda_{(4266)} = 0.7531 + 0.1005 \times [2.116 \times 10^{-10} \times 1944 \times (7.178 \times 10^9)^2 / (3 \times 14.18 \times 10^9 \times (273.15 + 143.0))] =$ 0.87 W m^{-1} K^{-1}

The mud log lists the lithology at these depths as predominantly clays and silts, consistent with the low conductivity estimates (lowered further by high temperature).

It should be stressed that this method was developed and calibrated using crystalline rocks, and its accuracy in sedimentary rocks is uncertain. In addition, the necessary logs are often unavailable. RHOB and DT are readily available for most wells, but PEF is rarer, and shear velocity very rare, especially in older wells. Consequently, other methods utilising empirical relationships are often required.

Houbolt and Wells (1980) derived a relationship for siliciclastic rocks with water-filled pores:

$$\lambda = 77 \times V/(a \times (c + T)) \, \text{W} \, \text{m}^{-1} \, \text{K}^{-1} \tag{4.62}$$

where V is acoustic velocity in kilometres per second, T = temperature in degrees Celsius, $a = 1.039$, $c = 80.031$.

Question: Recalculate the thermal conductivity values from the above example using Houbolt and Wells' (1980) empirical relationship.

Answer: From the previous example we know:

$V_{(3469)} = 4.443 \, \text{km} \, \text{s}^{-1},$ $T_{(3469)} = 107.6°\text{C}$
$V_{(4126)} = 3.735 \, \text{km} \, \text{s}^{-1},$ $T_{(4126)} = 135.2°\text{C}$
$V_{(4266)} = 3.171 \, \text{km} \, \text{s}^{-1},$ $T_{(4266)} = 143.0°\text{C}$

So from Equation (4.62):

$\lambda_{(3469)} = 77 \times 4.443/(1.039 \times (80.031 + 107.6)) =$
   $1.75 \, \text{W} \, \text{m}^{-1} \, \text{K}^{-1}$

$\lambda_{(4126)} = 77 \times 3.735/(1.039 \times (80.031 + 135.2)) =$
   $1.29 \, \text{W} \, \text{m}^{-1} \, \text{K}^{-1}$

$\lambda_{(4266)} = 77 \times 3.171/(1.039 \times (80.031 + 143.0)) =$
   $1.05 \, \text{W} \, \text{m}^{-1} \, \text{K}^{-1}$

These values are significantly higher than the values calculated in the previous example. Without independent verification, however, it is not possible to say which method gives the more realistic answer.

The intermediate step of determining mineralogy from well logs is highly recommended because it has the advantage that the mineralogy can be validated or repudiated against the known lithology. Demongodin et al. (1991) developed a method using four independent logs – gamma ray (GR), sonic velocity (DT), density (RHOB) and neutron porosity (NPHI). They assumed that the signal registered by each of these four logs is the sum of the individual signals from four common sedimentary components: quartz, calcite, clay and water. The log responses of pure quartz, calcite and water are reasonably well known, but those of clay can vary markedly depending on the specific mineralogy. The log values for the clay end-members were found by cross-plotting pairs of logs and identifying the quartz, calcite, clay and water vertices from the cloud of points. Once the individual responses of the four components were known, the mix that best reproduced the observed log responses at each depth was determined numerically. The thermal conductivity was then calculated using the geometric mean law (although the square-root mean law could also be used).

Question: A particular sequence is known to contain only quartz, clay and water. The responses of four logging tools to pure samples of the three individual components are as follows:

GR (API units) Quartz = 10, Clay = 180, Water = 0
NPHI (%)   Quartz = −2, Clay = 43, Water = 100
RHOB (g cm^{-3}) Quartz = 2.65, Clay = 3.10, Water = 1.00
DT (μs m^{-1}) Quartz = 165, Clay = 220, Water = 620

At a particular depth, the four logs have the following values:
    GR = 42.5, NPHI = 22.3, RHOB = 2.49, DT = 244
What mix of quartz, clay and water best approximates the observed log responses? Given the thermal conductivity of quartz is 6.97 W m^{-1} K^{-1}, clay is 1.85 W m^{-1} K^{-1} and water is 0.6 W m^{-1} K^{-1}, what is the bulk conductivity at the sampled depth (uncorrected for temperature)?

Answer: Through various numerical methods, the best fit to the observed logs is found with a mix consisting of 65% quartz, 20% clay and 15% water. Using a geometric mean, the bulk conductivity is therefore
$$\lambda = 6.97^{0.65} \times 1.85^{0.20} \times 0.60^{0.15} = 3.70 \text{ W m}^{-1} \text{ K}^{-1}$$
Using a square root mean:
$$\lambda = [(0.65 \times 6.97^{1/2}) + (0.20 \times 1.85^{1/2}) + (0.15 \times 0.60^{1/2})]^2 = 4.43 \text{ W m}^{-1} \text{ K}^{-1}$$
The average of the two estimates is $\lambda = 4.1 \pm 0.4$ W m^{-1} K^{-1}

The technique would yield spurious results in sediments containing significant amounts of minerals other than the four chosen (e.g. dolomite, feldspar or halite).

Methods such as that described above are simple to develop and employ, but more sophisticated log analysis should give results that are more accurate. In particular, much research has gone into determining porosity and clay content from well logs, and the more accurately these can be determined, the more accurate will be the conductivity estimate.

Asquith (1991) suggested that the volume of clay, V_{cl}, could be most accurately estimated with a combination of GR, SP (spontaneous potential), RHOB and NPHI logs. He suggested finding three independent estimates of clay volume. The first:

$$V_{cl(1)} = 0.330 \times [2^{(2 \times I_{GR})} - 1.0] \text{ for consolidated sediments} \tag{4.63}$$

$$V_{cl(1)} = 0.083 \times [2^{(3.7 \times I_{GR}) - 1.0]} \text{ for unconsolidated sediments} \tag{4.64}$$

where
$$I_{GR} = (GR - GR_{min})/(GR_{max} - GR_{min})$$

GR_{min} = GR minimum (recorded in clean sand)

GR_{max} = GR maximum (recorded in shale)

The second estimate of clay volume:

$$V_{cl(2)} = 1.0 - (PSP/SSP) \tag{4.65}$$

where PSP is SP deflection from shale line and SSP is SP deflection in thick clean sand.

The third and final estimate of clay volume:

$$V_{cl(3)} = (\phi_n - \phi_d)/(\phi_{nsh} - \phi_{dsh}) \frac{\phi_n - \phi_d}{\phi_{nsh} - \phi_{dsh}} \tag{4.66}$$

where ϕ_n is porosity calculated from NPHI log, ϕ_d is porosity calculated from RHOB log, ϕ_{nsh} is neutron porosity in adjacent shale and ϕ_{dsh} is density porosity in adjacent shale.

The actual volume of clay is taken to be the *minimum* of these three independent estimates ($V_{cl(1)}$, $V_{cl(2)}$ and $V_{cl(3)}$). Once the volume of clay is known, porosity can be calculated. There are three logs that relate to porosity – the so-called 'porosity logs'; DT, NPHI and RHOB. Each must be corrected for the effect of clay before a reliable porosity estimate can be obtained. The effective porosity, ϕ_e, is found from the sonic velocity log, DT:

$$\phi_e = \frac{DT - \Delta t_{ma}}{\Delta t_f - \Delta t_{ma}} \left[\times \frac{100}{\Delta t_{sh}} \right] - V_{cl} \left(\frac{\Delta t_{sh} - \Delta t_{ma}}{\Delta t_f - \Delta t_{ma}} \right) \tag{4.67}$$

where Δt_{ma} is interval transit time of the formation's clay-free matrix, Δt_f is interval transit time of pore fluid (fresh water = 612 μs m^{-1}, salt water = 607 μs m^{-1}), Δt_{sh} is interval transit time of adjacent shale and $[100/\Delta t_{sh}]$ is a compaction correction only used for unconsolidated sands.

Using the density log, RHOB, the effective porosity is given by:

$$\phi_e = \frac{\rho_{ma} - RHOB}{\rho_{ma} - \rho_r} - V_{cl} \left(\frac{\rho_{ma} - \rho_{sh}}{\rho_{ma} - \rho_f} \right) \tag{4.68}$$

where ρ_{ma} = density of the formation's clay-free matrix, ρ_f is density of fluid (fresh water = 1.0 g cm^{-3}, salt water = 1.1 g cm^{-3}) and ρ_{sh} is density of adjacent shale.

From the combination neutron-density log, effective porosity is

$$\phi_e = (\phi_{nc} + \phi_{dc})/2 \tag{4.69}$$

where

$\phi_{nc} = \phi_n - (V_{cl} \times \phi_{nsh})$

$\phi_{dc} = \phi_d - (V_{cl} \times \phi_{dsh})$

and all other variables are as defined for Equation (4.66).

Question: The log responses at 4397 m in the well Kalyptea
  Sidetrack 1 (see earlier examples) were as follows:

RHOB $= 2.50\,\mathrm{g\,cm^{-3}}$, DT $= 252\ \mu\mathrm{s\,m^{-1}}$, GR $= 115.3$ API units, NPHI $= 12.0\%$, SP $= 85$ mV

The log responses in an adjacent shale unit were as follows:

RHOB $= 2.70\,\mathrm{g\,cm^{-3}}$, DT $= 262\ \mu\mathrm{s\,m^{-1}}$, GR $= 178$ API units, NPHI $= 22.0\%$, SP $= 105$ mV

and in an adjacent pure sand unit:

GR $= 50$ API units, SP $= 70$ mV

Determine the clay volume and effective porosity.

Answer: Assuming a sandstone matrix, $\rho_m = 2.648\ \mathrm{g\,cm^{-3}}$, $\Delta t_m = 177\ \mu\mathrm{s\,m^{-1}}$ (Asquith and Gibson, 1982, pp. 66–7) and saline drilling mud:

Density porosity, $\phi_d = (2.648 - 2.50)/(2.648 - 1.10) = 9.56\%$

Shale density porosity, $\phi_{dsh} = (2.648 - 2.7)/(2.648 - 1.10) = -3.36\%$

Neutron porosity, $\phi_n = 16\%$ (Asquith and Gibson, 1982, Figure 30)

Shale neutron porosity, $\phi_{nsh} = 26.5\%$

Gamma-ray factor, $I_{GR} = (115.3 - 50)/(178 - 50) = 0.510$

Sandstone SP deflection, $SSP = 70 - 105 = -35$ mV

At 4397 m the sediments are consolidated so we use Equation (4.63) to find $V_{cl(1)}$:

$V_{cl(1)} = 0.330 \times (2^{2 \times 0.510} - 1.0) = 0.339$

We use Equation (4.65) to find $V_{cl(2)}$:

$V_{cl(2)} = 1.0 - ((85 - 105)/ - 35) = 0.429$

And Equation (4.66) to find $V_{cl(3)}$:

$V_{cl(3)} = (16.0 - 9.56)/(26.5 - -3.36) = 0.216$

Clay volume is the minimum of the three independent estimates, so:

$V_{cl} = V_{cl(3)} = 0.216$, or 21.6%

We can now estimate effective porosity from the three porosity logs. Firstly, DT using Equation (4.67):

$\phi_e = [(252 - 177)/(612 - 177)] - 0.216[(262 - 177)/(612 - 177)] = 0.130$, or 13.0%

Then RHOB using Equation (4.68):

$\phi_e = (2.648 - 2.50)/(2.648 - 1.1) - 0.216[(2.648 - 2.70)/(2.648 - 1.1)] = 0.103$, or 10.3%

Finally, NPHI using Equation (4.69):

$\phi_e = (\phi_{nc} + \phi_{dc})/2 = [(0.16 - (0.216 \times 0.265)) + (0.0956 - (0.216 \times -0.0336))]/2 = 0.103$, or 10.3%

Agreement between the logs is quite good, with an average $\phi_e = 11.2\%$. The logs suggest that the composition of the rock at 4397 m is 21.6% clay, 11.2% water and 67.2% sandstone minerals.

With the volume of clay and the effective porosity determined, the composition of the remaining portion of rock can be estimated. The methods for doing this generally involve correcting various logs for the effects of clay and porosity, then calculating the mineral mix most likely to produce the observed log response. The number of mineral components identified depends on the sophistication of the method. Five components, including clay and water, is common. We now have a multi-component mixture from which we can deduce bulk conductivity using either a geometric or a square-root mean mixing law.

Question: The temperature at 4397 m in Kalyptea Sidetrack 1 is 145°C. Subsequent analyses of the logs in the previous example indicated the remaining matrix to be 85% quartz, 10% calcite and 5% dolomite. What is the bulk conductivity of the rock?

Answer: The components of the rock and their conductivities are therefore as follows:

water (11.2%), $\lambda = 0.604$ W m^{-1} K^{-1} (at room temperature, see Table 4.3)
clay (21.6%), $\lambda = 2.3$ W m^{-1} K^{-1} (see Table 4.5)
quartz (57.1%), $\lambda = 7.69$ W m^{-1} K^{-1} (see Table 4.1)
calcite (6.7%), $\lambda = 3.59$ W m^{-1} K^{-1} (see Table 4.1)
dolomite (3.4%), $\lambda = 5.51$ W m^{-1} K^{-1} (see Table 4.1)

By the geometric mean:
$$\lambda = 0.604^{0.112} \times 2.3^{0.216} \times 7.69^{0.571} \times 3.59^{0.067} \times 5.51^{0.034} = 4.187 \text{ W m}^{-1}\text{ K}^{-1}$$

By the square-root mean:
$$\lambda = [(0.112 \times 0.604^{1/2}) + (0.216 \times 2.3^{1/2}) + (0.571 \times 7.69^{1/2}) + (0.067 \times 3.59^{1/2}) + (0.034 \times 5.51^{1/2})]^2 = 4.861 \text{ W m}^{-1}\text{ K}^{-1}$$

Using Sekiguchi's (1984) temperature correction, Equation (4.59), these estimates reduce to 3.06 W m^{-1} K^{-1} and 3.49 W m^{-1} K^{-1}, respectively. The average is 3.3 ± 0.2 W m^{-1} K^{-1}.

4.5. Shale Conductivity

An increasing body of evidence suggests that the vertical conductivity of shale does not vary with depth in the same way as other lithologies. The thermal conductivity of sedimentary rocks generally increases with depth through the effect of compaction. In many documented cases, however, the vertical conductivity of shale remains constant or decreases with depth and compaction (e.g. Blackwell, Wisian and Beardsmore, 1997).

Blackwell and Steele (1989a, 1989b) proposed a mechanism to explain this observation. Shale is composed of highly anisotropic sheet silicates (Table 4.5).

Table 4.5. Average Thermal Conductivity for a Number of Sheet Silicates, Parallel and Perpendicular to Main Cleavage Plane, and for a Mixed Aggregate

Mineral	Average Conductivity ($W\,m^{-1}\,K^{-1}$)		
	Parallel	Perpendicular	Aggregate
Muscovite	3.89	0.52	2.35
Phlogopite	4.01	0.48	
Biotite	3.14	0.52	2.02
Lepidolite		0.48	2.30
Phyrophyllite	6.17	1.15	4.50
Talc	11.5		2.97
Chlorite			2.52
Clinochlore	10.3	1.97	

Sources: Diment and Pratt (1988) and Williams and Anderson (1990).

Thermal conductivity parallel to the mineral sheets is often many times greater than conductivity perpendicular to the sheets. Very fine shale grains are deposited in random orientation in low-energy environments and, soon after deposition, porosity is generally on the order of 75–80% (Hamilton, 1976). Freshly deposited shale therefore has an average vertical conductivity, λ_{av}, somewhere between the maximum, λ_{max}, and minimum, λ_{min}, conductivities of the mineral. Sass et al. (1992) suggested the following relationship:

$$\lambda_{av} = \lambda_{max}^{2/3} \times \lambda_{min}^{1/3} \tag{4.70}$$

Water is very quickly expelled from compacting layers and porosity generally decreases to around 60% at depths of only 50 m (e.g. Faas and Crocket, 1983). Subsequent burial is associated with loss of water and decreasing porosity in the same way as for other lithologies. However, during compaction of shale there is also a progressive rotation of the clay platelets to a preferred horizontal orientation (Bennett, Bryant and Keller, 1981). This effectively reduces the vertical matrix conductivity as the bulk conductivity increases due to loss of water. The two mechanisms act against each other and the vertical conductivity of the shale may vary little with depth.

Question: A particular shale is composed almost entirely of phyrophyllite, with thermal conductivity along the platelets, $\lambda_{max} = 6.17\ W\,m^{-1}\,K^{-1}$, and across the platelets, $\lambda_{min} = 1.15\ W\,m^{-1}\,K^{-1}$. Porosity of the shale at 50 m depth is 60%. What is the difference in modelled vertical conductivity, λ_v, between the shale at 50 m and a sample compressed to 10% porosity, assuming that in the latter case all the grains have rotated to a preferred horizontal orientation?

Answer: The freshly deposited shale has randomly oriented grains, so the matrix conductivity can be estimated using Equation (4.70):

$$\lambda_{av} = \lambda_{max}^{2/3} \times \lambda_{min}^{1/3} = 6.17^{2/3} \times 1.15^{1/3} = 3.524 \text{ W m}^{-1}\text{K}^{-1}$$

Combined with 60% water using a square root mean, the vertical conductivity of the shale is:

$$\lambda_v = [(0.4 \times 3.524^{1/2}) + (0.6 \times 0.63^{1/2})]^2 = 1.51 \text{ W m}^{-1}\text{K}^{-1}$$

The compacted shale has vertical matrix conductivity equal to λ_{min}, so the bulk conductivity of the shale is

$$\lambda_v = [(0.9 \times 1.15^{1/2}) + (0.1 \times 0.63^{1/2})]^2 = 1.09 \text{ W m}^{-1}\text{K}^{-1}$$

Effective vertical thermal conductivity of the shale has decreased with compaction.

The values used in the example above are realistic and dramatically illustrate the possibility that vertical shale conductivity may actually decrease with compaction, opposite to the predictions of simple models. Applying a temperature correction enhances the effect.

4.6. Summary

Thermal conductivity is a measure of how easily heat is transmitted through a material. Under typical crustal conditions, phonon conduction is the dominant mode of heat transfer, although thermal radiation increases in importance with increasing temperature. Below about 600 K (about 300°C), the effect of thermal radiation can be safely ignored.

The thermal conductivity of a rock is a function of the conductivities and the geometrical relationships between its constituent minerals and fluids. A rock or formation composed of layers stacked perpendicular to the direction of heat flow is best modelled using a harmonic mean. If layers are aligned parallel to the direction of heat flow, the conductivity is best estimated with an arithmetic mean. A random aggregate of minerals and fluids can be modelled with either a geometric mean or a square-root mean model.

Thermal conductivity is closely related to lithology. Whenever possible, each lithology within the region of interest should be sampled for physical thermal conductivity measurements. Ideally, these involve steady-state measurements on core samples, but drill cuttings and transient methods can also be used with reduced accuracy and precision. The important parameter to be determined is the conductivity of the granular matrix of the rock.

Compaction reduces the amount of pore fluid in a rock and generally increases the bulk conductivity. Different lithologies compact at different rates. In the absence of porosity logs, compaction models must be constructed for each lithology under investigation. Once porosity is known, it can be combined with matrix conductivity to determine the bulk conductivity of each formation. This conductivity must then be corrected for *in situ* temperature conditions.

Electric well logs are an invaluable source of data about the formations penetrated by a well. Although no logging tool directly measures thermal conductivity, it can be estimated using a variety of different methods and well logs. The most reliable techniques are those that determine mineralogy and porosity, then combine the two to estimate conductivity.

The vertical thermal conductivity of shale does not follow the same trends as other lithologies. The vertical conductivity of other lithologies invariably increases as depth of burial and compaction increase. Due to clay platelet rotation, however, the vertical conductivity of shale may change very little, or even decrease, with increasing depth of burial and compaction.

CHAPTER FIVE

Thermal Maturity

> Do not the vast masses of basalt, the general appearances of mountain-ranges, the violent distortions and fractures of strata, *the great prevalence of metamorphic action* (which must have taken place at depths of not many miles, if so much), all agree in demonstrating that the rate of increase of temperature downwards must have been more rapid ... in geological antiquity than in present age?
>
> *On the Secular Cooling of the Earth* – Prof. William Thomson, 1862.

The usual objective of thermal modelling is to identify the time and depth of hydrocarbon generation. Hydrocarbons are by-products of the metamorphism of organic material (kerogen) within sediments – a gradual process involving the expulsion of volatiles, gases, liquids and oils during the chemical alteration of buried organic matter. The *thermal maturity* of a rock is a measure of the degree to which this metamorphism has progressed.

Before we can make predictions about the timing of petroleum generation, we need to know two things. We need to understand the precise response of organic detritus to changes in thermal conditions, and we also need to know the thermal history of potential petroleum source beds. This chapter addresses those two requirements.

5.1. The Generation of Hydrocarbons from Organic Matter

The chemistry of converting organic matter into hydrocarbons is a whole separate field of expertise and is far too complex to investigate in great detail here. But a few words are necessary to illustrate the importance of temperature on the organic metamorphic process.

5.1.1. Kerogen

The term 'organic matter', or OM, covers a wide range of material naturally deposited along with inorganic minerals in a typical sedimentary basin. The term includes material transported from onshore, as well as that originating in the oceans or seas much closer to the final point of deposition. Everything from woody cell walls, spore cases, leaf cuticles, pollen and resins to algae, bacteria and plankton may end up within sediment and be classified as OM. Each of these components is unique in chemical composition and the products and rates of decomposition are similarly unique. Thus, information is required on the *type* of OM present in a source rock before the effect of temperature can be properly assessed. This, in turn, requires a standard classification scheme.

Sedimentary organic matter can be divided into two broad categories: *bitumen* and *kerogen*. Bitumen is that fraction of sedimentary OM that is soluble in typical organic solvents such as chloroform, benzene or methanol–benzene mixtures. The group includes hydrocarbons and more complex compounds typically of high atomic weight (Durand, 1980). Kerogen is the insoluble fraction of OM left in the rock after washing with solvents, and it generally makes up the largest part. Sediments almost always contain some kerogen – sandstones generally contain the least, while coal consists almost entirely of it.

Hydrocarbons generate from kerogen, and thus it is the response of kerogen to changes in temperature that interests us. The type and *maturity* (degree of thermal degradation) of kerogen is most often characterised by its carbon (C), hydrogen (H) and oxygen (O) content. This information is generally presented on a Van Krevelen diagram (Van Krevelen, 1961), which plots the kerogen's H:C atomic ratio against its O:C atomic ratio (e.g. Figure 5.1).

Note: Hydrogen, carbon and oxygen are the major constituents of most kerogens, but other elements such as sulphur, nitrogen and iron are also significant components of some (e.g. Durand and Monin, 1980; Hunt, Lewan and Hennet, 1991).

Tissot et al. (1974) defined three broad types of kerogen based on where they plot on a Van Krevelen diagram, and their path through the diagram as they mature:

Type I – Hydrogen-rich, oxygen-poor. The type-specimen of these kerogens comes from the Eocene Green River Formation in Colorado. They are thought to have been deposited in large lakes as bacterial and vegetable waxes (Durand and Monin, 1980) or algal organic matter (Wignall, 1994), and they produce distinctive oils containing low proportions of aromatic compounds. Kerogens of this type are rare.

Type II – Intermediate H:C and O:C ratios. By far the great majority of source rocks globally contain Type II kerogens, which are thought to be derived from bacterially altered marine organic matter (e.g. phytoplankton) with some lipid-rich terrigenous material (e.g. leaf cuticles and spores). The type-specimen comes from a Lower Toarcian (Early Jurassic) shale unit in the Paris Basin, France, and the resulting oils are rich in naphthene and aromatic hydrocarbons (Wignall, 1994).

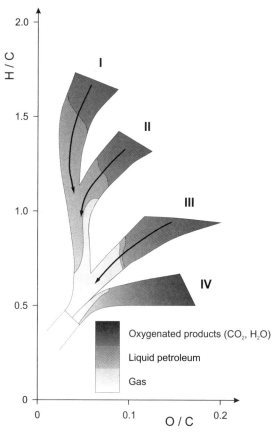

Figure 5.1. A Van Krevelen diagram showing the H:C and O:C ratios of kerogen types I–IV. The arrows show the evolution paths of increasing maturity. The shading indicates the products at different points along the evolution path. Modified after Van Gijzel (1982).

Type III – Hydrogen-poor, oxygen-rich. These kerogens are mostly derived from terrigenous woody organic matter and cuticular waxes, and are generally associated with near-shore deltaic environments. The type-sample is from Upper Cretaceous shale in the Douala Basin, Cameroon. Hydrocarbons from these kerogens are characterised by short *n*-alkanes (C_{20}–C_{35} range).

Tissot (1984) added a fourth kerogen type:

Type IV – Very low H:C, high O:C. These consist of highly reworked organic matter with no hydrocarbon generative potential.

Chemical reactions that involve the expulsion of part of the initial material usually alter the atomic ratios within the remaining kerogen, and thus change its position on a Van Krevelen diagram. The gradual maturation of kerogens over time therefore causes them to trace characteristic paths through the Van Krevelen diagram (Figure 5.1).

Question: What effect do simple reactions such as removal of water (H_2O), carbon dioxide (CO_2) and methane (CH_4) have on a kerogen's position on a Van Krevelen diagram?

Answer: Dehydration removes two hydrogen atoms for each oxygen atom. So the H:C ratio decreases at twice the rate as the O:C ratio and the kerogen moves downwards and to the left along a path of gradient $m = 2$.

Removing CO_2 from the kerogen increases the H:C ratio, and generally decreases the O:C ratio, causing the kerogen to follow a path upwards and to the left. The gradient depends on the initial O:C ratio. (Note, however, that if the initial O:C > 2 then the ratio increases after the reaction and the kerogen follows a path upwards and to the right, but an O:C ratio this high is rarely, if ever, observed.)

A similar effect is associated with a methane-releasing reaction. The O:C ratio increases and the H:C ratio generally decreases, so the reaction moves the kerogen downwards and to the right on the Van Krevelen diagram, at a gradient dependent on the initial H:C ratio. (Note that if H:C > 4 initially, the ratio increases with the reaction, but this initial ratio is rarely observed.)

5.1.2. Kinetics

The individual chemical steps required to convert kerogen into petroleum products are many and complex, and the great majority of such reactions remain unknown. However, this apparent obstacle does not restrict our ability to study the reaction system *as a whole*. The object of such studies is generally to determine the rate at which reactions proceed under different thermal conditions, with the ultimate aim of applying this knowledge to geological situations in order to model natural hydrocarbon generation. The concepts and methods employed in studies of this kind are those of *chemical kinetics*.

The general premise of chemical kinetics is that reactions proceed at a rate dependent on the absolute temperature and the amount of reactant available. The relationship is best described by the Arrhenius equation:

$$k(T) = A \times \exp(-E/RT) \tag{5.1}$$

where T is absolute temperature (K), R is the gas constant (8.314472 $J\,K^{-1}\,mol^{-1}$) and k is the rate of the reaction (s^{-1}):

$$(\partial N/\partial t) = -k \times N \tag{5.2}$$

where N is the amount of reactant remaining (moles).

There remain two undefined parameters in Equation (5.1). Parameter E is known as the *activation energy* and can be thought of as an energy barrier

$(\mathrm{J\,mol^{-1}})$ that must be overcome before the reaction can proceed. Parameter A is known variously as the *pre-exponential factor* or *frequency factor* and is possibly related to the vibrational frequency $(\mathrm{s^{-1}})$ of the reacting molecules (Schenk et al., 1997). Parameters E and A define the rate at which a reaction proceeds at any given temperature.

Question: Five moles ($N = 5$) of a particular reactant is heated to a temperature of 350°C. The reaction in question has an activation energy $E = 1.5 \times 10^5\,\mathrm{J\,mol^{-1}}$ and frequency factor $A = 10^{14}\,\mathrm{s^{-1}}$. At what rate does the reaction initially proceed? At what rate would it proceed at 150°C and 100°C?

Answer: 350°C $= 623.15$ K, so $RT = 5181\,\mathrm{J\,mol^{-1}}$
 From Equation (5.1):
  $k(623.15) = 10^{14} \times \exp(-150{,}000/5181) = 26.67\,\mathrm{s^{-1}}$
 and from Equation (5.2):
  $dN/dt = -k \times N = -133\,\mathrm{mol\,s^{-1}}$
 The reaction proceeds to completion within a fraction of a second.
  At 150°C $= 423.15$ K, $RT = 3518\,\mathrm{J\,mol^{-1}}$
 So
  $k(423.15) = 10^{14}\exp(-150{,}000/3518) = 3.04 \times 10^{-5}\,\mathrm{s^{-1}}$
 and
  $dN/dt = -k \times N = -1.52 \times 10^{-4}\,\mathrm{mol\,s^{-1}}$
 The reaction takes the better part of a day to complete.
  At 100°C the reaction rate is $k = 1 \times 10^{-7}\,\mathrm{s^{-1}}$ and the reaction takes many months to complete.

Unfortunately, the generation of hydrocarbons from kerogen cannot be written as precise chemical equations. One problem is that the precise reactions taking place during the process are largely unknown. But, also, the initial material (kerogen) and final product (a mixture of hydrocarbons) are generally so complex that they defy precise identification. So rather than deriving equations for molar amounts of real chemical reactants, hydrocarbon generation kinetics deal with a concept usually termed *hydrocarbon potential*. Rather than thinking of a particular kerogen as containing a certain amount of compound X that degrades through time to produce compound Y, we instead visualise the kerogen as initially having a certain *potential* to produce hydrocarbons, $\mathbf{M_0}$, which decreases as the hydrocarbon *yield*, $\mathbf{m}$, increases. The rate of hydrocarbon generation can now be written in a form similar to Equation (5.2):

$$-(\partial M/\partial t) = (\partial m/\partial t) = k \times (\mathbf{M_0} - \mathbf{m}) \tag{5.3}$$

where k is temperature dependent and has the form of Equation (5.1).

The technique works best when hydrocarbon generation is modelled as a weighted series of parallel reactions with the same frequency factor, but different activation energies. These energies have been determined for a large number of source rocks globally, generally utilising the technique of *pyrolysis*. Pyrolysis involves artificially forcing a source rock to release all of its potential hydrocarbon products, usually by increasing the temperature of a sample at a constant heating rate, r. The rate at which hydrocarbons are released from the sample is temperature dependent:

$$\frac{\partial \mathbf{m}}{\partial T} = \sum_{i=1}^{n} \mathbf{M}_{i0} \frac{A}{r} \left(-\frac{E_i}{RT} - \frac{A}{r} \mathbf{J}_i \right) \tag{5.4}$$

$$\mathbf{J}_i = \int_0^T \exp\left(-\frac{E_i}{RT} \right) dT$$

The task is to determine, for different kerogen types, the value of A and the hydrocarbon potential, $\mathbf{M}_{i0}$, across a range of activation energies. Generally, the activation energies fall in the range 160–340 kJ mol^{-1} (Schenk et al., 1997) and the frequency factor falls in the range $10^{13} - 10^{16}$ s^{-1}.

Many kinetic figures have been published for hydrocarbon systems. Most have been derived for individual source beds and specific products, and this is the recommended approach. Kinetic parameters determined for one source bed are not necessarily relevant to any other region, as particular kerogen mixes may be unique to particular localities. But, for illustration purposes, we include kinetic data derived for the generation of hydrocarbons from Type I and Type II kerogens and an Indonesian coal (Figure 5.2).

Pyrolysis can be performed in either an inert gas (open system) or in water (hydrous pyrolysis). The former is useful for information about bulk petroleum generation, whereas hydrous pyrolysis gives information specifically about oil generation, and arguably recreates natural conditions better than gas hydrolysis. For example, Hunt et al. (1991) published kinetic parameters for the generation of oil from four different Type II kerogens, varying in sulphur content (Table 5.1; also see the spreadsheet 'EASY%RO.XLS' available from the web site mentioned in the Preface). Note that the authors used just one activation energy rather than a range of energies, and that the values of E

Table 5.1. Kinetic Parameters for the Generation of Oil from Type II Source Rocks

Source	E (kJ mol^{-1})	A (s^{-1})	Sulphur Content (wt% kerogen)
Monterey shale	143.4	2.224×10^7	11.0
Phosphoria shale	178.7	1.338×10^{10}	9.0
Alum shale	201.3	4.899×10^{11}	7.4
Woodford shale	218.3	1.792×10^{13}	5.4

Source: Hunt et al. (1991)

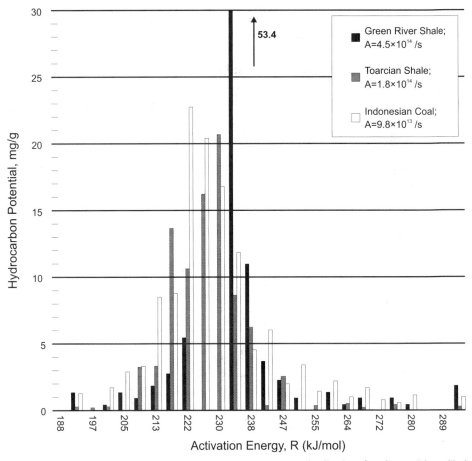

Figure 5.2. Hydrocarbon potential versus activation energy distribution for Green River Shale (Type I kerogen), Toarcian Shale (Type II kerogen) and an Indonesian coal (Modified after Schenk et al., 1997). Frequency factor, *A*, is also indicated for the three source rocks. Data were obtained by open-system pyrolysis.

and *A* are lower than those in Figure 5.2, meaning that the reactions proceed much faster and at lower temperature.

Reactions must proceed much faster in the laboratory than under geological conditions if they are to be observed. During a pyrolysis experiment, the heating rate [parameter *r* in Equation (5.4)] is many orders of magnitude higher than in nature. When geological heating rates and pyrolysis-derived kinetic parameters are substituted into Equation (5.4), hydrocarbon formation is predicted over a much lower and narrower temperature range than during the pyrolysis experiments (Figure 5.3).

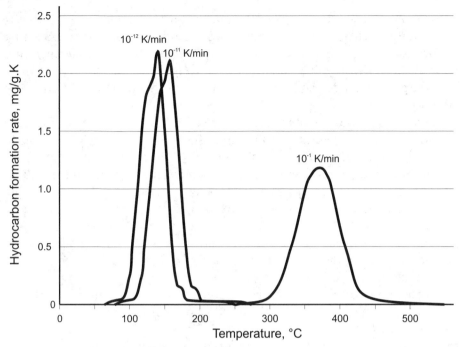

Figure 5.3. Calculated rate of hydrocarbon formation from a Toarcian shale (Figure 5.2) at three different heating rates. Note that petroleum forms at a lower and narrower temperature range at geological heating rates. Modified after Schenk et al. (1997).

EXAMPLE

In a particular pyrolysis experiment, hydrocarbon generation peaked at 435°C when the heating rate was 8×10^{-2} K min^{-1}. Petroleum was generated over a temperature range of 80°C. The experiment was repeated (with fresh material) using a slower heating rate (2×10^{-2} K min^{-1}) and hydrocarbon generation peaked at 420°C, spread over 75°C. At geological heating rates on the order of 10^{-12} K min^{-1}, peak hydrocarbon generation might be at temperatures as low as 160°C, and restricted to a range of 35°C.

In terms of chemical kinetics, we can now model the response of organic material to changes in thermal conditions. We have seen that the rate of hydrocarbon generation depends on heating rate. As a direct consequence, an accurate temperature history reconstruction becomes vital to an overall petroleum exploration strategy (Schenk et al., 1997).

5.2. Geochemical Indicators of Maximum Palaeotemperature

Constraining the present thermal state of a sedimentary basin is only the first step in modelling petroleum generation, because the thermal maturity of a rock is the culmination of its complete thermal history. The influence of time is roughly linear, while the temperature dependence is closer to exponential (e.g. Gretener and Curtis, 1982), so it is the *maximum* historic temperature that exerts the greatest influence on maturity. Conversely, the thermal maturity of a rock gives a crude estimate of the maximum temperature the rock has experienced.

Simple tectonic models predict progressive burial and greater temperature with time for any sedimentary formation. Assuming no secondary heat pulses, heat removal or intervening period of denudation, all present-day sedimentary formations should be at their maximum historic temperature. But are these assumptions valid? Of course they are not in all cases. While some basins develop along simple paths, many are exposed to tectonic forces subsequent to those from which they were born. Others pass close to hot spots, or develop aquifer systems that redistribute heat by non-conductive means. Each of these events could result in the present temperature of a formation being lower than at some time in the past.

The generation of oil generally requires that the temperature of the source sediment be maintained in the range 70–120°C for geologically significant periods of time (e.g. Hood, Gutjahr and Heacock, 1975). At cooler temperatures the reaction rate is too slow to produce economic deposits, while at higher temperatures the rate is too rapid to allow expulsion before the oil degrades. Present thermal conditions tell us nothing of possible past elevated temperatures. We must look within the sediment itself for clues to its maturity, and thus the maximum temperature that it has experienced.

The necessity of unravelling the thermal history of sedimentary rocks has given birth to a great number of methods for achieving that aim. The methods all rely on specific, non-reversible, temperature-dependent, chemical or physical processes that act on one or more components of the sediment. Collectively, the products of these processes are termed *palaeotemperature indicators*.

Note: The study of palaeotemperature indicators falls loosely under the heading of 'geochemistry'. While some techniques are obviously chemical in nature (requiring the separation and comparison of specific chemical species), others depend more on physical attributes such as reflectance and fluorescence. However, in keeping with convention, we will refer to all such techniques as being geochemical.

A number of techniques for estimating maximum palaeotemperature are examined below. The list is not exhaustive, but covers the main methods and a few others. Fission track thermochronology is discussed separately and in detail in Section 5.3.

Note: No palaeotemperature indicator is useful in all situations. All have advantages and limitations. Although this may seem self-apparent, there is an inclination to grow attached to one particular technique and to prefer it to all others. This is a perilous tendency if one aims to remain objective.

5.2.1. Vitrinite Reflectance (VR)

The most widely accepted indicator of thermal maturity is *vitrinite reflectance* (VR). Wide usage, however, does not preclude *flawed* usage, and it is always useful to take a fresh look at the underlying principles and potential pitfalls associated with the method. It would be easy to become lost in an in-depth discussion of coal petrology and chemistry, but entire books have already been written on the subject (e.g. Durand, 1980; Barker, 1996) and interested readers are referred to them. This section briefly covers the reflectance technique, how we use it to tell us maximum palaeotemperature, and what hazards are associated with it.

To understand VR we need to answer three basic questions: What is vitrinite? What is reflectance? What is the relationship between vitrinite reflectance and temperature? The first two questions are fairly simple to answer. The third requires a little more explanation.

Vitrinite is a class of *maceral* – the equivalent of a mineral in an inorganic rock, one of the microscopic constituents of coal and dispersed organic matter. Macerals can be broadly divided into three classes on the basis of morphology, reflectance in white light, and fluorescence:

Liptinite (Figure 5.4, Figure 5.5) – Often referred to as '*exinite*' in older texts. Derived from the waxy, lipid-rich (hence the name) and resinous parts of plants, such as spore cases, leaf cuticles, single-cell algae and resin globules. Liptinite macerals are the least reflective and, in all but the highest ranked coals, always fluoresce under ultraviolet light.

Vitrinite (Figure 5.6) – Derived from plant cell walls, cell contents or precipitated gels. In any coal sample, the reflectance of vitrinite lies somewhere between that of liptinite and inertinite, and *increases with the rank of the coal*. Some low-rank vitrinite macerals show a weak brownish fluorescence.

Note: See below for definitions of 'reflectance' and 'rank'.

Inertinite (Figure 5.7) – Derived from essentially the same plant material as vitrinite, but has been oxidised, altered or otherwise degraded (e.g. by reworking, changes in redox conditions or forest fires) in the early stages

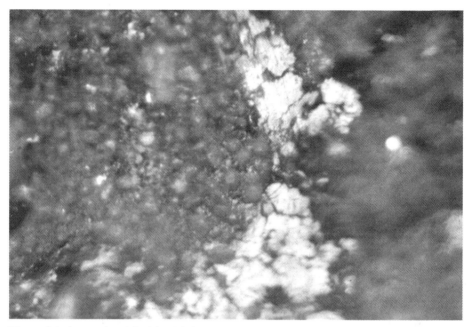

Figure 5.4. A sample of liptinite (globular material) from the well Yampi 1, offshore Western Australia, under reflected white light. Field of view is approximately 0.075 mm.

Figure 5.5. The same sample from Figure 5.4, under ultraviolet light. The liptinite fluoresces.

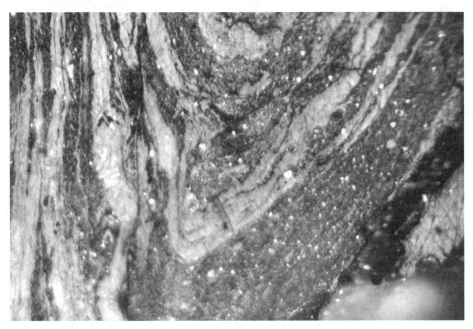

Figure 5.6. A sample of vitrinite (light-coloured bands) from Yampi 1, under reflected white light. Field of view is approximately 0.075 mm.

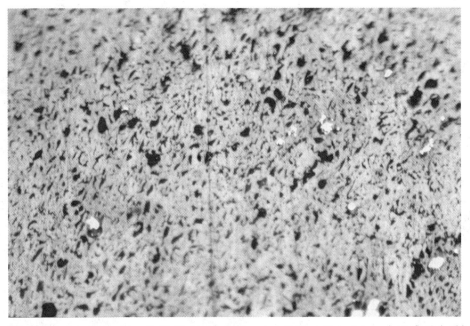

Figure 5.7. A sample of inertinite (light-coloured material) from Yampi 1, under reflected white light. Field of view is approximately 0.075 mm.

of coal formation. Inertinites are the most reflective of the macerals, but generally do not fluoresce.

Whitten and Brooks (1972) defined coal *rank* as 'the percentage of carbon in dry, mineral-free coal.' The percentage of carbon (and therefore the rank) increases with increasing burial time and temperature (e.g. LaPlante, 1974; Hood et al., 1975; Gretener and Curtis, 1982; Figure 5.8). It is the property of vitrinite in *italics* above that makes it a useful indicator of palaeotemperature. A higher reflectance implies that the coal is of a higher rank and has thus been raised to a higher temperature or buried for a longer time. Qualitatively, this is a simple relationship. Quantifying the relationship, however, is not so simple.

Quantifying reflectance is the first difficulty. *Reflectance* is a measure of the percentage of an incident beam of polarised white light that is reflected off a polished surface. Putting aside the technical difficulties of separating, mounting and polishing the vitrinite (standard procedures are set down in various parts of the world: e.g. Australian Standard AS2061-1989, 'Preparation of coal samples for incident light microscopy.'), and constructing instrumentation to measure the reflectance, there remains an intrinsic difficulty related to the nature of the maceral. Vitrinite is an anisotropic, *bireflectant* substance. Rotation through 360° under a reflecting light microscope reveals two identical reflectance maxima at 180° separation, with minima mid-way between the maxima. The magnitude of the anisotropy is related to sediment compaction and generally increases with increasing rank (Robert, 1988).

The anisotropy has led to the development of two different systems for recording VR. One system advocates recording the maximum reflectance for each vitrinite grain, while the other suggests recording only a single, randomly oriented reflectance for each grain. The first method (often quoted as $R_{v\,max}$) is more precise, but requires that the microscope stage be rotated through 360° for each measurement in order to identify the maxima. The second method

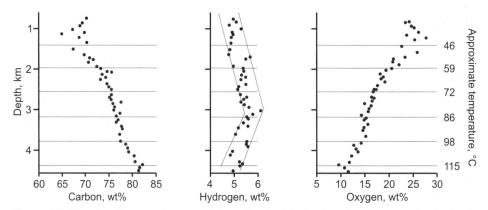

Figure 5.8. The changing atomic composition of coal with depth and temperature in the South Pecan Lake field of Louisiana. Modified after LaPlante (1974).

(referred to as R_o, $R_o\%$, $R_v\%$ or other variations) is less precise but much faster. For VR less than about 2%, R_o is statistically reliable if at least thirty grains are counted and averaged (Robert, 1988).

Note: When considering published values of vitrinite reflectance, pay particular attention to whether they are R_o or $R_{v\max}$. Exercise caution in situations where $R_o > 2\%$. The standard method of measuring vitrinite reflectance is by an oil emersion lens, hence the 'o' subscript.

R_o is related to $R_{v\max}$ by (Ting, 1978):

$$R_{v\max}(\%) = R_o(\%) \times 1.066 \qquad (5.5)$$

The exact nature of the correlation between vitrinite reflectance and maximum palaeotemperature is not immediately evident. Much effort has been expended over the years to clarify the relationship. For many years the most popular method for equating time, temperature and vitrinite reflectance was based on a concept developed by Lopatin in Russia in the 1970s (Lopatin, 1971, 1976), but popularised in the West by Waples (1980).

The thermal history of a rock is broken down into a number of discrete temperature intervals. The thermal maturity of the kerogen within the rock is then estimated using an equation of the following form:

$$\text{TTI} = \Sigma(\Delta t_i)(2^{n_i}) \qquad (5.6)$$

where Δt_i is the time (Ma) spent by the kerogen in the ith temperature interval, n_i is an integer weighting factor for the ith interval and TTI stands for time–temperature index, a term suggested by Lopatin and still in use.

Lopatin assumed that the process of petroleum production from organic material is a simple reaction that doubles its rate for every 10°C increase in temperature. Each temperature interval, therefore, covers 10°C and is assigned a weighting factor, n_i. Lopatin arbitrarily chose 100–110°C as the reference interval, $n = 0$. The weighting factor changes by 1 for each 10°C, in the negative direction for lower temperatures and the positive direction for higher temperatures.

Question: Given that the temperature interval 100–110°C has a weighting factor $n = 0$, what are the weighting factors of the intervals 50–60°C and 130–140°C?

Answer: Each 10°C interval alters the weighting factor by 1, so $n = -1$ for 90–100°C, $n = -2$ for 80–90°C, and $n = -5$ for 50–60°C. Similarly, $n = 1$ for 110–120°C, $n = 2$ for 120–130°C and $n = 3$ for 130–140°C.

Question: Organic matter is deposited within sediment at a surface temperature of 20°C. It is buried to a temperature of 100°C at a constant rate over a period of 40 Ma. Sedimentation then slows, and after another 30 Ma the temperature has increased to 130°C. How mature is the sediment, expressed in terms of TTI?

Answer: In the first period of burial, the sediment passes through eight temperature intervals in 40 Ma, spending 5 Ma in each. The initial interval (20–30°C) has a weighting factor $n = -8$. So from Equation (5.6), after 40 Ma the progressive TTI is

$$\text{TTI} = \Sigma(\Delta t_i)(2^{n_i}) = (\Delta t_i)\Sigma(2^{n_i}) = 5 \times (2^{-8} + 2^{-7} + 2^{-6} + 2^{-5} + 2^{-4} + 2^{-3} + 2^{-2} + 2^{-1}) = 4.98$$

The sediment then spends 10Ma in each of the next three temperature intervals, $n = 0$, 1 and 2. The total maturity is then

$$\text{TTI} = 4.98 + 10 \times (2^0 + 2^1 + 2^2) = 74.98$$

Note: There is no intrinsic reason to retain the 10°C interval or the 100–110°C reference interval. However, the convention is entrenched so all published TTI values should be calculated using these parameters.

McKenzie (1981) preferred to state the TTI function as a continuous integral:

$$\text{TTI} = \int_{t_s}^{t_f} 2^{[(T_t - 105)/10]} dt \tag{5.7}$$

where T_t is time-dependent temperature (°C), and t_s and t_f are start and finish times (Ma), respectively, for the interval under investigation.

Equation (5.7) is, effectively, a smoothed version of Equation (5.6).

Question: Use Equation (5.7) instead of Equation (5.6) to find TTI for the situation given in the previous example.

Answer: The thermal history should be broken into two segments. The first:

$$T_t = 20 + 2t, \qquad 0 \le t \le 40 \text{ Ma}$$

the second:

$$T_t = 60 + t, \qquad 40 \le t \le 70 \text{ Ma}$$

Substituting these into Equation (5.7) and solving the integral for the first segment:

$$\text{TTI} = \int [2^{(-8.5+0.2t)} dt = 2^{-8.5} \int [2^{0.2t}] dt = 2^{-8.5} \times [1/(0.2 \times$$
$$\ln 2)] \times [2^8 - 2^0] = 5.08$$
and for the second segment:
$$\text{TTI} = \int [2^{(-4.5+0.1t)} dt = 2^{-4.5} \int [2^{0.1t}] dt = 2^{-4.5} \times [1/(0.1 \times$$
$$\ln 2)] \times [2^7 - 2^4] = 71.41$$
So the total TTI $= 5.08 + 71.41 = 76.49$

A TTI value on its own has no physical meaning, being based wholly on a system of arbitrary temperature intervals and weighting factors. It does, however, express the complete thermal history of a rock as a simple numerical value, which may then be calibrated against physical maturity indicators. Waples (1980) presented such a calibration between TTI and the reflectance of vitrinite, using data from thirty one wells worldwide. His calibration was roughly linear on a log–log plot (Figure 5.9), and took the following form:

$$\ln(\text{TTI}) = A \times \ln(R_o) + B \tag{5.8}$$

where $A = 4.144$ and $B = 4.168$ (over the range $0.5 \leq R_o \leq 5.0\%$).

Note: This calibration was developed using TTI values calculated from Equation (5.6). The calibration is slightly different for TTI values calculated using Equation (5.7).

Question: According to Waples' (1980) calibration, what VR value should we expect for our kerogen in the previous example?
Answer: Given TTI $= 74.98$, Equation (5.8) tells us:
$$\ln(74.98) = 4.144 \times \ln(R_o) + 4.168$$
so
$$\ln(R_o) = [\ln(74.98) - 4.168]/4.144 = 0.036$$
$$R_o = 1.04\%$$

In practice, we generally use the method in reverse. That is, we measure vitrinite reflectance and use the results to calculate thermal maturity in terms of TTI. As with all empirical methods, a single calibration does not necessarily suit all geological situations. Waples' calibration is often accused of underestimating TTI for a given VR (e.g. Morrow and Issler, 1993). Issler (1984) proposed an alternative calibration, which appears to give better results for many deposits. Issler used fifteen wells on the Scotian Shelf of eastern Canada to derive a relationship that took the same form as Equation (5.8) but with $A = 6.184$ and $B = 6.115$ (over the range $0.29 \leq R_o \leq 0.98\%$). Horváth et al. (1988) proposed yet another calibration, over a greater range of $R_o(0.27–3.0\%)$. Their calibration had $A = 8.278$, $B = 6.992$ for TTI < 292 ($R_o < 0.853\%$), and

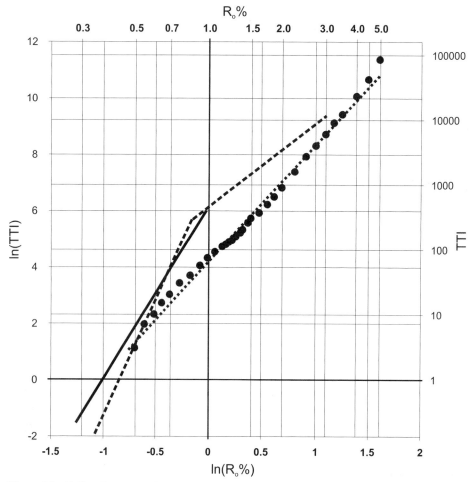

Figure 5.9. Calibration curves for time–temperature index (TTI) versus vitrinite reflectance (R_o%). Data points are from Waples (1980) and the dotted line is the least-squares best fit. Solid line is the calibration of Issler (1984), and dashed line is that of Horváth et al. (1988).

$A = 2.936$, $B = 6.143$ for TTI > 292 ($R_o > 0.853$%), and gave similar results to Issler's (1984) for mid-range values. These calibration curves are also shown on Figure 5.9.

Question: Using McKenzie's (1981) formulation of TTI [Equation (5.7)], calculate the equivalent R_o values for the above example using the calibrations of Issler (1984) and Horváth et al. (1988).

Answer: Given $TTI = 76.49$,

according to Issler, $A = 6.184$ and $B = 6.115$, so:

$\ln(76.49) = 6.184 \times \ln(R_o) + 6.115$

$R_o = 0.75$%

and according to Horvath et al., $A = 8.278$ and $B = 6.992$, so:

$\ln(76.49) = 8.278 \times \ln(R_o) + 6.992$

$R_o = 0.73\%$

Both of these are significantly lower than Waples' (1980) estimate in the previous example.

These sorts of calibrations are of use for forward modelling only. That is, TTI is calculated for a thermal history model and compared with TTI derived from observed VR. The thermal history model is adjusted and the process repeated until the model adequately explains the observed reflectance values. The process is often slow and painstaking. However, it is possible to invert the procedure and estimate the maximum temperature experienced by a sample of vitrinite, T_{max}, directly from its reflectance.

Barker and Pawlewicz (1986) suggested a relationship based on a compilation of over 600 $R_o(\%) - T_{max}(°C)$ data pairs from regions with well-understood thermal histories:

$$\ln(R_o) = 0.0078 \times T_{max} - 1.2 \tag{5.9}$$

Question: Vitrinite reflectance is measured at 1.05% in 80-Ma-old sediments. What maximum temperature does this suggest?

Answer: According to Barker and Pawlewicz's (1986) relationship, Equation (5.9), time plays no role, and T_{max} is a function only of R_o:

$$T_{max} = [\ln(R_o) + 1.2]/0.0078 = [\ln(1.05) + 1.2]/0.0078 = 160°C$$

This kind of relationship is very simplistic, however, and fails to take into account the large scatter introduced by the role that time plays in the maturation process. A more elaborate inversion of R_o data is possible if we assume a simple burial model, with burial depth proportional to the square root of time, $\sqrt{t}$, and thermal gradient proportional to $1\sqrt{t}$ (see Chapter 7 for an explanation of why these relationships are assumed). Burial temperature then increases linearly with time. Using the nomenclature of Equation (5.7), and setting $t_s = 0$, the temperature at time t is

$$T_t = T_s + (T_f - T_s)t/t_f \tag{5.10}$$

Substituting this into Equation (5.7) and solving the integral we get

$$\text{TTI} = 2^{[(T_s-105)/10]} \times [2^{[(T_f-T_s)/10]} - 1] \times 10t_f/[(T_f - T_s) \times \ln(2)] \tag{5.11}$$

The '−1' in the square parentheses is negligible in most situations, and can be ignored. Equation (5.11) then reduces to an iterative solution for T_f:

$$T_f = P \times \ln(Q \times (T_f - T_s)) \tag{5.12}$$

where

$$P = [10/\ln(2)] = 14.427$$

$$Q = \text{TTI} \times 2^{(105/10)}/(P \times t_f)$$

Equation (5.12) converges rapidly for any initial estimate of $T_f >$ $[T_s + 2\exp(T_s/P)/Q]°\text{C}$.

Question: Using Issler's (1984) calibration and assuming a simple burial history, a constant surface temperature $T_s = 15°\text{C}$ and a linear increase in temperature with time, what present temperature, T_f, is required to account for the reflectance in the example above?

Answer: The first step is to convert VR to TTI using Issler's coefficients in Equation (5.8):

ln(TTI) = 6.184 × ln(1.05) + 6.115

Implying TTI = 612

Substituting TTI = 612, $T_s = 15°\text{C}$ and $t_f = 80$ Ma into Equation (5.12):

$$P = 14.427, \qquad Q = 767.89$$

To converge, our initial estimate of T_f must be greater than $15 + 0.0074$, so we will begin with an estimate of $T_f = 15.008°\text{C}$ (although a higher estimate will converge faster). Subsequent iterations of Equation (5.12) give $T_f = 26.19°\text{C}$, $130.69°\text{C}$, $164.39°\text{C}$, $168.08°\text{C}$, $168.43°\text{C}$, $168.46°\text{C}$ and $168.47°\text{C}$ (obviously converging). If present temperature is at or about $168°\text{C}$, then a simple burial history is supported by the vitrinite reflectance measurement. A temperature presently less than $168°\text{C}$ would imply that the kerogen has experienced somewhat higher temperatures in the past. A present temperature higher than $170°\text{C}$ may imply that the material has been heated rapidly and recently.

The style of vitrinite reflectance analysis described above has waned in popularity since the development of chemical kinetic models. The increasing reflectance of vitrinite with time and temperature can be modelled as the result of a number of parallel chemical reactions that act to eliminate water, carbon dioxide, methane and higher hydrocarbons from the maceral (Burnham and Sweeney, 1989). Different reactions proceed at different rates for a given temperature and it is this spectrum of reaction rates that leads to the observed gradual increase in reflectance with temperature and time. This is analogous to the kinetic model of petroleum generation discussed in Section 5.1.2.

The Lawrence Livermore National Laboratory in California carried out much of the development in the area of VR kinetics in the late 1980s (e.g. Burnham and Sweeney, 1989). The result was a method known as EASY%Ro (Sweeney and Burnham, 1990). This is now the most popular method of modelling VR, so we will describe it here in some detail, reproducing to a large degree Sweeney and Burnham's (1990) Appendix I.

The basic principles of kinetic modelling are the same as those presented in Section 5.1.2 and will not be reproduced here. For the purposes of modelling VR, EASY%Ro uses a set of twenty reactions with a common pre-exponential factor $A = 1.0 \times 10^{13}$ s^{-1}, and stoichiometric (or weighting) factors and activation energies as listed in Table 5.2.

Question: At what rate does reaction 12 proceed at 100°C and 200°C?

Answer: From Equation (5.1) we know the reaction rate:

$$k_{12} = A \exp(-E_{12}/RT)$$

At 100°C, $T = 100 + 273.15 = 373.15$ K, $E_{12} = 234 \times 10^3$ J mol^{-1}, $A = 1 \times 10^{13}$ s^{-1}, $R = 8.314$ J K^{-1} mol^{-1}

so

$$k_{12} = 1 \times 10^{13} \exp[-234 \times 10^3/(8.314 \times 373.15)] = 1.749 \times 10^{-20} \text{ s}^{-1}$$

At 200°C, $T = 473.15$ K, so

$$k_{12} = 1 \times 10^{13} \times \exp[-234 \times 10^3/(8.314 \times 473.15)] = 1.466 \times 10^{-13} \text{ s}^{-1}$$

Table 5.2. **Stoichiometric Factors and Activation Energies used in EASY%Ro**

Reaction Number i	Stoichiometric Factor f_i	Activation Energy E_i, (kJ mol^{-1})
1	0.03	142
2	0.03	151
3	0.04	159
4	0.04	167
5	0.05	176
6	0.05	184
7	0.06	192
8	0.04	201
9	0.04	209
10	0.07	218
11	0.06	226
12	0.06	234
13	0.06	243
14	0.05	251
15	0.05	259
16	0.04	268
17	0.03	276
18	0.02	285
19	0.02	293
20	0.01	301

Source: Sweeney and Burnham's (1990) Table 1.

Although the actual figures may be difficult to comprehend, the important point is that the reaction proceeds approximately ten million times faster at 200°C than at 100°C.

In practice, EASY%Ro breaks the thermal history of a source bed into segments of constant heating rate. For example, the thermal history of a particular bed might be broken into an initial period of rapid heating at $1°C\,Ma^{-1}$ for 20 Ma, followed by a longer period of slower heating at $0.2°C\,Ma^{-1}$ for 80 Ma. The jth heating period is defined as that extending from time t_{j-1} to t_j, and raises the temperature from T_{j-1} to T_j. Thus, the heating rate during the jth period, H_j, is

$$H_j = (T_j - T_{j-1})/(t_j - t_{j-1})\ \mathrm{K\,s^{-1}} \tag{5.13}$$

We now define a parameter I_{ij}, which is related to the rate of the ith reaction during the jth heating period:

$$I_{ij} = T_j A \times \left[1 - \frac{(E_i/RT_j)^2 + a_1(E_i/RT_j) + a_2}{(E_i/RT_j)^2 + b_1(E_i/RT_j) + b_2} \right] \times \exp(-E_i/RT_j) \tag{5.14}$$

where $a_1 = 2.334733$, $a_2 = 0.250621$, $b_1 = 3.330657$ and $b_2 = 1.681534$.
Furthermore, we define $\Delta I_{i0} = 0$ and

$$\Delta I_{ij} = \Delta I_{i,j-1} + (I_{ij} - I_{i,j-1})/H_j \tag{5.15}$$

Note: The heating rate, H_j, is the same for all twenty parallel reactions (i.e. all values of i).

Defining $\mathbf{w}_{i0}$ as the initial amount of material available for reaction i, and $\mathbf{w}_i$ as the amount of *unreacted* material remaining, then the proportion of material remaining at the end of the jth time interval is

$$(\mathbf{w}_i/\mathbf{w}_{i0})_j = \exp(-\Delta I_{ij}) \tag{5.16}$$

The total converted fraction of all twenty reactants, $\mathbf{F}$, is

$$\mathbf{F} = \sum_i f_i[1 - (\mathbf{w}_i/\mathbf{w}_{i0})] \tag{5.17}$$

recalling that f_i is the stoichiometric factor of the ith reaction given in Table 5.2.
Parameter $\mathbf{F}$ has been calibrated directly with vitrinite reflectance:

$$R_o = \exp(-1.6 + 3.7\mathbf{F}) \tag{5.18}$$

Although the process looks complicated, it only involves arithmetic and is easily adapted to computer programs or, as Sweeney and Burnham (1990) themselves illustrated, spreadsheets. (See the spreadsheet 'EASY%RO.XLS' on the web site mentioned in the Preface.)

Question: Take the earlier example where a source bed is heated at a constant rate from 15°C to 168.46°C in 80 Ma. What is the predicted R_o according to EASY%Ro?

Answer: This is just a single period of heating, so from Equation (5.13):

$$H_1 = (168.46 - 15)/(80 \times 3.1557 \times 10^{13}) = 6.0787 \times 10^{-14} \text{ K s}^{-1}$$

We should now calculate I_{i0} and I_{i1} for all twenty reactions, using Equation (5.14) and noting $T_0 = 15°C = 288.15 \text{ K}$, and $T_1 = 168.46°C = 441.61 \text{ K}$. We do not present all the workings here, but the solution for **F** is:

$$\mathbf{F} = 0.516$$

so, from Equation (5.18):

$$R_o = \exp(-1.6 + 3.7 \times 0.516) = 1.36\%$$

This is considerably higher than $R_o = 1.05\%$ calculated using Lopatin's method.

EASY%Ro is calibrated in the range $0.3\% \leq R_o \leq 4.5\%$, and for heating rates in the range of about 10^{-15}–10^{-5} K s^{-1}, which cover most geological problems.

Morrow and Issler (1993) compared the results for different VR models and concluded that at low maturation levels (VR < 0.9%) EASY%Ro may slightly overestimate maturity, and the simpler TTI calibration of Issler (1984) may be more accurate in this range. Future adaptations to EASY%Ro are therefore likely to focus on improving the calibration of the technique in these lower ranges, and on extending the range of activation energies to higher values of R_o.

Note: In spite of subsequent attempts to modify EASY%Ro (e.g. Suzuki, Matsubayashi and Waples, 1993), it remains the most popular method of modelling VR.

There are certain limitations to VR that are not related to calibration errors. They relate, instead, to the complex nature and origin of vitrinite. The samples used to calibrate the different models are generally extensively studied and well understood, but the same cannot always be said for samples collected and analysed as part of routine basin history studies. If the quality of reflectance data is poor, then no amount of modelling can constrain the thermal history.

Another limitation is that vitrinite reflectance can only be measured for samples containing vitrinite. This obviously restricts the range of formations for which the method will work. Vitrinite is detrital plant material, usually transported with other sediment to the final deposition point. It may suffer diagenetic oxidation to inertinite (i.e. become *subhydrous* – having relatively low H:C ratio) during transport, thus forming a deposit barren of vitrinite. Carbonates, which tend to form in clear water with little external sediment input, are also generally barren of vitrinite.

Identifying vitrinite under a microscope also introduces a potential source of error. The level of experience of the operator determines how accurately vitrinite is distinguished from other macerals, but there is always an unavoidable subjective bias. If the material retains its original cellular structure then the task is simplified, but when all macerals are broken up and dispersed through the sample then liptinite and inertinite may be mistaken for vitrinite, with adverse effects on reflectance values.

Core or hand specimens are the preferred source for reliable vitrinite samples. Drill cuttings may be contaminated by material from higher up the well, depressing average reflectance values.

All the above problems are serious, but perhaps the *most* serious problem with VR arises from the fact that vitrinite is derived from many different sources, not all of which have identical chemical properties. In particular, there are major differences in the chemistry of terrestrial- versus marine-derived vitrinite (e.g. Wilkins et al., 1992). The reflectance of vitrinite of marine origin is often significantly lower than for a thermally equivalent terrestrial vitrinite. This effect, known as *reflectance suppression*, has been attributed to differences in original plant matter (e.g. Thomas, 1982; Price and Barker, 1985), and is due to the marine vitrinite remaining relatively *perhydrous* (i.e. having a high H:C ratio). All of the commonly used vitrinite models were calibrated using terrestrial vitrinite and often underestimate maximum temperature when used with marine material (George, Smith and Jardine, 1994).

> *Note:* If marine-influenced vitrinite is used to model the thermal history of a basin, thermal maturity is likely to be underestimated.

Khavari Khorasani and Michelsen (1993) published a kinetic model for the reflectance of marine-influenced vitrinite (included on the 'EASY%RO.XLS' spreadsheet available on the web site mentioned in the Preface). However, their recommended values predict reflectance *greater* than the terrestrial model in many situations, contrary to observation.

The importance placed on VR as a tool for constraining thermal history should reflect the quality of the data. Assessment of quality should take into account the source (e.g. core vs. cuttings), preservation (e.g. coal fragments vs. dispersed organic matter), depositional environment (terrestrial vs. marine) and the experience of the operator making the reflectance measurements.

5.2.2. *Fluorescence Alteration of Multiple Macerals (FAMM®)*

Australia's CSIRO Petroleum has developed a technique aimed at overcoming some of the problems associated with traditional vitrinite reflectance. The technique is called 'Fluorescence Alteration of Multiple Macerals', or *FAMM®* (registered trademark of the Commonwealth Scientific and Industrial Research Organisation; Wilkins et al., 1992), and specifically addresses the problems of vitrinite reflectance suppression and enhancement. The method involves measuring the fluorescence emission from polished maceral samples in response to a laser beam (488 nm radiation at ~0.1 mW at the sample surface; Cooper et al., 1998) focussed to about 1–2 μm diameter. The intensity of the emitted red (625 nm) fluorescence generally varies with time and the effect is quantified by measuring the intensity at the beginning (I_i) and end (I_f) of a specified time period. Presently, a period of 400 s is used, although 700 s was used earlier in the development of the technique.

A number of measurements are made on a single sample, focussing the laser at different maceral fragments. All maceral types are targeted, which gives the technique its name. All data from the sample are plotted on log–log axes, and this is referred to as a 'fluorescence alteration diagram' (Figure 5.10). The abscissa (*x*-axis), I_f, is mainly controlled by maceral type, with inertinite hav-

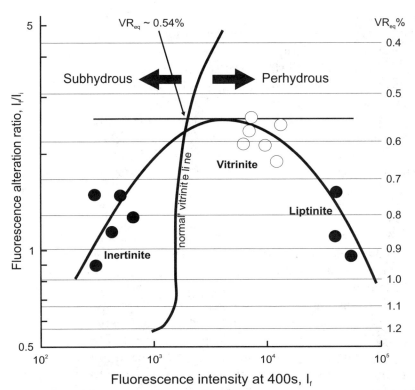

Figure 5.10. Fluorescence alteration diagram illustrating the procedure for deriving VR_{eq} for samples containing perhydrous vitrinite.

ing the lowest fluorescence, and liptinite the highest. The ordinate (y-axis), I_f/I_i, is mainly controlled by the thermal maturity, decreasing as maturity increases.

To determine the equivalent vitrinite reflectance (VR_{eq}) of the sample using the fluorescence alteration diagram, a best-fit second-order polynomial curve is placed through the complete data set. The technique has been calibrated such that the curve maximum indicates VR_{eq} for samples with $VR_{eq} < {\sim}0.9\%$. For samples with $VR_{eq} > 0.9\%$, the polynomial function is often relatively flat or concave upwards. Then, VR_{eq} is estimated from the average fluorescence alteration ratio of the vitrinite population.

All fluorescence alteration diagrams include a standard 'J-curve' indicating where 'normal' (*orthohydrous* – having a normal H:C ratio) vitrinite plots (originally calibrated using extensively studied coals from eastern Australia, with subsequent calibrations developed for other geologic provinces). Perhydrous grains (suppressed reflectance) plot to the right of the J-curve while subhydrous grains (enhanced reflectance) plot to the left. The technique inherently corrects the value of VR_{eq} for the state of the vitrinite. For example, in a population of perhydrous vitrinite, $VR_{eq} > R_o$.

EXAMPLE

Cooper et al. (1998) presented a comparison of maturity estimates using three methods for samples from the Otway Basin, Australia. In the Parker River/Blanket Bay area, traditional VR yielded a maturity estimate of $R_o = 0.53\%$, FAMM® suggested $VR_{eq} = 0.57\%$ and fission track analysis suggested an equivalent $FT - R_o = 0.53\%$. All were in close agreement. However, in the Beach Forest region, where the organic sedimentary material showed signs of early diagenetic oxidation, $R_o = 0.76\%$, $VR_{eq} = 0.49\%$ and $FT - R_o = 0.53\%$. Here, FAMM® was in close agreement with the independent fission track estimate, while the traditional VR appeared to overestimate maturity.

The FAMM® technique has many advantages over traditional VR analyses. Firstly, all macerals (liptinite, vitrinite and inertinite) are used, which has the dual advantage of removing the subjective element of the analysis and also making the method applicable to a greater range of samples (e.g. many marine shales, where vitrinite is sparse or non-existent). The exception to this rule is for samples that contain perhydrous vitrinite and are more mature than about $VR_{eq} = 0.9\%$. For those samples, the polynomial intersection method overestimates VR_{eq} (Wilkins et al., 1998) and vitrinite must be identified to determine VR_{eq} from the fluorescence alteration ratio. If no vitrinite can be identified, an approximate VR_{eq} can be ascertained by estimating the likely position of vitrinite between the inertinite and liptinite populations.

The current disadvantage of the FAMM® method stems from its relatively recent development – access to the technique remains limited. Only two laboratories in the world, in Sydney and Beijing, are currently equipped to make the necessary measurements, whereas there are any number of avenues for obtaining traditional VR data. FAMM® is still evolving, as evidenced by the recent change in measurement time, but consistently proves to be a valuable tool in overcoming many of the problems with traditional VR.

Note: The results of a FAMM® analysis are in units of equivalent R_o. These values can be used in maximum temperature modelling in place of traditional VR.

5.2.3. Thermal Alteration Index (TAI)

Alteration in the colour and opacity of liptinite (spore, pollen, and leaf cuticle material) with temperature, when viewed in transmitted light, was recognised as early as 1948 (Schopf, 1948), but it was not until the 1960s that the phenomenon was applied to petroleum exploration (Gutjahr, 1966; Correia, 1967; Staplin, 1969). The *thermal alteration index* (TAI) is an arbitrary scale (from 1 to 4) that ranks the gradual change in colour and opacity of liptinite with temperature. TAI 1 is equivalent to the lowest levels of organic maturity and is assigned to liptinite particles showing a greenish-yellow hue. TAI and maturity increase as the colour of liptinite progresses through pale yellow, yellow, amber, deep red-brown, and finally opaque black at TAI 4.

EXAMPLE

The liptinite sample illustrated in Figure 5.4 has, in reality, a greenish-yellow hue and could be assigned a maturity of TAI 1. This is consistent with the shallow depth (750 m below sea floor) and low temperature (about 55°C) from whence the sample was extracted.

The application of TAI to detailed petroleum exploration is of limited value. The technique is subjective in that the analyst must discriminate between colours on a standard scale. More importantly, however, the entire oil generation window falls within a narrow range of TAI values ($\sim 2.0 - 2.7$) and poor resolution of the index limits the precision of palaeotemperature estimates to about ±20°C. The technique is widely used, though, and provides a simple and useful initial indicator of the general level of organic maturity. Below TAI 2, no oil can be expected, while above TAI 3.5, thermal conditions are in the gas zone.

According to Staplin (1982), errors in maximum temperatures derived from this method result largely from a failure to standardise laboratory techniques, and not from inherent shortfalls in the process.

5.2.4. Conodont Alteration Index (CAI)

In many ways, the *conodont alteration index* (CAI) is analogous to TAI. With CAI, it is the colouration of conodont fossils that reflects the maximum palaeo-temperature. Epstein, Epstein and Harris (1977) first suggested the method after studying material from the Appalachians. The present index (e.g. Nicoll and Gorter, 1984; Nicoll and Foster, 1994; Helsen, David and Fermont, 1995) is, like TAI, an arbitrary numerical scale, with eight levels related to the gradual change in colour and opacity of conodonts with temperature. A value of CAI 1 is equivalent to the lowest maturity, while CAI 8 represents the highest temperatures. Intermediate values, such as CAI 2.5, are acceptable if a conodont appears to fall between two indices. The CAI scale is shown in Table 5.3.

The CAI method is relatively cheap and fast, does not require sophisticated instruments, and gives an initial indication of the thermal maturity of the host sediment. Unfortunately, CAI suffers from the same two problems as TAI. The first is the broad temperature range associated with each index. The uncertainty in maximum palaeotemperature ranges from about ±20°C to ±50°C. Uncertainty of that magnitude is too high for detailed thermal history reconstruction.

The second problem is the subjectivity of the method. Human operators must interpret the colour of the conodont and this interpretation will vary

Table 5.3. Conodont Alteration Index (CAI) and Corresponding Palaeotemperature

CAI	Colour	Palaeotemperature (°C)	Thermal Maturity
1.0	Pale yellow	50–80	Immature
1.5	Very pale brown	50–90	Early-mature
2.0	Brown to dark brown	60–140	Mature
2.5		85–180	Late-mature
3.0	Very dark grey	110–200	Over-mature
3.5		150–260	Over-mature
4.0	Light black	190–300	Partially carbonised
4.5		230–240	Partially carbonised
5.0	Dense black	300–400	Carbonised
5.5	Dark grey–black	310–420	Early volatilisation of carbon
6.0	Grey	350–435	Volatilisation of carbon
6.5	Grey-white	425–500	Late volatilisation of carbon
7.0	Opaque white	480–610	Carbon fully volatilised
7.5	Semi-translucent	> 530	Partially recrystallised
8.0	Transparent	> 600	Recrystallised

Source: Data from Helsen et al. (1995).

somewhat from person to person. One person's 'very dark grey' (CAI 3.0) is another's 'light black' (CAI 4.0), and the difference in interpreted palaeotemperature could be 100°C or more. Helsen et al. (1995) attempted to remove the subjective element of interpretation by using a computer to analyse the colour of conodonts in terms of red, green and blue components. They found that the intensities of all three colours decrease almost linearly with increasing CAI up to CAI 5.0, then increase rapidly in the range CAI 5.0–8.0. The results suggest that an objective method of determining CAI might be obtained after adequate calibration.

Deaton, Nestell and Balsam (1996) suggested a similar means to remove the subjective element from the CAI method. They published the results of a comparison between conventional CAI and the reflectance spectrum of white light off conodonts. They found that the reflectance increases almost linearly with wavelength in the range 550–800 nm, and that the slope of the reflectance versus wavelength plot varies with CAI according to the following formula:

$$\text{CAI} = -95.818(slope) + 623.996(slope)^2 + 4.558 \tag{5.19}$$

where *slope* is in units of percent per nanometre. The uncertainty in the CAI estimate is about ±0.2 with 95% confidence.

Question: A number of conodont fragments were separated from a potential source rock and subjected to reflectance spectrum analyses. The condonts were found to have the following average reflectances at the specified wavelengths:
  48.1% @ 550 nm
  50.8% @ 600 nm
  54.3% @ 650 nm
  56.9% @ 700 nm
  59.2% @ 750 nm
  61.9% @ 800 nm
What equivalent CAI level and maximum palaeotemperature do these results suggest?

Answer: The average slope of the reflectance versus wavelength data is 0.0553% nm^{-1}. So, from Equation (5.19):

$$\text{CAI} = (-95.818 \times 0.0553) + (623.996 \times 0.0553^2) + 4.558$$
$$= 1.2 \pm 0.2$$

This represents a maximum paleaotemperature in the range of about $50 - 85°C$.

An objective method of classifying CAI is a step towards developing a new maturity scale of much greater precision than conventional CAI. Such a scale has the potential to be a very useful tool, because conodonts are found in carbonates, while vitrinite is not (e.g. Mory, Nicoll and Gorter, 1998). Current drawbacks include the large number of pristine conodonts required for statistical accuracy (30–100), and the necessity for standardised preparation of the samples (Deaton et al., 1996). Further work will allow direct correlation of conodont optical properties against other maturity indicators, and remove the need for the original CAI scale.

5.2.5. *Clay Mineralogy*

There are four principal groups of clay minerals (kaolinite, smectite, illite and chlorite) and these include over a dozen individual species. Many of these species may be present within a clay-rich rock when it is first deposited. However, there is typically a downward trend of mineral simplification as diagenetic processes increasingly enrich units in illite and chlorite at the expense of smectite and kaolinite. In general, burial appears to favour the retention of illite over other clay minerals, and both the relative abundance and degree of order in the illite crystal lattice increase with depth (Chamley, 1989, Chapter 15). Smectite typically disappears totally at a depth in the range 1000–2500 m. These observations have given rise to a couple of methods of using illite to estimate maximum palaeotemperature. The first is based on the degree of order in the illite crystal lattice, and the second on the ratio of illite to smectite in the clay.

The abundance of illite in a material is best measured using x-ray diffraction. Illite shows up on an x-ray spectrum as a characteristic peak at a wavelength around 1 nm (Figure 5.11). As temperature increases, the degree of order within the illite crystal lattice also increases, leading to a progressive narrowing of the 1 nm peak. Kübler (1967) defined an *index of crystallinity* as the width of the 1 nm illite peak (at half-height) in millimetres on a standard x-ray spectrum. The index, in which lower numbers imply higher temperature, has proven a reliable and practical tool for estimating the degree of low-level metamorphism that a unit has undergone. It is most often used to determine the depth of the transition between diagenesis and metamorphism (the anchizone), which occurs between indices 4.0 and 2.5. This is equivalent to vitrinite reflectance on the order of $R_{v\,\mathrm{max}} = 2.3–3.3\%$ (Kisch, 1980).

Illite crystallinity index is, however, a poor tool for determining absolute palaeotemperature. Robert (1988) was critical of the technique because the actual width of the illite peak depends on the characteristics of the apparatus used, and is thus laboratory dependent. Also, the index shows poor correlation with other palaeotemperature indicators at low maturity (Kisch, 1980). It is, however, one of the few methods that work at high maturity levels, which makes it useful in situations where other methods may fail.

The relative proportion of smectite in clay decreases with respect to illite with increasing time and temperature. Temperature is by far the most important factor controlling the reaction rate, but the ratio of the chemical activities of potassium/sodium (K/Na) in the clay also exerts some control. Pytte and

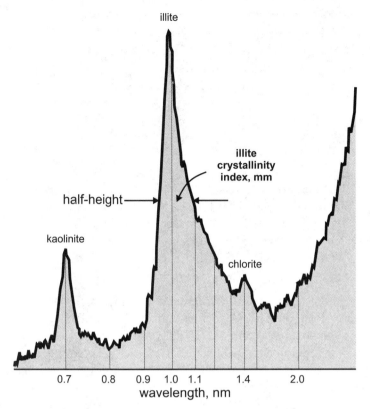

Figure 5.11. A typical x-ray diffraction spectrum of a sample of clay. Note the characteristic peaks denoting the abundance of kaolinite (~0.7 nm), illite (~1 nm) and chlorite (~1.4 nm). Illite crystallinity index is the width on the plot (in millimetres) of the illite peak at half its height.

Reynolds (1989) modelled the clay transformation in terms of chemical kinetics and derived the following sixth-order relationship:

$$-\partial S/\partial t = S^5 \times (\text{K/Na}) \times A \exp(-E/RT) \tag{5.20}$$

where

$S = $ mole fraction of smectite in illite–smectite mixed layers
$(\text{K/Na}) = 74.2 \times \exp(-2490/T)$
$A = 5.2 \times 10^7 \text{ s}^{-1}$
$E = 138 \text{ kJ mol}^{-1}$
$R = $ the gas constant $= 8.314472 \text{ J K}^{-1} \text{ mol}^{-1}$
$T = $ temperature (K)

Solving the differential equation, and inserting all known constants, Equation (5.20) can be written as follows:

$$S^4 = \frac{S_0^4}{1 + (1.543 \times 10^{10}) \times t \times S_0^4 \exp(-19{,}110/T)} \tag{5.21}$$

where S_0 is the mole fraction of smectite in the initial mixed-layer clay and t is time (s) spent by the clay at more than 90% of peak temperature.

Question: The mole fraction of smectite in an illite–smectite mixed-layer clay is initially 80%. The clay is placed in a kiln for a period of 12 days, after which time a sample is recovered and analysed. It is now found to contain just 23% smectite. At what temperature was the clay baked?

Answer: The variables in Equation (5.21) are

$S = 0.23$

$S_0 = 0.8$

$t = 12$ days $= 1.037 \times 10^6$ s

Substituting and rearranging gives

$\exp(-19{,}110/T) = [(S_0^4/S^4) - 1]/[1.543 \times 10^{10} \times t \times S_0^4] =$
$145.37/6.554 \times 10^{15} = 2.218 \times 10^{-14}$
$-19{,}110/T = \ln(2.218 \times 10^{-14}) = -31.44 \text{ K}^{-1}$
$T = -19{,}110/-31.44 = 607.8 \text{ K} = 335°\text{C}$

Bethke and Altaner (1986) examined the problem in a slightly different way. They treated the smectite layers differently depending on the number of adjacent illite layers, and found that a smectite layer adjacent to a single illite layer converts to illite at a lower energy than a smectite layer totally surrounded by either smectite or illite. They derived a weighted kinetic relationship for the reactions of the three different smectite types:

$$(\partial P_S/\partial t) = -P_S \times (X_0 k_0 + X_1 k_1 + X_2 k_2) \qquad (5.22)$$

where

P_S = fraction of smectite with respect to total smectite–illite
$X_i = N_i/N$ = fraction of smectite with i adjacent illite layers/total smectite
$N_i = N_{i0} \exp(-k_i t)$
N_{i0} = initial fraction of Type i smectite/total smectite
$k_i = A \exp(-E_i/RT)$ = reaction rate for smectite with i adjacent illite layers

Bethke and Altaner (1986) assumed a constant pre-exponential factor for all three smectite types, $A = 10^{-3} \text{ s}^{-1}$. They found the activation energy for Type 0 smectite (smectite layer with no adjacent illite layer) was $E_0 = 77.5 \text{ kJ mol}^{-1}$; for Type 1 smectite, $E_1 = 75.5 \text{ kJ mol}^{-1}$, and for Type 2 smectite, $E_2 = 100.5 \text{ kJ mol}^{-1}$. They provided a graphical solution to Equation (5.22) that illustrated the change in the proportion of illite in a smectite–illite mixed layer as temperature increases (Figure 5.12).

Question: A particular clay sample is exhumed from a formation that was deposited 25 Ma ago. The present temperature of the formation is 80°C. Assuming an average surface temperature of 15°C, what is the expected illite content in relation to the total smectite–illite?

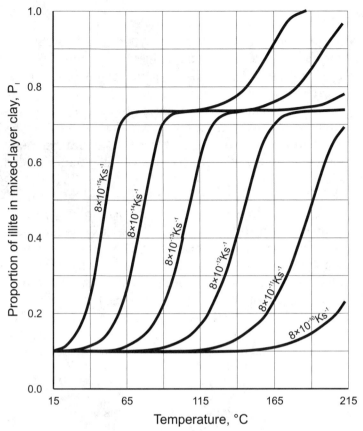

Figure 5.12. The proportion of illite in smectite–illite mixed layers calculated for different heating rates. Modified after Bethke and Altaner (1986).

Answer: We need to determine the average heating rate:

average heating rate = change in temperature / change in time

$$dT/dt = (80°C - 15°C)/25 \text{ Ma} = 65 \text{ K}/(25 \times 3.1557 \times 10^{13} \text{ s})$$

$$= 8.24 \times 10^{-14} \text{ K s}^{-1}$$

From Figure 5.12, assuming a heating rate of 8.24×10^{-14} K s^{-1} and a present temperature of 80°C:

$$P_I \approx 0.55$$

We expect approximately 55% of the smectite–illite mixed layers to be illite.

The main difficulty with smectite–illite geothermometers is that clay mineralogy is affected by many influences other than temperature. In particular, fluid interactions play a large role in altering clay chemistry. One can never be certain that the clay being analysed is truly representative of the thermal history of the host bed.

5.2.6. Pyrolysis (RockEval)

Pyrolysis is a general term meaning 'annihilation by fire', and refers to a number of different processes. The petroleum industry generally uses the term to describe a procedure whereby a small sample of rock is heated in an inert gas (e.g. argon). Heating rate is kept constant while hydrocarbons 'boiled' from the rock are carried with the inert gas to a chromatograph. The chromatograph records the volume of hydrocarbons through time, where time is directly proportional to temperature (owing to the constant heating rate). The results of such a procedure, and the effects of increasing maturity, are predictable in a general sense.

A rock sample containing organic material holds two sources of hydrocarbons. Firstly, there are those hydrocarbons that have naturally generated within the rock and remain within the pore spaces. These hydrocarbons are generally expelled from the sample and are detected by the pyrolysis instrument at temperatures below about 200°C. Plotting the amount of detected hydrocarbon versus temperature therefore reveals a distinct peak in the range 100–200°C (Figure 5.13). This peak is usually referred to as the S1 peak.

The second source of hydrocarbons in the rock is kerogen material that has not yet matured. We know from Section 5.1.2 (in particular, Equation 5.1) that reactions proceed at a rate dependent on temperature. If the temperature is

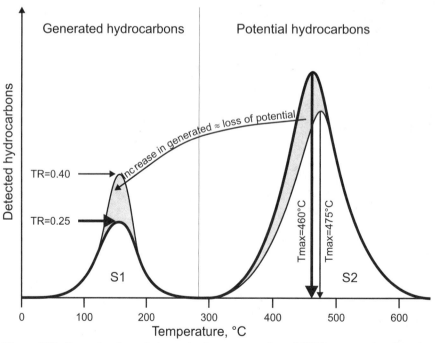

Figure 5.13. Example of pyrolysis results for two samples of different maturity. Sample 1 (thick line) is less mature than sample 2 (thin line). Sample 2 yields higher Tmax and TR values. An increase in generated hydrocarbons (amplitude of S1) is offset by a decrease in potential hydrocarbons (amplitude of S2).

raised to a sufficient level, reactions that normally take millions of years will run to completion within an hour or less. The kerogens generate all of their potential hydrocarbons, which are subsequently detected by the pyrolysis apparatus. These hydrocarbons show up on the pyrolysis graph as a second distinct peak, generally around 400–500°C (Figure 5.13), referred to as the S2 peak.

Of all the thermal maturity indicators, perhaps the most misleading (from the point of view of thermal history reconstruction) is the pyrolysis expression, *Tmax*. Tmax is the temperature corresponding to the summit of the S2 peak. In spite of its name, Tmax is only loosely related to the maximum palaeotemperature to which organic material has naturally been subjected.

The greater the thermal maturity of the original rock, the lower the volume of potential hydrocarbons still trapped within the kerogen. Also, and more importantly, the activation energy of the remaining material will be higher, requiring higher temperatures to mobilise it during pyrolysis, and resulting in a higher Tmax. Thus, two aspects of the pyrolysis plot relate to the effects of increasing maturity. Firstly, the location of the S2 peak moves progressively to the right, and, secondly, the amplitude of the S1 peak increases relative to the S2 peak (Figure 5.13). Both of these effects have been correlated with TTI and other maturity indicators – Tmax directly, and the amplitudes in terms of a *transformation ratio* (TR), defined as the height of S1 divided by the sum of the heights of S1 and S2:

$$TR = S1/(S1 + S2) \tag{5.23}$$

The specific results of pyrolysis strongly depend on both the heating rate and the recording equipment. A faster heating rate gives the reactions less time to proceed, so they require higher temperatures. In other words, a higher heating rate will result in a higher value of Tmax for the same sample.

Note: Pyrolysis results from different laboratories are mutually comparable only if a standardised pyrolysis technique is employed. There does exist such a technique, widely used amongst those in the petroleum industry, known as RockEval.

Using *RockEval* parameters, thermal maturity equivalent to $R_o = 0.6\%$ is characterised by TR ≈ 0.1 and Tmax $\approx 440°$C, while $R_o = 1.4\%$ is equivalent to TR ≈ 0.4 and Tmax $\approx 470°$C.

5.2.7. *Fluid Inclusion Microthermometry (FIM)*

During diagenetic crystallisation of minerals such as calcite, feldspar and quartz, small amounts (usually $< 10~\mu$m in diameter) of pore fluid are sometimes trapped within the growing crystal. Usually, these fluid inclusions are to be found along the irregular contact between the original sedimentary grain and the mineral overgrowth (Lisk and Eadington, 1994). Fluid inclusion analyses are generally used to constrain the timing of oil migration in relation to mineral diagenesis, but they are also able to give some indication of past temperature conditions. The fluid is generally trapped as a single phase, but commonly separates into liquid and vapour components upon reduction to surface conditions, forming an observable bubble within the inclusion. This reversible process is behind the principal of *fluid inclusion microthermometry* (FIM).

Upon reheating the inclusion, the fluid reverts to a single phase at a particular temperature, T_h, which is the homogenisation temperature (Roedder, 1981). Thus T_h is the *minimum* temperature at which the inclusion was originally trapped. Accuracy of the temperature estimate is approximately $\pm 1°C$ at below $100°C$ and $\pm 2°C$ at above $100°C$ (Lisk and Eadington, 1994). Although not an indicator of *maximum* palaeotemperature, the technique is included here because of the useful information it can yield.

In a limited way, FIM provides information on temperature conditions at particular times in the history of a basin. It is typical to find more than one generation of fluid inclusions within a given sample. Careful interpretation of the location of the inclusions with respect to the microstructure of the sediment (Figure 5.14) yields relative time–temperature information.

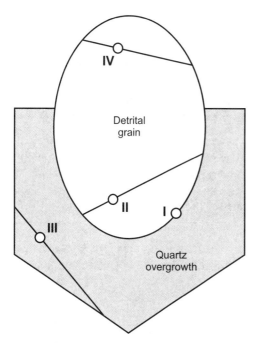

Figure 5.14. The location of fluid inclusions in relation to the timing of fluid migration. I – Fluid trapped during crystallisation of quartz overgrowths. II – Fluid trapped during healing of fractures prior to quartz overgrowth. III – Fluid trapped during healing of fractures during or after quartz overgrowth. IV – Cannot be used for timing in detrital quartz. Modified after Lisk and Eadington (1994).

Question: In a sample of sandstone obtained from a drill core taken from a depth of 1750 m, three distinct populations of fluid inclusions are discovered. The first population is found at the boundary of detrital grains and quartz overgrowths and has $T_h = 91°C$. The second population is from healed fractures within detrital grains and has $T_h = 74°C$, and the third population is from healed fractures within the quartz overgrowths and has $T_h = 52°C$. Present-day temperature at 1750 m is 65°C. What do these data tell us about the history of the sandstone?

Answer: There is an obvious timing relationship between the three populations. The second population must predate the first, while the third must postdate both. The first two events occurred at temperatures greater than 74°C and 91°C, respectively. Both tell us that palaeoconditions were hotter than at present. This could mean either higher heat flow or a greater depth of burial in the past. The third population of inclusions developed at a temperature greater than 52°C. Although not conclusive, the implication is that the conditions cooled (through lower heat flow or denudation) prior to the third event, then warmed to present levels.

The results of FIM can be tainted if the inclusion was originally a two-phase mixture (e.g. pore water and gas; pore water and oil; oil and gas). In particular, the initial presence of a gas phase increases the amount of vapour at room temperature and also increases T_h. In a population of inclusions exhibiting a significant spread in T_h values, the minimum value probably corresponds to an inclusion that initially contained only pore fluid, with no vapour phase. We can be confident that the fluid contained an equilibrium amount of dissolved gas at entrapment temperature, thus the minimum T_h recorded for a given population of fluid inclusions is likely be the actual temperature of entrapment (Roedder, 1981).

Note: $T_{h(min)}$ = equilibrium temperature = entrapment temperature

---

5.2.8. Molecular Biomarkers

The chemical reactions that take place during the thermal evolution of kerogen are many and complex. Such reactions control the evolution of reflectance, fluorescence, colour and numerous other physical properties. They also control the generation of volatiles and petroleum products. Most of these reactions cannot be investigated individually, but a small number can and have been studied in detail and provide an independent indication of thermal maturity.

As already described in Section 5.1.2, the rate of most chemical reactions depends on temperature and can be approximated by the Arrhenius equation:

$$k = A \times \exp(-E/RT) \tag{5.24}$$

where
  k = the rate of the reaction (s^{-1})
  A = pre-exponential factor (s^{-1})
  E = activation energy $(J\,mol^{-1})$
  R = the gas constant = 8.314472 $(J\,K^{-1}\,mol^{-1})$
  T = absolute temperature (K)

If a clear relationship exists between two molecules, where one is the product of a reaction involving the other, then the effect of temperature on the reaction can be investigated. Once the reaction is understood, the relative concentrations of the parent and product molecules within a sample tell us something about the maturity of that sample. Such molecules are then referred to as *molecular biomarkers*.

MacKenzie and coworkers carried out much of the early investigation into molecular biomarkers in the early 1980s. MacKenzie, Lewis and Maxwell (1981) presented the results of a study into the role played by temperature in the isomerisation of a number of chiral centres in several hydrocarbons (pristane, steranes and hopanes), and the aromatisation of steroid hydrocarbons. For their study, they used samples of Toarcian shale from the Paris Basin, the type-specimen of Type II kerogens (see Section 5.1.1). They were able to reproduce the isomerisation and aromatisation products observed in mature Toarcian shale by heating immature shale in the laboratory. Furthermore, they observed that the reaction rates depended on temperature in a way predicted by Equation (5.24). Although they did not quantify the kinetic parameters of the reactions, they concluded that the studies 'suggest it may be possible to distinguish the thermal histories of sediment sequences' using such markers.

Note: A chiral centre (from the Greek, '*cheir*', meaning 'hand') in a hydrocarbon is a carbon atom with four different chemical groups attached to it. There is no axis of symmetry through the atom. Chiral carbons can therefore be separated into two configurations, R and S. In order to decide which configuration applies to a particular chiral centre, the groups attached to the chiral carbon are numbered from 1 to 4, in the order of those groups that contain atoms of higher atomic number, and then to decreasing sizes of the hydrocarbon chains (Figure 5.15). After priority is determined, the molecule is drawn to depict the group of lowest priority, no. 4, pointing into the plane of the page. If the priority of the remaining groups increases clockwise, the molecule has an R configuration. If the priority increases counter-clockwise, it has an S configura-

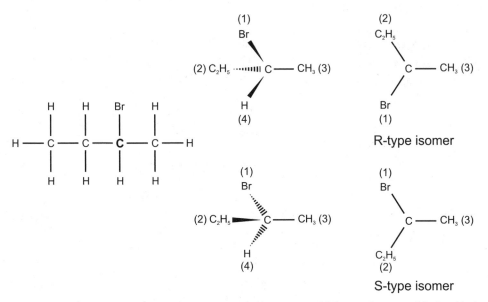

Figure 5.15. The figure on the left shows the general structure of 2-bromo butane, with the chiral carbon highlighted. The middle figures are the two possible geometric configurations of the groups around the chiral carbon, with the groups numbered in order of decreasing priority. The right-hand figures illustrate the isomers with the lowest priority group (the single hydrogen) pointing into the page. The top isomer is therefore (*R*)-2-bromo butane, and the bottom is (*S*)-2-bromo butane.

tion. If the chiral centre is part of a ring system, it is more convenient to label the configurations α and β. A chiral centre has an α-configuration if the lowest priority group points into the page when the molecule is drawn in the conventional manner, and a β-configuration if the lowest priority group points out of the page.

MacKenzie and McKenzie (1984) went a step further and calculated the kinetic parameters (E and A) for three reactions: the isomerisation of the C-20 chiral carbon in steranes ($E = 91$ kJ mol^{-1}, $A = 0.006$ s^{-1}), the isomerisation of the C-22 chiral carbon in hopanes ($E = 91$ kJ mol^{-1}, $A = 0.016$ s^{-1}) and the conversion of steroid hydrocarbons from mono-aromatic to tri-aromatic ($E = 200$ kJ mol^{-1}, $A = 1.8 \times 10^{14}$ s^{-1}).

Isomerisation is a reversible reaction, but the rate of the reaction in one direction generally exceeds the reverse reaction by a factor γ. For steranes and hopanes, the rate of conversion from the *R*-form to the *S*-form exceeds the reverse conversion by factors of $\gamma = 1.174$ and $\gamma = 1.564$, respectively. If the initial compound contains only *R*-isomers, it can be shown that after a certain time the proportion of *S*-isomers in the total sample is

$$P_S = [\gamma/(1 + \gamma)][1 - \exp(-(1 + \gamma)k_I t)] \tag{5.25}$$

where

$P_S = (S\text{-isomers})/(S\text{-isomers} + R\text{-isomers})$
$t = \text{time (s)}$
$k_I = A \times \exp(-E/RT) = $ the isomerisation rate of the forward (R to S)
conversion

Question: A compound initially containing only R-isomers of sterane is found to have 28% S-isomers of sterane after being buried at constant temperature for 5 Ma. At what temperature, T, was it most likely held?

Answer: We are told that 28% of the sterane is now S-type, so $P_S = 0.28$. We also know $\gamma = 1.174$ for steranes, and $t = 5$ Ma $= 1.5778 \times 10^{14}$ s. From this we can calculate k_I using Equation (5.25):

$P_S = [\gamma/(1 + \gamma)][1 - \exp(-(1 + \gamma)k_I t)]$
$0.28 = [1.174/2.174][1 - \exp(-2.174 \times k_I \times 1.5778 \times 10^{14})]$
$k_I = \ln[1 - (0.28 \times 2.174)/1.174]/(-2.174 \times 1.5778 \times 10^{14}) = 2.1307 \times 10^{-15} \text{ s}^{-1}$

We are now able to calculate T using Equation (5.24), and remembering that for sterane isomerisation $E = 91 \text{ kJ mol}^{-1}$ and $A = 6 \times 10^{-3} \text{ s}^{-1}$:

$k_I = A \exp(-E/RT)$

$2.1307 \times 10^{-15} = 0.006 \times \exp(-91{,}000/(8.314 \times T))$

$1/T = \ln[2.1307 \times 10^{-15}/0.006][-8.314/91{,}000]$
$\quad\quad = 0.00262 \text{ K}^{-1}$

So:

$T = 1/0.00262 = 382 \text{ K} = 109°C$

Note: The maximum value for $P_S = \gamma/(1 + \gamma)$. This means that the concentration of S-isomers of steranes and hopanes cannot exceed about 0.54 and 0.61, respectively.

The aromatisation reaction is irreversible so the concentration of the product, P_A, is

$$P_A = 1 - \exp(-k_A t) \tag{5.26}$$

where

$P_A = \text{product}/(\text{reactant} + \text{product})$
$k_A = \text{the aromatisation rate}$

Question: What concentration of aromatised steroid hydrocarbons would we expect to find in the sample from the above example?

Answer: We can determine k_A using Equation (5.24) and the kinetic parameters given in the text for the reaction:

$$k_A = A \times \exp(-E/RT) = 1.8 \times 10^{14} \exp(-200{,}000/(8.314 \times 382))$$
$$= 8.06 \times 10^{-14} \text{s}^{-1}$$

Equation (5.26) then gives us the solution for P_A at $t = 5$ Ma $= 1.5778 \times 10^{14}$ s:

$$P_A = 1 - \exp(-k_A t) = 0.999997$$

That is, the aromatisation reaction has effectively progressed to completion.

Note: Unlike isomerisation products, the maximum concentration of aromatisation products is 1.0.

Sedimentary basins tend not to remain at constant temperature, so in real situations the equations given above must be integrated over time and temperature to estimate the concentrations of product in a particular sample. This is done in a way similar to that presented for kerogen maturation in Section 5.1.2.

The *methylphenanthrene index* (MPI; Radke and Welte, 1983) was the first maturity indicator developed using aromatic compounds and is still widely used (e.g. Tupper and Burckhardt, 1990). Phenanthrene, P, is made up of three benzene rings. A phenanthrene molecule with a methyl (OH) group attached to its nth carbon is called n-methylphenanthrene, or n-MP (Figure 5.16). The MPI is defined as

$$\text{MPI} = \frac{1.5 \times (2\text{-MP} + 3\text{-MP})}{\text{P} + 1\text{-MP} + 9\text{-MP}} \tag{5.27}$$

MPI has been calibrated against vitrinite reflectance in the range $0.65 \leq R_o \leq 1.35\%$ for Type III kerogen source rocks (Radke and Welte, 1983):

$$R_o = 0.6(\text{MPI}) + 0.4\% \tag{5.28}$$

A later calibration for Type II, III and IV kerogens in the range $0.3 \leq R_o \leq 1.7\%$ (Boreham, Crick and Powell, 1988):

$$R_o = 0.7(\text{MPI}) + 0.22\% \tag{5.29}$$

was found to be less reliable for some localities (Tupper and Burckhardt, 1990).

The molecular biomarkers described above are probably the most commonly used, but many more have been investigated. These include the following:

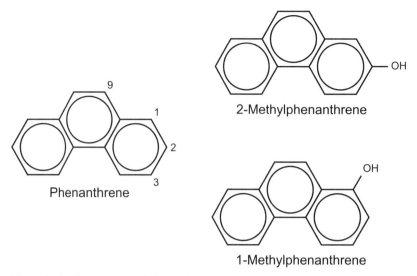

Figure 5.16. The structure of phenanthrene and methylphenanthrene, indicating the numbers of the carbon atoms relevant to MPI. The 2- and 3-methylphenanthrene isomers are more stable than 1- and 9-methylphenanthrene.

$5\beta(H),14\alpha(H),17\alpha(H)$-steranes are converted to $5\alpha(H),14\beta(H),17\beta(H)$-steranes with increasing maturity (MacKenzie et al., 1980).

Note: Chiral carbons in ring-molecules are labelled $\alpha(H)$ on a standard molecular representation if a hydrogen bond points into the page, and $\beta(H)$ if a hydrogen bond points out from the page. Steranes have three chiral carbons in rings, numbered 5, 14 and 17. Biological organisms only synthesise $(\beta\alpha\alpha)$ steranes. The more stable $(\alpha\beta\beta)$-isomers appear progressively with thermal maturity.

The ratio of $(6R, 10S)$-pristane to $(6R, 10R)$-pristane to $(6S, 10S)$-pristane starts at 1:0:0 and approaches 2:1:1 with increasing maturity (Tissot and Welte, 1984).

Note: Pristanes have two chiral carbons, numbered 6 and 10. Biological organisms only synthesise $(6R, 10S)$-pristane (meso-pristane). The other isomers appear progressively with thermal maturity.

The ratio of deoxophylloerythroetioporphyrin (DPEP) to etioporphyrin approaches zero with increasing maturity (Tissot and Welte, 1984). Different types of porphyrins approach zero at different rates.

> *Note:* Porphyrins are a class of water-soluble, nitrogenous biological pigments. When combined with metals and proteins, porphyrins form substances such as chlorophyll and haemoglobin. Porphyrins have been identified in even the oldest sediments, making them very useful biomarkers.

17β(H),21β(H)-hopanes are converted to the more stable 17α(H),21β(H)-hopanes with increasing maturity (MacKenzie et al., 1980).

The ratio of 17β(H),21α(H)-hopanes (moretanes) to 17α(H),21β(H)-hopanes decreases markedly at a greater depth than the previous reaction (MacKenzie et al., 1980).

> *Note:* Living organisms always synthesise ($\beta\beta$)-hopanes and sometimes moretanes. ($\alpha\beta$)-Hopanes are *only* generated in sediment through isomerisation. The moretane to ($\alpha\beta$)-hopane ratio initially drops at shallow depths owing to the conversion of ($\beta\beta$)-hopanes to ($\alpha\beta$)-hopanes. There is then a period of relative stability in the ratio before moretanes are converted to ($\alpha\beta$)-hopanes at higher temperature.

17α(H)-trisnorhopane is progressively converted to 18α(H)-trisnorhopane with increasing maturity (Peters and Moldowan, 1993).

Methyladamantane index (MAI) – the ratio of 1-methyladamantane to (1-methyladamantane + 2-methyladamantane) increases from 0.5 at VR $\approx$ 0.9% with increasing maturity (Chen et al., 1996).

Methyldiamantane index (MDI) – the ratio of 4-methyldiamantane to (1-methyldiamantane + 3-methyldiamantane + 4-methyldiamantane) increases from 0.3 at VR $\approx$ 0.9% with increasing maturity (Chen et al., 1996).

> *Note:* Diamondoid hydrocarbons such as methyladamantane (MA) and methyldiamantane (MD) are rigid, three-dimensionally fused ring alkanes with a diamond-like structure. The former has only two isomers, 2-MA and 1-MA, in order of increasing stability. MD has three isomers, 2-MD, 1-MD and 4-MD, in order of increasing stability.

Figure 5.17 summarises the results of the above discussion by comparing the ranges of maturity covered by each of the biomarker methods.

Molecular biomarkers are commonly found in most sediment, making them widely available for maturity studies. Shales of all ages yield sufficient amounts of biomarkers to determine a number of isomeric and other ratios discussed above. However, the precision of the maturity estimate derived from biomarkers is

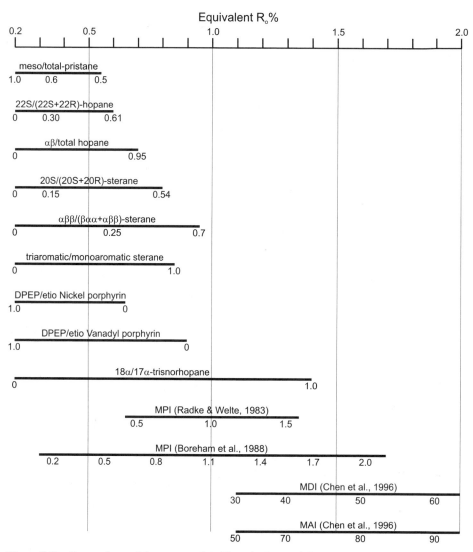

Figure 5.17. Comparison of the range and R_o% equivalence of the molecular biomarkers discussed in the text. Ranges and values shown are approximate only, and the solid lines are *not* intended to represent linear scales.

generally low. Also, all values may vary with heating rate, lithology and organic source (Peters and Moldowan, 1993). For these reasons, biomarkers are a useful indicator of the *general* degree of maturity (e.g. early oil generation zone, diagenetic zone, etc.), but are inadequate for more detailed thermal modelling.

5.3. Fission Track Thermochronology (FTT)

Only one phenomenon has so far been identified that preserves information on the *timing* of past thermal events – the annealing of spontaneous fission tracks within certain minerals. All of the palaeotemperature indicators discussed so far (with the exception of FIM) only yield information about the cumulative thermal stress imposed on particular components of the sediment. The actual time–temperature path followed by the material is not uniquely constrained. In general, maximum temperature can only be deduced by making assumptions about the burial history. Fission track thermochronology can constrain the absolute timing of certain events.

Fission track thermochronology (FTT, sometimes referred to as 'fission track analysis') is based on the observation that the crystal lattices of some minerals preserve the scars of spontaneous nuclear fission events. Uranium-238 (the most common isotope of uranium, with a half-life of 4.5×10^9 years) is the major source of such events. During fission, the nucleus of a uranium atom splits with such force that the two products leave a trail of damage, or *fission track*, through the host crystal. Apatite, zircon and sphene commonly contain uranium and are prone to fission track damage.

The initial length of a fission track is relatively constant for a given mineral (e.g. close to 16 μm for apatite; Gleadow, Duddy and Lovering, 1983) but gradually heals, or *anneals*, as broken chemical bonds are repaired. Annealing begins at each end of the fission track and moves inwards, decreasing the track length at a rate primarily dependent on temperature (e.g. Green, 1986; Moore, Gleadow et al., 1986). The temperature dependence is such that the annealing rate increases approximately 10-fold for a 10°C increase in temperature (Fleischer, Price and Walker, 1975; Gleadow et al. 1986; Green et al., 1989b). Other factors affecting the annealing rate include mineral composition (see Section 5.3.3) and the angle between the fission track and the *c*-axis of the crystal (annealing is most rapid when the fission track lies perpendicular to the *c*-axis; Green et al., 1986).

Fission tracks begin to anneal as soon as they form. However, new tracks continue to form within the crystal at a rate dependent only on the amount of uranium present. The distribution of fission track lengths within a single crystal is therefore related to the thermal history experienced by that crystal, and it is this observation that underpins FTT.

Each mineral anneals at a different rate for a given temperature. Tracks remain in some minerals at much higher temperatures than in others, so different minerals are useful for FTT under different thermal conditions. Table 5.4 gives an indication of the temperatures required to achieve the same level of annealing in the three commonly studied minerals. Zircons retain fission tracks for much longer, and at higher temperatures, than either sphene or apatite. An FTT analysis is generally performed only on the most appropriate mineral for the age and burial history of the sample being assessed. That is, in a metamorphic or igneous sequence, one of the higher temperature minerals may be chosen, whereas in a sedimentary basin setting, apatite is commonly used. The remainder of this section concentrates primarily on apatite FTT.

Table 5.4. Temperature Required to Anneal Fission Tracks in 1 hour at 1 atmosphere Pressure

Mineral	Temperature Required to Anneal Indicated Proportion of Tracks		
	0%	50%	100%
Apatite[a]	275–400°C	322°C	375–530°C
Sphene	520°C	620°C	637°C
Zircon		700°C	

[a]Apatite data derived from more than one primary source
Source: Data from Fleischer et al. (1975).

5.3.1. Choice and Preparation of Samples

> *Note:* Preparation and measurement methods are similar for all minerals, so the details for apatite FTT may be regarded as typical of the general technique.

The lithologies most likely to yield apatite grains are medium to coarse sandy units and acidic igneous bodies. Fine-grained units, basalt and carbonates are poor sources of apatite for FTT purposes.

The required quantity of raw material depends on the concentration of apatite within the formation being sampled, but generally 1–2 kg are necessary to yield a sufficient number of apatite grains. As with other methods, sampling from a single depth in a sequence is of limited value, and considerably more information is obtained by analysing samples from a number of different stratigraphic depths within a section.

> *Note:* Sampling can be performed at a number of different depths within a single well, or from surface outcrops at a number of different stratigraphic levels within a region. When sampling from wells, it is common to assemble each FTT sample from regular portions of drill chips collected over depth intervals of 100 m or more. In contrast, sufficient amounts of outcrop material can be collected from precise stratigraphic levels.

Consolidated samples must be crushed to a medium-to-coarse grain size (1–2 mm) prior to separation of apatite grains by heavy liquid flotation and magnetic differentiation, or other means – manual identification and separation may be required if the necessary equipment is unavailable. The apatite grains are then mounted in epoxy resin on separate glass slides for each depth interval. The fossil fission tracks, known as *spontaneous* tracks, are revealed by grinding and polishing to expose internal grain surfaces, and are enlarged by etching with 5 M nitric acid.

The number and density of spontaneous fission tracks revealed on the polished surface are functions of both the concentration of uranium within the grain and the time that the grain has been at a relatively low temperature. A high concentration of uranium naturally produces a large number of spontaneous tracks, but long time periods at low temperature also generate a high density of tracks because old tracks do not anneal. The concentration of uranium in the grain must be determined in order to resolve this ambiguity.

The most popular way to measure uranium concentration in an apatite grain is the so-called 'external detector method' (Hurford and Green, 1982). A low-uranium muscovite sheet is held against the polished and etched slide and used as an external monitor of uranium-235 fission events induced by an applied neutron bombardment. After irradiation, the detector is removed from the apatite slide and etched to reveal the *induced* fission tracks. The density of induced fission tracks is directly proportional to the total concentration of uranium in the grain (Figure 5.18).

Note: A source of neutrons sufficient to induce fission tracks is generally found only within controlled nuclear reactors (e.g. the Australian Atomic Energy Commission's HIFAR reactor at Lucas Heights, New South Wales).

5.3.2. Analysis and Interpretation

Moore et al. (1986) and Green (1986) detailed the standard method of counting fission tracks, and Gleadow et al. (1986) and Green et al. (1989a, 1989b) discussed the principles of interpreting fission track length distribution. The reader wanting detailed background information should pursue those references. Only a summary of the process is presented here.

Conventional analyses use only fully etched *confined tracks* (Lal, Rajan and Tamhane, 1969; Laslett et al., 1982) lying horizontally within a polished surface containing the *c*-axis of the grain. Confined tracks lie predominantly within the plane of the slide and are characterised by a tapered point at both ends. A rounded or flat point on a track indicates that it exits the polished surface, and is thus not an entire track. Ideally, the lengths of at least one hundred tracks should be measured on each slide, spread across at least twenty grains. The number of apatite grains recovered from a particular sample affects the reliability of the final result. The quality of the yield can be roughly assessed using Table 5.5. Poor apatite yields (i.e. fewer than five apatite grains recovered from a particular stratigraphic sample) give highly uncertain results.

Fission tracks retain both time and magnitude information about the thermal conditions experienced by the apatite grains since they were last fully annealed. The primary purpose of fission track thermochronology is to determine an absolute date for a specific thermal event. A parameter known as the

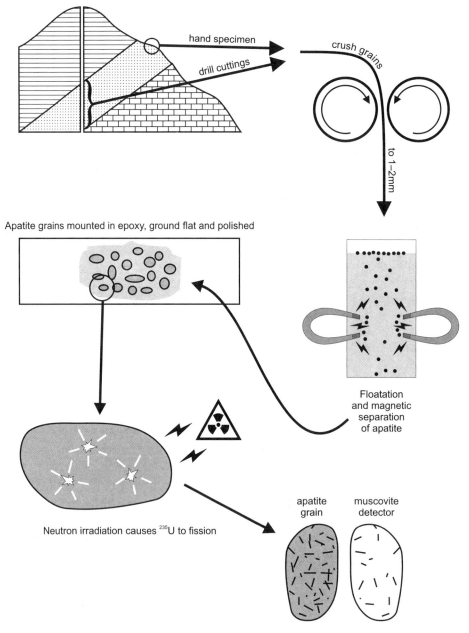

hand specimen

drill cuttings

crush grains

to 1–2mm

Apatite grains mounted in epoxy, ground flat and polished

Floatation
and magnetic
separation
of apatite

Neutron irradiation causes ^{235}U to fission

apatite
grain

muscovite
detector

Etching reveals ^{238}U and ^{235}U tracks in apatite, but
only the induced ^{235}U tracks in muscovite detector

Figure 5.18. Preparation of apatite grains for fission track analyses, illustrating the external detector method for determining the concentration of uranium.

Table 5.5 Quality of Apatite Yields

Number of Grains	Yield Quality
< 5	Very poor
5–10	Poor
10–15	Fair
15–20	Good
~ 20	Very good
> 20	Excellent

apparent fission track age, t_a (Ma), can be determined from the following (Price and Walker, 1963; Naeser, 1967):

$$t_a = \frac{1}{\psi_D} \ln\left[1 + \frac{\psi_D \phi \sigma I \rho_s}{\psi_f \rho_i}\right] \tag{5.30}$$

where

ψ_D = total decay constant of $^{238}U = 1.55125 \times 10^{-4}$ (Ma^{-1})

ψ_f = spontaneous fission decay constant of ^{238}U (Ma^{-1})

I = isotopic ratio of $^{235}U/^{238}U = 7.2527 \times 10^{-3}$

σ = thermal neutron fission cross section for $^{235}U = 580.2 \times 10^{-24}$ (cm^2)

ϕ = thermal neutron fluence (cm^{-2})

ρ_s/ρ_i = spontaneous/induced track density ratio

A glass standard of known uranium concentration (e.g. NBS standard glass – SRM612) is included with each batch of apatite grain mounts sent for irradiation. The glass is mounted with a muscovite detector in a manner identical to that for the apatite mounts. The density of fission tracks later revealed on the muscovite detector, ρ_D, is used to determine the value of ϕ:

$$\phi = B \times \rho_D \tag{5.31}$$

where B is a calibration factor characteristic of the glass and measuring technique.

Note: Fission track density must be measured over the same area of both the apatite grain and the muscovite detector. Furthermore, ρ_i must be doubled because spontaneous tracks revealed on the etched surface of a grain are the product of fission events both above and below the current grain surface, while induced events recorded on the muscovite detector come only from below the surface. That is, the measured ρ_s/ρ_i ratio must be halved to get an indication of the true ratio.

Only ψ_f remains unknown, and must be calibrated against a sample of known age. In practice, most of the constants are grouped into a single calibration constant, ζ (e.g. Hurford and Green, 1983), so that track densities are the only variables:

$$t_a = \frac{1}{\psi_D} \ln\left[1 + \frac{\rho_s \rho_D}{\rho_i} \zeta \psi_D\right] \qquad (5.32)$$

Question: The fission track laboratory at Latrobe University, Australia, has calibrated its technique with a value $\zeta = 352.7 \pm 3$ yr cm^{-2} (e.g. Beardsmore and O'Sullivan, 1995). The calibration is only valid when the correct units are used for ρ ($\times 10^6$ cm^{-2}) and Ψ_D (Ma^{-1}). The laboratory examined a population of apatite grains separated from drill cuttings taken from a well in the Browse Basin, Western Australia. The grains revealed fission track densities $\rho_D = 1.378 \times 10^6$ cm^{-2}, $\rho_s = 6.294 \times 10^5$ cm^{-2} and $\rho_i = 3.618 \times 10^6$ cm^{-2}. What is the apparent fission track age, t_a, for the sample?

Answer: In terms of the correct units for the calibration:

$\rho_D = 1.378$

$\rho_s/\rho_i = 0.5 \times 0.6294/3.618 = 0.08698$ (see Note above for explanation of the 0.5 factor)

$\Psi_D = 1.55125 \times 10^{-4}$

From Equation (5.32):

$t_a = 1/(1.55125 \times 10^{-4}) \times \ln[1 + 0.08698 \times 1.378 \times 352.7 \times 1.55125 \times 10^{-4}] = 6446.4 \ln[1 + 0.00656] = 42.1$ Ma

---

There will usually be a spread in the apparent ages determined for a number of apatite grains from a single population. This is due to natural variation in the distribution and decay of uranium. But two grains will also yield different ages if they are from different sources. It is therefore important to determine whether a group of individual apatite grains represent a single population (and thermal history), or a mixture from different sources (and different thermal histories). The chi-squared statistical test (Galbraith, 1981) is used for this purpose.

The chi-squared test determines the probability that all grains belong to a single age population. Less than 5% probability indicates a significant spread in single grain apparent ages; this is either due to their derivation from more than one source, or due to differential annealing in grains of different compositions (see Section 5.3.3).

The apparent age of an apatite grain is rarely the same as the actual length of time since the grain was last fully annealed (Gleadow et al., 1986; Green, 1986). The *length distribution* of the fission tracks holds the key to interpreting the thermal history of the grain. In rocks that have not been heated above 50°C subsequent to rapid cooling, the mean length of all apatite fission tracks is between 14 and 15 μm, with standard deviation around 1 μm. Such a simple length distribution can arise only from a rapid and permanent cooling event – a rapidly uplifted basement terrane, for example. In most apatite grains, the length distribution is more complex.

Above 50°C an apatite fission track begins to anneal. Annealing results in shorter spontaneous tracks, reduced spontaneous track density, and a reduction in the apparent fission track age (Gleadow et al., 1986; Green et al., 1989a, 1989b). Above a critical temperature (dependent on the chemical composition of the apatite grain), fission tracks effectively anneal instantaneously, and apparent fission track age is reduced to zero, resetting the fission track clock. The length of a partially annealed track reflects the integrated temperature history of the apatite grain since the track formed. New tracks form regularly throughout geological time, so the length distribution represents a complete thermal record since the last total annealing event (e.g. Green et al., 1989b).

Thermal history reconstruction from fission track length distribution is a developing field. Laslett et al. (1987) published kinetic models of annealing that are still considered applicable today. However, models are continuously being calibrated and refined, and this is yet another field in which advances in computing technology allow advances in the complexity of models. One new avenue being investigated is the statistical analysis of *projected* tracks (Grivet et al., 1993) – tracks that are not wholly confined within the plane of the grain surface and have some part of their original length missing.

5.3.3. Limitations to FTT

Fission track age determination is a relatively precise and reliable tool. Different laboratories are able to arrive at consistent results for the same sample. The current limitations to FTT lie in the accuracy of thermal histories reconstructed from fission track length distributions. Most interpretations until recently have followed the method described by Green et al. (1989b), based on extrapolation of empirical laboratory annealing data obtained from Durango apatite (Green et al., 1986; Laslett et al., 1987). However, this early model is an oversimplification and ignores certain facts. In particular, it has been convincingly shown (e.g. Green et al., 1985, 1986; Sieber, 1986) that fission track annealing rate varies with the chemical composition of the host mineral. Chlorine-rich apatites are more resistant to annealing than fluor-apatites.

The composition of apatite from Durango, Mexico (chlorine content = 0.5 wt%) is generally considered 'typical' (Sieber, 1986), and many models are calibrated using samples of the same. When such models are used to interpret fission track results from apatite of unknown composition, errors of unknown magnitude are introduced. There are numerous examples in the literature of apatite composition varying significantly from the Durango standard. For example, apatites from the Otway Basin in Australia are generally chlorine-rich with respect to Durango (Figure 5.19; Green et al., 1986), while those from the Sverdrup Basin, in the Canadian Arctic Archipelago, are rela-

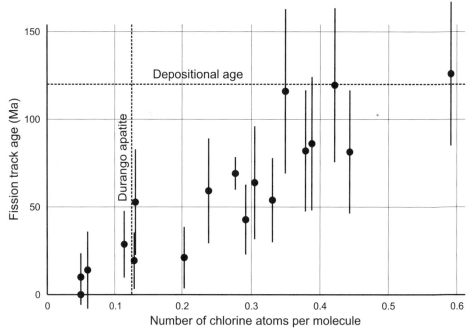

Figure 5.19. Relationship between apparent fission track age and composition for a sample from the Otway Basin, Australia. The relationship suggests that chlorine-rich apatite is more resistant to annealing than fluor-apatite. The composition of Durango apatite is shown for reference. Modified after Green et al. (1986).

tively chlorine-depleted (Collins, 1996). Green et al. (1986) and Boudreau, Love and Hoatson (1993) found that apatite composition could vary markedly, even between samples derived from the same borehole.

Modern fission track interpretation has moved away from empiricism towards kinetic models. The variation in activation energy with apatite composition argues strongly against using a 'standard' apatite for calibration, and many laboratories now determine apatite composition as a necessary and routine part of fission track analyses. Early published results may be questionable if the composition of the sample was unknown.

Another obvious limitation to FTT is that a track-bearing mineral must be present in sufficient quantities within the region or formation under investigation. Surprisingly, apatite has been found to be a relatively common trace mineral within many medium- to coarse-grained sand units. Carbonate and shale sequences, on the other hand, rarely host apatite, and thus the method is unsuitable in many localities.

Note: Marine sediments may, in fact, contain a large amount of apatite in the form of residual bone and shell material. However, the typical grain size of biogenic apatite is far too fine for FTT purposes.

As with any analytical technique, FTT is limited by the quality of the samples. This goes beyond the mere necessity that the samples contain measurable numbers of apatite grains. It is sometimes difficult to even be sure of the origin of the sample. The quantity of material required for a typical FTT study demands the use of drill cuttings over core in most down-hole situations. The quality of drill cuttings is less reliable than core, and if samples are retrieved from a centralised storage facility the analyst must accept that they were, in fact, sourced from the location stated. Results of analyses can sometimes raise serious doubts as to validity of this assumption.

EXAMPLE

Beardsmore and O'Sullivan (1994) raised concerns about the integrity of drill cuttings reputedly from the well Yampi 1 on the North West Shelf of Western Australia. The FTT on apatite retrieved from the cuttings strongly suggested that the sampled sequence had not experienced temperatures greater than 90°C since deposition. However, the ambient temperature of the sampled formation is in excess of 140°C, implying that fission tracks should be completely annealed, regardless of apatite composition. It appeared likely that the samples did not represent the *in situ* formation. This conclusion was supported by a distinct bimodal age distribution within the apatite grains taken from the drill cuttings. A strong peak of ages in the range 200–500 Ma contrasted with another strong peak of ages at less than 10 Ma. Given the ambient temperature, it is likely that the apparently very young grains were from the sampled unit while the older grains were from elsewhere. The origin of such contamination is contentious.

5.4. Deciding which Thermal Maturity Indicators to Use

The experts and aficionados of each of the techniques discussed above are often guilty of exaggerating the capabilities of their particular method. Fission track thermochronology is particularly prone to this sort of overenthusiasm because it is an excellent tool – often the *only* tool – for determining the *date* of a particular event. However, as a *palaeotemperature* indicator it is not necessarily more accurate than any other method. It is strongly recommended, therefore, that thermal maturity be independently estimated using several of the techniques discussed in this chapter.

When it comes to determining which techniques are most appropriate for any given situation, a number of factors need to be addressed. These include, but are not necessarily limited to, the lithology to be examined, the present-day temperature, the required precision in the palaeotemperature estimate, and (of course) fiscal limitations. Figure 5.20 tabulates the relevant parameters pertaining to each technique, to help in the decision process.

Many of the techniques are relevant only for particular lithologies. For example, if the formation to be tested is predominantly shale, then fission track thermochronology is unlikely to provide any useful data. However, vitrinite reflectance, FAMM®, TAI, illite crystallinity, pyrolysis and biomarkers all offer themselves as possible sources of palaeotemperature estimates. An assessment of core or cuttings will reveal whether the relevant sediment components are present for each of the techniques. If the particular shale under investigation is barren of kerogen material, for example, then illite crystallinity might be the only technique worth considering, in spite of its shortfalls.

Often it is possible to be selective as to which formations are sampled in a particular section. The decision about which palaeotemperature techniques to use then becomes one of required precision. The greater precision of some of the techniques (VR, FAMM®, Pyrolysis, FTT) explains their popularity. The lower precision of the other techniques means they are sufficient only for a general estimate of thermal maturity. However, the less precise techniques have certain other advantages, the major one being that palaeotemperature estimates can be made at the same time as other studies are carried out. For example, CAI and TAI can be estimated during a biostratigraphy study.

Some of the techniques are automatically ruled out under certain conditions. For example, it is probably a wasted effort to measure VR on marine-influenced samples, CAI on samples younger than Triassic, or FTT on samples drawn from a formation presently at 150°C.

Recall that FTT is the only technique able to give an estimate of the absolute *age* of an event. The age estimate becomes more reliable as the maximum temperature is constrained. Therefore, FTT, VR (FAMM® if suppression is suspected) and pyrolysis are mutually advantageous. Maximum temperature estimates from several sources help constrain the age of the last cooling event, which in turn helps to refine the maximum temperature estimate. Through an iterative process, a thermal history ideally emerges that is consistent with all available data.

Consistency between data sets is not always encountered. In fact, it is very common that palaeotemperature estimates from different techniques are mutually inconsistent. Beardsmore and O'Sullivan (1995) presented estimates of maximum palaeotemperature derived from the same well using four different methods (CAI, FAMM®, VR, FTT). The four techniques produced contradictory results (Figure 5.21). Foland (1997) also found inconsistencies between FTT, VR, FAMM® and TAI data sets from around the world. These results highlight the danger of using highly empirical procedures to model complex phenomena. The greater the number of independent indicators that give consistent results, the greater the confidence one may have in the subsequent interpretations. The information and confidence to be gained by including a number of different maturity techniques in a study of one region far outweigh the capabilities of any one method (e.g. Cooper et al., 1998).

	Sediment component analysed	Main rock type	Equipment required	Approximate maturity range (VR%)	Palaeotemperature precision	Maximum measurable palaeotemperature (from Equation 5.9)	Time information	Limitations
VR	vitrinite		reflecting light microscope, photometer	0.3–5.0	±5°C	>350°C	none	reflectance suppression in marine deposits
FAMM®	various macerals	siltstone, shale	laser fluorescence microprobe	0.4–1.2		175°C		small number of laboratories set up for technique
TAI	palymorphs		transmitting light microscope, non-colourblind operator	0.3–2.4	±20°C or higher	260°C		subjective, poor temperature resolution
CAI	conodonts	carbonates		0.3–5.0+	±20–50°C	>600°C		subjective, Cambrian-Triassic only, poor resolution
Illite crystallinity	illite	shale	xray diffraction instrument	0.4–5.5	±10°C	300°C		clay composition changes due to fluids
Pyrolysis	organic matter	all sediments	GC-MS[a] with pyrolysis inlet, or RockEval tool	0.6–1.4	±5°C	200°C		must use standard heating rate, results vary for sediment type
Fluid inclusions	fluid inclusions in calcite, f'spar, quartz grains	sandstone, limestone	microscope, heating stage	n/a[b]	±2°C[b]	>1000°C	relative timing of different inclusions	only gives temperature at times of fluid migration
Biomarkers	extracts and pyrolysates	all sediment	GC-MS[a]	0.3–2.0	±15°C	240°C	none	most indices are poorly calibrated to VR
FTT	apatite	sandstone	microscope, thermo-nuclear reactor	0.3–1.0	±5°C	130°C	absolute age of last cooling event	composition of apatite must be determined

Figure 5.20. Comparison of palaeotemperature indicators. Note that temperature and maturity ranges are only approximate, the equipment and limitations lists are not exhaustive. [a]GC-MS = gas-chromatograph mass spectrometer. [b]Fluid inclusions yield a *minimum* palaeotemperature estimate for a particular event.

199

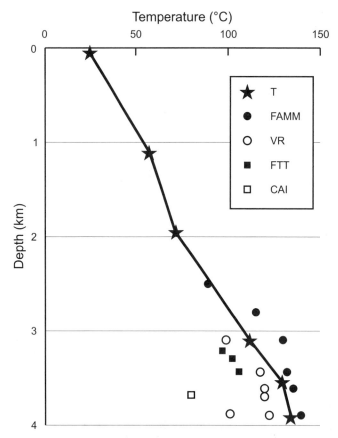

Figure 5.21. Horner-corrected bottom-hole temperature, T, and maximum palaeotemperatures deduced from a number of techniques for the well Ashmore Reef 1, on the North West Shelf, Australia. The lack of agreement highlights the danger of relying too heavily upon a single indicator of palaeotemperature.

Note: Geochemical temperature indicators should never be viewed as infallible. When quoted or used in modelling, appropriate error margins should be clearly noted.

Figure 5.22 shows the approximate correlation between VR, TTI, TAI, CAI, Tmax and petroleum generation windows. Note that most of the calibrations between techniques are imprecise and that the illustrated relationships are only approximate.

5.5. Summary

The thermal maturity of a rock is a measure of the degree to which organic metamorphism has progressed, and gives a crude indication of the maximum temperature the rock has experienced. The dependence of maturity on time is

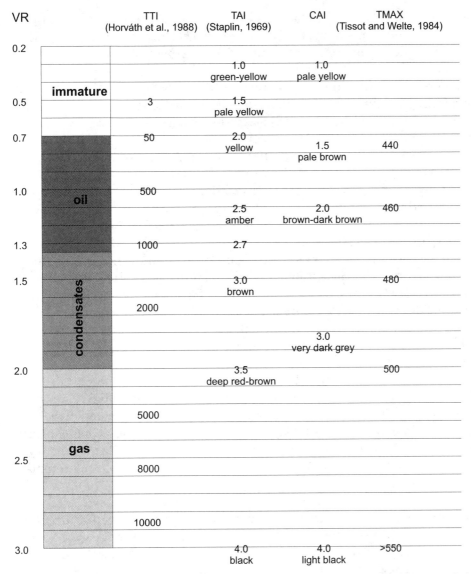

Figure 5.22. Approximate correlation between vitrinite reflectance (VR), time–temperature index (TTI), thermal alteration index (TAI), conodont alteration index (CAI) and Tmax from RockEval.

roughly linear, while the temperature dependence is closer to exponential. There are many methods for quantifying thermal maturity, each of which relies on a specific, non-reversible, temperature-dependent, chemical or physical process that acts on one or more components of the sediment.

Vitrinite reflectance, VR, is arguably the most widely used thermal maturity indicator. Vitrinite is a microscopic constituent of coal and dispersed organic matter, and its reflectance in white light increases with the rank of the coal. The increasing reflectance of vitrinite with time and temperature can be modelled as a number of parallel chemical reactions. Different reactions proceed at different rates for a given temperature, leading to the observed gradual increase in reflectance with temperature and time. The reflectance of marine-influenced vitrinite is often significantly lower than for a thermally equivalent terrestrial vitrinite.

Fluorescence Alteration of Multiple Macerals, FAMM®, was developed specifically to address the problem of vitrinite reflectance suppression and enhancement. The method involves measuring the intensity of red fluorescence emitted from polished maceral samples in response to a laser beam. The fluorescence intensity generally varies with time (from I_i to I_f over 400 s) and data are plotted on log–log axes of I_f/I_i versus I_f. The ratio I_f/I_i is mainly controlled by the thermal maturity, decreasing as maturity increases. The FAMM® technique removes the subjective element of VR analyses and is applicable to a greater range of samples (e.g. many marine shales, where vitrinite is sparse or non-existent).

The gradual change in colour and opacity of some varieties of liptinite and conodonts with increasing thermal maturity has given rise to two maturity indices: thermal alteration index, TAI, and conodont alteration index, CAI. Both indices are numeric scales where higher numbers indicate greater maturity. They share the advantage of being very simple and easy to measure, and also share the drawbacks of being subjective and of low precision.

Kübler (1967) defined an index of illite crystallinity as the width of the 1 nm illite peak (at half-height) in millimetres on a standard x-ray diffraction plot. As temperature increases, the peak progressively narrows, representing increasing order within the illite crystal lattice. Lower numbers therefore imply higher temperatures. Similarly, the amount of illite in a smectite–illite mixed layer increases with temperature and time, and gives an indication of maximum palaeotemperature.

Pyrolysis is a procedure whereby a small sample of rock is heated at a constant rate in an inert gas. The volume of hydrocarbons 'boiled' from the rock is recorded as temperature increases. There are generally two distinct peaks in detected hydrocarbons. The first, in the range 100–200°C, are those that have already generated naturally and remain within the pore spaces of the rock. The second, around 400–500°C, are those that generate from the kerogen due to the high temperature of pyrolysis. The second peak represents the hydrocarbon potential of the rock under investigation. The greater the thermal maturity of the original rock, the lower the amplitude of the second peak (with respect to the first) and the greater the temperature at which the second peak appears.

If a clear relationship exists between two molecules, where one is the product of a reaction involving the other, then the relative concentrations of the parent and product molecules within a sample tell us something about the thermal maturity of that sample. Such systems are referred to as molecular biomarkers, and are commonly found in sediment. The precision of the maturity estimate derived from biomarkers is generally low, so while biomarkers are a useful indicator of the *general* degree of maturity, they are inadequate for detailed thermal modelling.

Fission track thermochronology, FTT, is based on the observation that the trail of damage left by the products of a fission event in a crystal lattice heals at a rate dependent on temperature. As existing fission tracks anneal, new tracks continue to form at a rate dependent on the amount of uranium present within the lattice. The thermal history of the grain is reflected in the length distribution of the fission tracks.

Which thermal maturity techniques are most appropriate for a given situation depends on the lithology to be examined, the present-day temperature, the required precision in the palaeotemperature estimate, and fiscal limitations. In general, the greater the number of independent indicators that give consistent results, the greater the confidence one may have in subsequent interpretations.

PART THREE

Modelling Techniques

CHAPTER SIX

Heat Flow

A perfectly complete geothermic survey would give us data for determining an initial epoch in the problem of terrestrial conduction. At the meeting of the British Association in Glasgow in 1855, I urged that special geothermic surveys should be made for the purpose of estimating absolute dates in geology.

On the Secular Cooling of the Earth – Prof. William Thomson, 1862.

Parts One and Two detailed how to estimate or measure the basic parameters required for geothermal investigations. For present heat flow, these parameters are heat generation, thermal gradient and thermal conductivity. If we wish to project heat flow into the past, additional information about palaeotemperature is required. The task now is to combine these different data sets into a coherent heat flow model.

The specific method chosen to combine the physical data sets and best determine heat flow depends on the quality of the data and the precision required in the result. The equation for steady-state conductive heat flow was given in Section 1.3, and is repeated here:

$$\mathbf{Q}_0 = \mathbf{Q}_d + \int A(z)\partial z = \lambda_d \left[\frac{\partial T}{\partial z}\right]_d + \int A(z)\partial z \tag{6.1}$$

where $\mathbf{Q}_0$ is the surface heat flow, $\mathbf{Q}_d$, λ_d and $[\partial T/\partial z]_d$ are the heat flow, thermal conductivity and thermal gradient, respectively, at depth, d; and $\int A(z)\partial z$ is the integral of volumetric heat generation from the surface to d.

6.1. Product Method

Heat flow at depth, d, $(\mathbf{Q}_d)$ is the product of two values: thermal gradient × thermal conductivity. The same rule applies over a depth interval. The average heat flow over an interval is the product of the average thermal gradient × average thermal conductivity over the same interval. However, even this simple rule must be followed with due care and with a thought for the physical processes at work.

The average thermal gradient is calculated using only the temperatures at the top and bottom of the interval (Figure 6.1). The average thermal conductivity is generally the harmonic mean of the conductivities of all horizontally bedded layers within the depth interval. There may be a number of temperature data within a specified depth interval, but it is important to note that the average thermal gradient is *not* found by linear regression through all temperature points. It follows that the accuracy of the heat flow estimate relies strongly on the accuracy of only two temperature estimates.

Question: A particular well is 3.7 km deep (Figure 6.1). From the surface down, it penetrates 1.1 km of sandstone ($\lambda = 3.6$ W m^{-1} K^{-1}), 1.1 km of marl ($\lambda = 2.8$ W m^{-1} K^{-1}), 1.0 km of shale ($\lambda = 1.8$ W m^{-1} K^{-1}) and 0.5 km of sandstone ($\lambda = 4.3$ W m^{-1} K^{-1}). Average surface temperature is 10°C, corrected bottom-hole temperature is 85°C. The temperatures at 1.1 km, 2.2 km and 3.2 km are 25°C, 45°C and 78°C, respectively. Uncertainty in temperature and conductivity is ±0.1°C and 0.1 W m^{-1} K^{-1}, respectively. What is the average heat flow in the well?

Answer: We calculate average gradient using only the surface and bottom-hole temperatures:

$$(\partial T / \partial z)_{av} = (85 \pm 0.1 - 10 \pm 0.1)/3.7 = 20.27 \pm 0.05°\text{C km}^{-1}$$

Average conductivity is the harmonic mean conductivity of the four layers:

$$\lambda_{av} = 3.7/((1.1/3.6 \pm 0.1) + (1.1/2.8 \pm 0.1) + (1.0/1.8 \pm 0.1) + (0.5/4.3 \pm 0.1)) = 2.70 \pm 0.11 \text{ W m}^{-1} \text{K}^{-1}$$

Heat flow is the product of $(\partial T / \partial z)_{av} \times \lambda_{av}$:

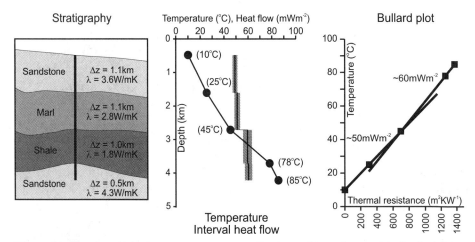

Figure 6.1. Stratigraphic column, temperature profile, interval heat flow and Bullard plot for an example well. Shaded area around interval heat flow represents uncertainty range.

$\mathbf{Q}_{av} = 20.27 \pm 0.05 \times 2.70 \pm 0.11 = 54.7 \pm 2.4 \text{ mW m}^{-2}$
For comparison, linear regression through all the temperature data indicates a best-fit gradient of $21.25°\text{C km}^{-1}$, about 5% higher than the true average.

A single, average vertical heat flow estimate is a good way to get a feel for the 'background' heat flow in a region, but any vertical variation in heat flow is masked. Applying the product method to each successive temperature interval may identify variation in heat flow with depth. Interval heat flow is found by multiplying the interval thermal gradient by the interval thermal conductivity, yielding a series of heat flow estimates down the hole.

Question: Determine interval heat flow for each of the four temperature intervals in the previous example (Figure 6.1).

Answer: $\mathbf{Q}_i = (\partial T / \partial z)_i \times \lambda_i$

First interval: $z = 0$–1.1 km, $T = 10$–25°C, $\lambda = 3.6 \pm 0.1$ $\text{W m}^{-1}\text{K}^{-1}$

  $(\partial T / \partial z) = (25 - 10)/(1.1 - 0) = 13.64 \pm 0.18°\text{C km}^{-1}$
  $\mathbf{Q}_1 = 13.64 \times 3.6 = 49.1 \pm 2.0 \text{ mW m}^{-2}$

Second interval: $z = 1.1$–2.2 km, $T = 25$–45°C, $\lambda = 2.8 \pm 0.1 \text{ W m}^{-1}\text{K}^{-1}$

  $(\partial T / \partial z) = (45 - 25)/(2.2 - 1.1) = 18.18 \pm 0.18°\text{C km}^{-1}$
  $\mathbf{Q}_2 = 18.18 \times 2.8 = 50.9 \pm 2.3 \text{ mW m}^{-2}$

Third interval: $z = 2.2$–3.2 km, $T = 45$–78°C, $\lambda = 1.8 \pm 0.1$ $\text{W m}^{-1}\text{K}^{-1}$

  $(\partial T / \partial z) = (78 - 45)/(3.2 - 2.2) = 33.00 \pm 0.20°\text{C km}^{-1}$
  $\mathbf{Q}_3 = 33.00 \times 1.8 = 59.4 \pm 3.7 \text{ mW m}^{-2}$

Fourth interval: $z = 3.2$–3.7 km, $T = 78$–85°C, $\lambda = 4.3 \pm 0.1$ $\text{W m}^{-1}\text{K}^{-1}$

  $(\partial T / \partial z) = (85 - 78)/(3.7 - 3.2) = 14.00 \pm 0.40°\text{C km}^{-1}$
  $\mathbf{Q}_4 = 14.00 \times 4.3 = 60.2 \pm 3.1 \text{ mW m}^{-2}$

The four intervals do not yield the same value for heat flow. The upper two are around 50 mW m^{-2} and the deeper two are around 60 mW m^{-2}.

The product method is able to delineate heat flow variation with depth, but is very sensitive to data quality and spacing. Small errors in closely spaced temperature data can introduce large errors in thermal gradient and interval heat flow. Narrow depth intervals may have apparent heat flow very different from intervals immediately above and below. Interpreting a real heat flow anomaly from those that are relics of poor temperature or conductivity data can be difficult. Every effort should be made to ensure that all data are as accurate as possible to reduce the likelihood of spurious heat flow anomalies.

High-precision temperature logs (see Section 3.1.1) allow individual formations to be matched with precise thermal gradient measurements. This is useful
for a couple of reasons. Firstly, a best estimate of heat flow can be obtained
using only those formations for which accurate conductivity data are available.
Secondly, this accurate heat flow estimate can then be applied to poorly
sampled formations to determine their thermal conductivity. This is an indirect
way of assessing the *in situ* thermal conductivity of shale layers (Blackwell et
al., 1997), for example, for which conventional methods are often inadequate.

Question: Several samples from a particular carbonate sequence are
collected and found to have a consistent thermal conductivity of
3.07 ± 0.04 W m^{-1} K^{-1}. A precision temperature log over the
same formation indicates a relatively constant thermal gradient
of $22.5 \pm 0.1°$C km^{-1}. A shale sequence lies immediately above
the carbonate, but no conductivity measurements were possible
on the friable unit. The temperature log, however, indicates a
thermal gradient of $48.0 \pm 0.2°$C km^{-1} across the shale. What is
the heat flow in the well and the conductivity of the shale formation?

Answer: The heat flow in the carbonate layer is

$\mathbf{Q} = \lambda \times \partial T / \partial z = 3.07 \pm 0.04 \times 22.5 \pm 0.1 = 69.1 \pm$
    1.2 mW m^{-2}

Assuming heat flow remains constant across the boundary, the
conductivity of the shale unit is found from

$69.1 \pm 1.2 = \lambda_{\text{shale}} \times 48.0 \pm 0.2$

or

$\lambda_{\text{shale}} = 69.1 \pm 1.2/48.0 \pm 0.2 = 1.44 \pm 0.03$ W m^{-1} K^{-1}

6.2. Bullard Plots

A more coherent and demonstrable calculation of heat flow is based on the
concept of *thermal resistance*. Thermal resistance, R, is defined as the integral
of the reciprocal of thermal conductivity, λ, over the depth range, z:

$$R = \int (1/\lambda) \partial z \tag{6.2}$$

In SI units it is expressed in metres squared kelvin per watt (m^{2}K W^{-1}). As
the name suggests, the thermal resistance of a material is a measure of how
effectively it retards the flow of heat (e.g. Cull, 1982; Gallagher, 1990).
From Equation (6.1):

$$\mathbf{Q}_d = \lambda_d \times (\partial T / \partial z)_d,$$

or

$$\mathbf{Q}_d = (\lambda/\partial z)_d \partial T = (1/\partial R)_d \partial T = (\partial T/\partial R)_d \tag{6.3}$$

Also from Equation (6.1):

$$\mathbf{Q}_0 = \mathbf{Q}_d + \int A(z)\partial z$$

So:

$$(\partial T/\partial R)_d = \mathbf{Q}_0 - \int_0^d A(z)\partial z \tag{6.4}$$

Equation (6.3) states that at any depth, d, heat flow is equal to the gradient of temperature with respect to thermal resistance. Equation (6.4) states that this gradient will change with depth only if heat is generated in, or removed from, the section. A graph of temperature against thermal resistance is known as a *Bullard plot*, after Sir Edward Bullard (Bullard, 1939).

The format of the thermal conductivity data set controls how thermal resistance is best calculated. For conductivity data defined by individual formations or layers, each of thickness, Δz_i, and conductivity, λ_i, Equation (6.2) can be written

$$R = \sum_i \left(\frac{\Delta z_i}{\lambda_i}\right) \tag{6.5}$$

However, for thermal conductivity derived from well logs, Δz is generally constant (usually 0.1524 m; 6 in), and

$$R = \Delta z \cdot \sum_i \left(\frac{1}{\lambda_i}\right) \tag{6.6}$$

Question: Present the data from the previous example on a Bullard plot. Interpret the results.

Answer: The problem is defined as a sequence of four distinct layers, each with characteristic thermal conductivity, so we use Equation (6.5) to calculate R at each temperature datum. First temperature datum (surface, $d = 0$):

$T_0 = 10°C$

$R_0 = 0 \ \mathrm{m^2\,K\,W^{-1}}$

Second temperature datum ($d = 1.1$ km $= 1100$ m):

$T_{1100} = 25°C$

$R_{1100} = (1100/3.6) = 305.6 \ \mathrm{m^2\,K\,W^{-1}}$

Third temperature datum ($d = 2.2$ km $= 2200$ m):

$T_{2200} = 45°C$,

$R_{2200} = (1100/3.6) + (1100/2.8) = 698.4 \ \mathrm{m^2\,K\,W^{-1}}$

Fourth temperature datum ($d = 3.2$ km $= 3200$ m):

$T_{3200} = 78°C$

$$R_{3200} = (1100/3.6) + (1100/2.8) + (1000/1.8) =$$
$$1254.0 \text{ m}^2 \text{ K W}^{-1}$$

Fifth temperature datum ($d = 3.7$ km $= 3700$ m):

$$T_{3700} = 85°C$$
$$R_{3700} = (1100/3.6) + (1100/2.8) + (1000/1.8) + (500/4.3) =$$
$$1370.2 \text{ m}^2 \text{ K W}^{-1}$$

The data are plotted in Figure 6.1. Heat flow is calculated from the gradient of the line on the Bullard plot. The data depict steady-state heat flow close to 50 mW m^{-2} in the top 2.2 km of the section, increasing to around 60 mW m^{-2} In the deeper section. The result suggests that heat is being removed at a rate of about 10 mW m^{-2} around 2.2 km depth (remembering that heat is flowing upwards).

Figure 6.1 illustrates an important point. It is fundamentally wrong to assume a constant thermal gradient across a section containing thermal conductivity contrasts (i.e. in all real situations). However, points on a Bullard plot should lie on a straight line for a section supporting a constant heat flow. In practice, errors in thermal conductivity and temperature data cause points on a Bullard plot to be somewhat scattered, and linear regression is required to find a best estimate of heat flow. Uncertainty in the final estimate is assessed statistically from the degree of scatter in the points (e.g. Kreyszig, 1983, p. 986).

Note: In a steady-state, conductive heat regime with negligible heat production, all points on a Bullard plot fall along a straight line, the gradient of which is equal to the surface heat flow, $\mathbf{Q}_0$. This is not true for a temperature-depth plot, which is generally non-linear.

6.3. Non-Linear Bullard Plots

Purely conductive, steady-state vertical heat flow with no internal heat production produces a straight line on a Bullard plot. Any deviation from these conditions may introduce non-linearity to the plot. A sufficient number of high-precision temperature measurements coupled with accurate conductivity data may reveal either gradual or sharp increases or decreases in heat flow with depth.

6.3.1. Reasons for Non-Linear Bullard Plots

A number of factors may be responsible for a non-linear Bullard plot, not all of which arise from actual thermal conditions. Any systematic error in estimating thermal conductivity, for example, translates directly into a systematic error in calculated heat flow.

If sediment compaction is underestimated, then porosity may be modelled higher than its true value. Modelled thermal conductivity will therefore be lower than true conductivity, and the error will generally increase with depth. Thermal resistance will be progressively overestimated with depth and the resulting Bullard plot will have a convex-up shape. That is, heat flow will appear to decrease with depth. The converse is also true. Too-high compaction coefficients result in an apparent increase in heat flow with depth. Fortunately, the magnitude of the error is generally small, and in most cases the percentage error in the heat flow estimate is less than that in the compaction coefficient.

A larger error results from a poor initial porosity estimate. Too large an initial porosity yields a modelled heat flow that is too low, and vice versa. The magnitude of the percentage error in heat flow is about the same as the percentage error in initial porosity. The spreadsheet, BULLARD_ERROR.XLS (available from the web site mentioned in the Preface) can be used to investigate the magnitude of errors arising from poor compaction models.

Question: A very thick sandstone sequence (2000 m) has matrix conductivity, $\lambda_m = 4 \text{ W m}^{-1} \text{ K}^{-1}$ and fluid conductivity $\lambda_w = 0.6 \text{ W m}^{-1} \text{ K}^{-1}$. Heat flow in the region is 60 mW m^{-2}. True initial porosity is 65% and true compaction coefficient [assuming Sclater and Christie-type (1980) compaction] is $5 \times 10^{-4} \text{ m}^{-1}$. What is the effect on modelled heat flow if the initial porosity is overestimated by 15%, and the compaction coefficient is overestimated by 20%?

Answer: To a certain degree the two errors cancel each other, but the overall effect is that surface heat flow is modelled about 10 mW m^{-2} lower than the true value. Heat flow appears to increase with depth and reaches its true value at about 1000 m.

High heat generation results in a decrease in heat flow with depth – a result similar to poor compaction modelling. Typical heat generation values in sediments, however, are insufficient to produce an observable effect on heat flow over the depth range of a typical exploration well. Ore-grade proportions of radiogenic material are required (e.g. Beardsmore, 1996). In hard rocks, where compaction is insignificant and radiogenic material more abundant, high heat generation is the most probable explanation for an observed decrease in heat flow with depth.

It may sometimes happen that an individual bed or formation appears to deviate from an apparently linear heat flow profile. This is only likely to become apparent with the resolution of precision temperature logs. Dog-legs in heat flow profiles are not physically sustainable in a steady-state conductive situation. Fluid advection can affect heat flow within individual beds (Section 6.3.5), but another likely explanation is an error in the conductivity estimate for the unit in question. This may arise from errors in matrix conductivity, porosity or fluid properties. In the latter case, most compaction models assume water- or brine-filled pores, so the modelled conductivity of oil and gas reservoirs (both have conductivities less than water) is higher than their true conductivity. They therefore appear to have higher heat flow.

Abrupt and sustained changes in vertical heat flow at specific depths generally indicate the introduction or removal of heat from the system. The most common cause of such an effect in a sedimentary setting is fluid migration (Section 6.3.5). Other causes may include heat refraction (Section 6.4) or diagenetic/metamorphic processes (Section 2.3), although the effect of the latter is generally negligible.

Note: It is a widely held misconception that high heat production can raise heat flow within an individual unit without affecting surrounding formations. In reality, heat generated within a unit must flow into adjacent units. The general effect is that heat flow is raised in *all* layers above the generating zone. Another common error is confusion between high temperature gradient and high heat flow. A relatively high gradient is expected in low-conductivity shale units, but this does not imply an increase in heat flow within the unit.

6.3.2. Climatology

Steady-state heat flow requires constant boundary conditions. Transient effects such as climate change, sedimentation and erosion alter the surface boundary condition, and can have an appreciable effect on the slope of a Bullard plot.

The disturbance to thermal gradient caused by periodic changes in surface temperature has already been discussed in Section 3.3. The effect of such temperature fluctuations on near-surface heat flow is directly proportional to the effect on gradient:

$$(\partial \mathbf{Q}_\theta / \partial z) = \lambda \times T_0 \exp(-\varepsilon z)(-\varepsilon)[\sin(\omega t - \varepsilon z) + \cos(\omega t - \varepsilon z)] \qquad (6.7)$$

Equation (6.7) describes the departure, $\mathbf{Q}_\theta$, from a mean value of heat flow at a particular depth, z, and time, t, resulting from a surface heating cycle with amplitude T_0 (half of the peak–trough amplitude) and frequency $\omega(\omega = 2\pi/P; P = \text{period})$. The thermal conductivity of the medium is λ and the thermal diffusivity, κ, is included in ε, where $\varepsilon = (\pi/P\kappa)^{1/2}$.

Heat flow at depth z is undisturbed by the surface temperature fluctuation when $(\partial \mathbf{Q}_\theta / \partial z) = 0$:

$$\sin(\omega t - \varepsilon z) = -\sin(\pi/2 - \omega t + \varepsilon z) \qquad (6.8)$$

$$t_{min} = (3\pi/4 + \varepsilon z)/\omega \quad \text{or} \quad (7\pi/4 + \varepsilon z)/\omega$$

Maximum disturbance to heat flow occurs when

$$\sin(\omega t - \varepsilon z) = \sin(\pi/2 - \omega t + \varepsilon z) \tag{6.9}$$

$$t_{max} = (\pi/4 + \varepsilon z)/\omega \quad \text{or} \quad (5\pi/4 + \varepsilon z)/\omega$$

Onshore heat flow data are only reliable for depths greater than 30 m if measurements are expected to be within 1% of undisturbed values. Values obtained closer to the surface will likely be disturbed by diurnal and seasonal heating cycles.

Long-period climatic changes affect the heat flow profile to greater depths and times. The change in heat flow, $\Delta \mathbf{Q}$, due to a permanent change in surface temperature, T_0, is

$$\Delta \mathbf{Q} = -\lambda \times T_0[(\pi\kappa t)^{-1/2}\exp(-z^2/4\kappa t)] \tag{6.10}$$

and the effect of more than one event is found by simple addition:

$$\Delta \mathbf{Q} = \Sigma \Delta \mathbf{Q}_i \tag{6.11}$$

where $\Delta \mathbf{Q}_i$ is the deviation in heat flow due to the ith event.

Question: Around 15,000 years ago the average surface temperature was 5°C and a sequence of basalt layers accommodated a stable heat flow of 54 mW m^{-2}. An ice age began at that time and lowered the average surface temperature to 0°C. The ice age ended 10,000 years ago and the average surface temperature rose to 10°C. What is the present surface heat flow and present heat flow at 500 m depth, given that the thermal conductivity and diffusivity of basalt are $\lambda = 1.8$ W m^{-1} K^{-1} and $\kappa = 7.9 \times 10^{-7}$ m^2 s^{-1}, respectively?

Answer: For the first event:

$T_0 = -5$ K

$t = 15,000 \times 3.15567 \times 10^7 = 4.7335 \times 10^{11}$ s

At the surface, the exponential term disappears from Equation (6.10):

$\Delta \mathbf{Q}_{surf} = 1.8 \times 5[(1.17479 \times 10^6)^{-1/2}] = 0.00830$ W m^{-2}

$\Delta \mathbf{Q}_{500\,m} = 1.8 \times 5[(1.17479 \times 10^6)^{-1/2}\exp(-0.16714)] = 0.00703$ W m^{-2}

For the second event:

$T_0 = 10$ K

$t = 10,000 \times 3.15567 \times 10^7 = 3.15567 \times 10^{11}$ s

$\Delta \mathbf{Q}_{surf} = -1.8 \times 10[(7.83193 \times 10^5)^{-1/2}] = -0.02034$ W m^{-2}

$\Delta \mathbf{Q}_{500\,m} = -1.8 \times 10[(7.83193 \times 10^5)^{-1/2}\exp(-0.25070)] = -0.01583$ W m^{-2}

So present surface heat flow, from Equation (6.11), is

$\mathbf{Q}_{\text{surf}} = 0.0540 + 0.00830 - 0.02034 = 0.0420 = 42.0 \text{ mW m}^{-2}$
and present heat flow at 500 m is
$\mathbf{Q}_{500\,\text{m}} = 0.0540 + 0.00703 - 0.01583 = 0.0452 = 45.2 \text{ mW m}^{-2}$
Heat flow is significantly (22%) lower than steady state at the surface but increases with depth.

When average surface temperature is used as the top boundary condition in regions of recent glaciation, it is observed that surface heat flow estimates increase with the depth of the bottom temperature determination (Crain, 1968). This is attributed to the propagation into the ground of a surface temperature rise, reducing near-surface temperature gradient.

6.3.3. *Sedimentation*

Sediment settling through a column of water attains the temperature of the water at the sediment–water interface where it is deposited. Previously deposited material moves downwards, relative to this interface, at a rate equal to the sedimentation rate (ignoring the effects of compaction). Heat is physically carried with the grains as they move downwards, and if this downward advection is significant it can reduce the observed surface heat flow. The magnitude and depth of the effect depend on the rate and duration of sedimentation.

Von Herzen and Uyeda (1963) developed an idealised model that assumes a horizontal depositional surface at depth $z = 0$ m and temperature $T = 0°\text{C}$. The surface is initially in thermal equilibrium with an underlying stratum of constant thermal conductivity, λ, thermal diffusivity, κ, and heat generation, A_0. Sedimentation of material identical to the underlying stratum commences at time $t = 0$ s and progresses at a constant rate, U (m s^{-1}). The heat transfer equation is

$$\frac{\partial^2 T}{\partial z^2} - \frac{U}{\kappa} \times \frac{\partial T}{\partial z} - \frac{1}{\kappa} \times \frac{\partial T}{\partial t} = -\frac{A_0}{\lambda} \tag{6.12}$$

Carslaw and Jaeger (1959, p. 387) gave a solution for $T(t, z)$, which Von Herzen and Uyeda (1963) used to calculate the effect on surface heat flow for the case where $A_0 = 0$. They calculated the ratio of modified to initial surface heat flow ($\mathbf{Q}^*/\mathbf{Q}$) in terms of elapsed time and sedimentation rate (Figure 6.2):

$$\frac{\mathbf{Q}^*}{\mathbf{Q}} = 1 - \text{erf}(X) - \frac{2X}{\sqrt{\pi}} e^{-X^2} + 2X^2 \, \text{erfc}(X) \tag{6.13}$$

where $X = 0.5 \times U \times \sqrt{(t/\kappa)}$, and erf( ) and erfc( ) denote the error function and its compliment [see Equation (3.29), or the spreadsheet 'FUNCTIONS.XLS' available from the web site mentioned in the Preface].

Although the model is largely unrealistic, the results can be used to draw general conclusions. For normal rates of sedimentation of 10^{-6}–10^{-5} m yr^{-1}, the effect on surface heat flow is insignificant. It is only when sedimentation

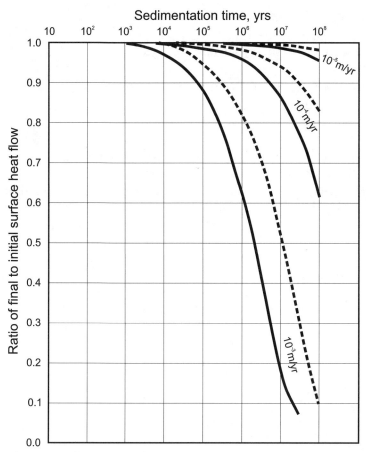

Figure 6.2. Effect of a constant sedimentation rate on surface heat flow. Values indicate rate of sedimentation. Solid lines are for sediment with thermal diffusivity $\kappa = 2 \times 10^{-7} \mathrm{m^2\,s^{-1}}$, dashed lines for $\kappa = 1.2 \times 10^{-6}$ m^2 s^{-1}. Modified after Von Herzen and Uyeda (1963).

rates reach relatively high levels $(> 10^{-4}$ m yr$^{-1})$ that there are noticeable effects on the near-surface heat flow. These high rates of sedimentation are only found in a limited number of environments.

Question: What is the effect on surface heat flow of sedimentation at a rate of 5×10^{-4} m yr^{-1} for 1 Ma, given that the diffusivity of the sediment is $\kappa = 8 \times 10^{-7}$ m^2 s^{-1}?

Answer: Converting all the parameters into SI units:

$U = 1.584 \times 10^{-11}$ m s^{-1}

$t = 3.1557 \times 10^{13}$ s

$\kappa = 8 \times 10^{-7}$ m^2 s^{-1}

Solving for X in Equation (6.13):

$X = 0.5 \times U \times \sqrt{(t/\kappa)} = 0.04976$

$\mathrm{erf}(X) = 0.05610$

erfc(X) = 0.94390
Solving for the rest of Equation (6.13):
  Q*/Q = 1 − 0.05610 − 0.05601 + 2 × 0.00248 × 0.94390 =
    0.89257 = 89.3%
Surface heat flow is reduced by almost 11% from steady-state
conditions.

Not all sedimentation is gradual. Turbidity flows and landslides can
result in sudden thick deposits of sediment at water temperature. Von
Herzen and Uyeda (1963) also examined this problem and produced a
second set of heat flow curves based on the thickness of the deposit and
the time since deposition (Figure 6.3). They found that the thermal effect of

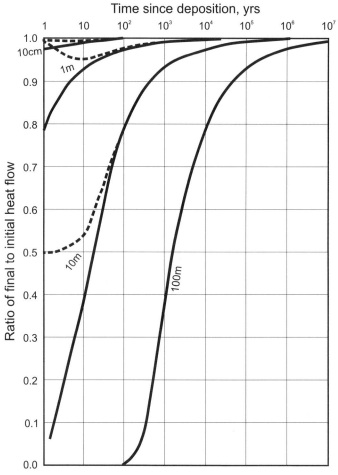

Figure 6.3. Effect on surface heat flow after sudden deposition of a thickness of sediment. Labels
indicate the thickness of the layer deposited. Dashed lines indicate effect at < 1 m depth, solid lines
indicate effect at 10 m depth. Modified after Von Herzen and Uyeda (1963).

a 1 m thick deposit largely dissipates within 10 years, but a 10 m thick deposit can significantly affect surface heat flow for up to 1000 years, and a 100 m thick deposit for up to 100,000 years. This result has implications for the accuracy of many shallow heat flow measurements made with deep-sea probes (Section 3.1.2) near continental slopes, where turbidity flows may be expected.

6.3.4. Erosion

The thermal effect of eroding the surface of the Earth is the opposite to that of sedimentation. Erosion results in an upward movement of rocks in relation to a reference point on the surface. Surface heat flow increases by an amount dependent upon the elapsed time and rate of erosion. The equation of heat transfer is the same as Equation (6.12), but U now has a negative value. The effect on surface heat flow is again determined using Equation (6.13).

Note: Erosion very rarely takes place in a region that can be modelled as a horizontal plain. There is often significant topographic relief in the region of erosion, and observed heat flow should be corrected for the effect of topography (Section 6.4.2) before the effect of erosion can be examined.

6.3.5. Groundwater Migration

The assumption of steady-state vertical heat conduction may be largely invalidated in the presence of hydrothermal processes, because moving fluids carry heat by advection. A fluid moving laterally through a permeable body repeatedly divides and recombines around grains within the matrix. This process provides a mechanism for mixing and diffusion, the main effect of which is to reduce the thermal gradient within the rock without affecting the net vertical heat flow. The thermal conductivity of the rock is effectively increased. If normal mixing models are used to construct a Bullard plot, then a laterally flowing aquifer will appear as a layer with heat flow lower than those layers above and below.

If there is a net vertical motion of the fluid with respect to the surface, then the effect can be quantified by considering the ratio of fluid-transported heat to conducted heat, called the Peclet number, P_e:

$$P_e = \phi \rho_f c_f w_z \partial z / \lambda \tag{6.14}$$

where ϕ is porosity of the rock; ρ_f, c_f are density $(\mathrm{kg\,m^{-3}})$ and thermal capacity $(\mathrm{J\,kg^{-1}\,K^{-1}})$ of the fluid; w_z is vertical flow rate of the fluid $(\mathrm{m\,s^{-1}})$, ∂z is the vertical distance over which the fluid flows (m) and λ is the thermal conductivity.

The vertical thermal conductivity of the aquifer effectively decreases by $\lambda \times P_e$ (positive values of w_z imply downward fluid flow):

$$\lambda_e = \lambda - \lambda \times P_e \tag{6.15}$$

Therefore, vertical fluid flow can be modelled as an effective increase or decrease in thermal conductivity.

Question: Water $(\rho_f = 1100 \text{ kg m}^{-3}, c_f = 4100 \text{ J kg}^{-1} \text{K}^{-1})$ flows through a 20 m thick aquifer with 20% porosity at a vertical rate of 15 mm yr^{-1} upward relative to the surface. By how much is the effective thermal conductivity of the aquifer increased?

Answer: The relevant parameters are

   $\phi = 0.2$

   $\rho_f = 1100 \text{ kg m}^{-3}$

   $c_f = 4100 \text{ J kg}^{-1}\text{K}^{-1}$

   $w_z = -15 \times 10^{-3}/3.1557 \times 10^7 = -4.7533 \times 10^{-10} \text{ m s}^{-1}$

   $\partial z = 20 \text{ m}$

The effective conductivity, λ_e, is found using Equation (6.15):

   $\lambda_e = \lambda - \lambda \times P_e = \lambda - \phi\rho_f c_f w_z \partial z = \lambda - 0.2 \times 1100 \times 4100 \times$
      $-4.7533 \times 10^{-10} \times 20 = \lambda + 8.575 \times 10^{-3} \text{ W m}^{-1} \text{ K}^{-1}$

That is, conductivity effectively increases by 8.575×10^{-3} W m^{-1} K^{-1}.

The heat is being transported by advection, so it does not necessarily conduct into the overlying layers. In a steady-state situation, however, the overlying layers will be in equilibrium with the aquifer, and the thermal effect can be modelled in a different way. When fluids move vertically within a sustained thermal gradient, the rate of heat transport varies with depth:

$$(\partial \mathbf{Q}_f/\partial z) = \phi\rho_f c_f w_z (\partial T/\partial z) \tag{6.16}$$

where $\mathbf{Q}_f$ = vertical heat flow (W m^{-2}), $\partial T/\partial z$ = thermal gradient (K m^{-1}) and the other symbols are as defined for Equation (6.14).

The left-hand side of Equation (6.16) reads 'rate of change in heat flow with depth' and the right-hand side shows that this is proportional to the thermal gradient, porosity, fluid thermal properties and fluid velocity. The effect of an upward flow of fluid (negative value of w_z) is to decrease heat flow with depth. This is the same effect as a heat source. Similarly, a downward fluid flow has the same effect as a heat sink. This means that vertical fluid flow can be modelled in terms of heat generation.

Question: Given that the average thermal gradient across the aquifer in the above example is 25°C km^{-1}, what value of heat generation would produce the equivalent effect and by how much does heat flow increase across the aquifer?

Answer: From Equation (6.16):

$$\partial Q/\partial z = 0.2 \times 1100 \times 4100 \times 4.7533 \times 10^{-10} \times 0.025 = -1.072 \times 10^{-5} \text{ W m}^{-2} \text{m}^{-1}$$

This is the same effect as a heat source of 1.072×10^{-5} W m^{-3}, or 10.7 μW m^{-3}. The heat flow decreases downwards over the 20 m interval by

$$1.072 \times 10^{-5} \times 20 \text{ m} = 0.214 \text{ mW m}^{-2}$$

We can test this result against the one calculated in the previous example. The change in vertical heat flow is equal to the increase in conductivity multiplied by the thermal gradient:

$$\partial Q = 8.575 \times 10^{-3} \times 0.025 = 0.214 \text{ mW m}^{-2}$$

which is the same result.

6.3.6. Deep Flow of Hot Fluids

Bruce et al. (1996) demonstrated how a short-term (< 1 Ma) flow of hot fluid along an aquifer within a sedimentary sequence might have a significant effect on the thermal history of the formations immediately above the aquifer. Fluids expelled from deep sedimentary formations (e.g. due to heating by a nearby intrusive body) may reach temperatures in excess of 200°C. If such temperatures are maintained within a thin aquifer for a thousand years or more, temperatures may be significantly raised for several hundred metres above the aquifer. Petroleum production during the period of elevated temperature may be equivalent to millions of years of generation at equilibrium temperature. Even if temperatures revert to equilibrium once hot fluids cease to flow, the event will be recorded by the organic maturity of the nearby sediments.

6.4. Non-Vertical Heat Flow

Heat does not always flow vertically in the Earth. In fact, this is only the case for conformable horizontal beds with a horizontal surface boundary with no lateral temperature variation. Non-horizontal surface boundaries, surface boundaries with laterally varying temperature, non-horizontal contacts between formations and formations with laterally varying thermal conductivity all distort heat flow away from vertical. These conditions describe a large portion of real-world situations, so an understanding of the factors leading to non-linear heat flow paths is important to determine under what conditions a one-dimensional model is sufficient. This section examines the two main factors that cause heat flow to deviate from vertical – basement relief and surface topography – and also looks at the effect of a salt dome on nearby heat flow.

Note: For most situations, numerical models are the best way to investigate heat flow paths. Only for a small number of ideal cases do analytical solutions exist.

6.4.1. Basement Relief

Heat is like most forms of energy. When it flows, it follows a path of least resistance. In practice, this means that heat escaping the interior of the Earth preferentially flows through regions of higher thermal conductivity. In an area of undulating, high-conductivity basement rocks beneath a blanket of low-conductivity sediment, heat is refracted away from regions of thick sediment cover and preferentially channelled through thinly covered areas. An exact analytical solution for the magnitude of the refraction due to an arbitrary basement shape is impossible, but some solutions have been found for geometrical approximations to real situations.

Bullard, Maxwell and Revelle (1956) considered the two-dimensional problem of a sinusoidal series of parallel ridges of amplitude, 2l (measured from crest to trough) and wavelength, **w**, covered by sediments (Figure 6.4). They

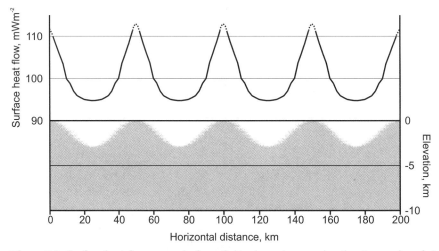

Figure 6.4. Surface heat flow anomaly for a finite element approximation to a series of sinusoidal ridges (amplitude = 3 km, wavelength = 50 km, $\lambda_r = 4$ W m^{-1} K^{-1}, $\lambda = 2$ W m^{-1} K^{-1}). Basal heat flow is 100 mW m^{-2} and surface temperature is 0°C. Maximum heat flow exceeds minimum by about 18.2 mW m^{-2}, very close to 18.8 mW m^{-2} predicted by Bullard et al. (1956).

found that when the ratio $\mathbf{l}/\mathbf{w}$ is small, heat flow through the crests exceeds that through the troughs by approximately $\partial\mathbf{Q}$, given by

$$\frac{\partial\mathbf{Q}}{\mathbf{Q}} = \frac{2\pi\mathbf{l}(\lambda_r - \lambda)}{\mathbf{w}\lambda} \tag{6.17}$$

where $\mathbf{Q}$ is the mean heat flow and λ_r and λ are the thermal conductivities of the underlying rock and sediment, respectively.

> *Note:* Equation (6.17) was originally published with the denominator $\mathbf{w}\lambda_r$, but was subsequently revised, with Bullard's permission, by Von Herzen and Uyeda (1963).

Von Herzen and Uyeda (1963) examined the problem of a semi-ellipsoid packet of sediment of depth $\mathbf{l}$ within a basement succession. The basin (Figure 6.5) can be visualised in plan view as an ellipse with long-axis length $2\mathbf{n}$ and short-axis length $2\mathbf{m}$. Von Herzen and Uyeda found the surface heat flow above the packet of sediment, $\mathbf{Q}_s$, to have a constant ratio to the mean heat flow, $\mathbf{Q}$:

$$\frac{\mathbf{Q}_s}{\mathbf{Q}} = \frac{\lambda/\lambda_r}{1 + (\lambda/\lambda_r - 1)\mathbf{F}} \tag{6.18}$$

where

$$\mathbf{F} = \mathbf{lmn} \times \int_n^\infty \frac{dx}{(x^2 - \mathbf{n}^2 + \mathbf{l}^2)^{3/2}(x^2 - \mathbf{n}^2 + \mathbf{m}^2)^{1/2}} \tag{6.19}$$

Equation (6.19) simplifies for certain values of $\mathbf{l}$, $\mathbf{m}$ and $\mathbf{n}$. When $\mathbf{n} = \infty$, the model is an infinite trough with elliptical cross section, and $\mathbf{F} = \mathbf{m}/(\mathbf{m} + \mathbf{l})$. The largest values of $\mathbf{Q}_s/\mathbf{Q}$ are obtained when $\mathbf{m} = \mathbf{n}$, describing a circular basin. Then:

$$\mathbf{F} = \frac{\mathbf{lm}^2}{(\mathbf{m}^2 - \mathbf{l}^2)^{3/2}} \cdot \left[\frac{-\pi}{2} + \arctan\frac{\mathbf{l}}{(\mathbf{m}^2 - \mathbf{l}^2)^{1/2}} + \frac{(\mathbf{m}^2 - \mathbf{l}^2)^{1/2}}{\mathbf{l}}\right] \tag{6.20}$$

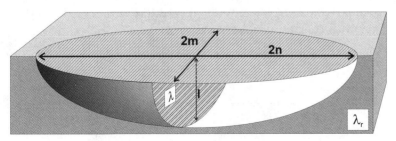

Figure 6.5. An ellipsoidal basin of length, $2n$, width, $2m$, and depth, l. Thermal conductivity of the underlying rocks is λ_r, and thermal conductivity of the sediment fill is λ.

It is interesting to note that heat flow is constant across the basin, although the base of the sediment is curved. The reduction in heat flow above the sediments is more pronounced for higher values of l/m and λ_r/λ.

Question: What is the heat flow above a package of sediment ($\lambda = 2.3$ W m^{-1} K^{-1}) that forms a circular basin 50 km wide and 8 km deep, given the mean heat flow in the basement ($\lambda_r = 4.2$ W m^{-1} K^{-1}) is 60 mW m^{-2}?

Answer: **F** is found using Equation (6.20) for a circular basin. In this case, **m** $= 25$ (2**m** $= 50$), **l** $= 8$:

$\mathbf{lm}^2/(\mathbf{m}^2 - \mathbf{l}^2)^{3/2} = 8 \times 25^2/(25^2 - 8^2)^{3/2} = 5000/13{,}287.53 = 0.37629$

$\mathbf{l}/(\mathbf{m}^2 - \mathbf{l}^2)^{1/2} = 8/(25^2 - 8^2)^{1/2} = 8/23.68544 = 0.33776$

$(\mathbf{m}^2 - \mathbf{l}^2)^{1/2}/\mathbf{l} = (25^2 - 8^2)^{1/2}/8 = 23.68544/8 = 2.96068$

So:

$\mathbf{F} = 0.37629 \times [(-\pi/2) + \arctan(0.33776) + 2.96068] = 0.64557$

We can now calculate the heat flow in the basin, $\mathbf{Q}_s$, using Equation (6.18):

$\mathbf{Q} = 60, \qquad \lambda = 2.3, \qquad \lambda_r = 4.2$

$\mathbf{Q}_s/60 = (2.3/4.2)/[1 + ((2.3/4.2) - 1) \times \mathbf{F}] = 0.77352$

The expected heat flow above the sediments is

$\mathbf{Q}_s = 60 \times 0.77352 = 46.41 \text{ mW m}^{-2}$

Of course, real basins do not conform to strict geometric shapes. Even so, the above models can often provide approximations to true basin geometry, and an estimate of the mean basement heat flow can be derived from the observed basin heat flow. Analytical solutions such as these are also useful for testing numerical models.

6.4.2. Topography

In mountainous terrain, heat has further to flow to reach the surface beneath a peak compared with in a valley. This suggests that, although heat flow at depth may be homogeneous, heat will preferentially flow into the valleys and surface heat flow will vary with elevation.

Note: It is vital to be aware of the effect of topography when comparing heat flow measurements from different elevations.

Lees (1910) investigated this problem, and his topographic correction remains the most commonly used. He developed a set of equations to model the distortion in the geothermal field beneath idealised mountain ranges. Lees' mountains consist of long, straight ranges of uniform height, H, lying on a horizontal plain. The mountain ranges are composed of the same material as the rocks beneath the plain and have thermal conductivity, λ, and radiogenic heat production, A.

The correction process begins by determining the height, h, such that

$$[h(1 + \beta h)]/[H(1 + \beta H)] = 0.5 \tag{6.21}$$

where $\beta = A/[2\lambda(\alpha - \alpha')]$.

The mountain range is defined such that its summit is at $x = 0$, $z = -H$, and its elevation varies along a perpendicular section according to

$$z(1 - \beta z) + Hd(1 + \beta H)\frac{z + H + d}{x^2 + (z + H + d)^2} = 0 \tag{6.22}$$

where x is the horizontal axis (origin directly beneath the summit); z is the vertical axis (origin at the surface of the plain, increases downwards) and d is the length of the straight line from the summit to a point at height h on the mountainside (d must be greater than H for the correction to be valid).

The equation describing the temperature field beneath the mountain range and plain is then

$$T = T_0 + \alpha z - \frac{A}{\lambda} \cdot \frac{z^2}{2} + (\alpha - \alpha')Hd(1 + \beta H)\frac{z + H + d}{x^2 + (z + H + d)^2} \tag{6.23}$$

where α is the thermal gradient beneath the plain away from the mountain range, α' is the decrease in air temperature with altitude, T_0 is the temperature at the surface of the plain and T is the temperature at point (x, z).

The thermal gradient at the summit of the range is given by

$$\frac{\partial T}{\partial z} = \alpha + \frac{A}{\lambda}H - (\alpha - \alpha')H(1 + \beta H)/d \tag{6.24}$$

If the rocks produce no heat ($A = 0$), Equation (6.24) reduces to

$$\frac{\partial T}{\partial z} = \alpha(1 - H/d) + \alpha'H/d \tag{6.25}$$

Remember that the shape of the mountain range must be such that $d > H$.

Question: Imagine a granitic mountain range ($\lambda = 3.2$ W m^{-1} K^{-1}, $A = 2.8$ μW m^{-3}) of moderate height ($H = 2$ km), 8 km wide at about half its height. Assume a gradient in air temperature of 5°C per kilometre height, and a geothermal gradient of 30°C km^{-1} away from the mountain range. By how much is the thermal gradient reduced beneath the summit of the mountain range?

Answer: The answer is found by inserting the relevant values (in SI units) into Equation (6.24):

$H = 2000$ m

$\lambda = 3.2 \text{ W m}^{-1} \text{ K}^{-1}$

$A = 2.8 \times 10^{-6} \text{ W m}^{-3}$

$\alpha = 0.030 \text{ K m}^{-1}$

$\alpha' = 0.005 \text{ K m}^{-1}$

$\beta = A/[2\lambda(\alpha - \alpha')] = 2.8 \times 10^{-6}/[2 \times 3.2 \times (0.030 - 0.005)] = 1.75 \times 10^{-5} \text{ m}^{-1}$

From Equation (6.21):

$[h(1 + \beta h)]/[H(1 + \beta H)] = 0.5$

$[h(1 + h \times 1.75 \times 10^{-5})]/[2000(1 + 2000 \times 1.75 \times 10^{-5})] = 0.5$

$1.75 \times 10^{-5} \times h^2 + h - 1035 = 0$

$h = [-1 + (1^2 - 4 \times 1.75 \times 10^{-5} \times -1035)^{0.5}]/[2 \times 1.75 \times 10^{-5}] = 3.56 \times 10^{-2}/3.5 \times 10^{-5} = 1016.9 \text{ m}$

This is approximately half the height of the mountain range, and we know that the width of the range at that height is 8 km. So a point at height h on the mountainside is about 4000 m horizontally from the summit. Parameter d can be found from the hypotenuse of a right-angle triangle with base = 4000 m and height = (2000 − 1016.9) = 983.1 m:

$d = (4000^2 + 983.1^2)^{0.5} = 4119 \text{ m}$

Now, from Equation (6.24):

$\partial T/\partial z = 0.030 + (2.8 \times 10^{-6} \times 2000/3.2) - (0.030 - 0.005) \times 2000 \times (1 + 1.75 \times 10^{-5} \times 2000)/4119 = 0.0192 \text{ K m}^{-1}$

At the peak of the mountain range, thermal gradient is reduced to $19.2°\text{C km}^{-1}$. This is 36% lower than on the plain.

These equations are built into the spreadsheet, 'LEES.XLS', which is available on the web site mentioned in the Preface, and may be used to investigate the effect of topography on temperature and gradient.

6.4.3. Salt Domes

Diapiric salt columns make up a significant part of the sedimentary fill of some basins. The thermal properties and geometry of such columns are different to all other lithologies commonly present in a basin, and they can strongly influence the geothermal field. Special attention should, therefore, be paid to salt bodies when building thermal models of a basin. They act to increase surface heat flow in at least three ways. Firstly, salt has a higher thermal conductivity than most other lithologies. Secondly, salt bodies generally resemble vertical dykes more than horizontal layers. Thirdly, salt may transport heat via advection if its rate of ascent is significant. We will investigate each of these points in a little more detail.

Halite (rock salt) has a thermal conductivity around $5.5 \text{ W m}^{-1} \text{ K}^{-1}$ (see Table 4.1). This is more than twice the mean conductivity of other sediments. The only other common lithology with a comparable conductivity is dolomite (about 4.7 W m^{-1} K^{-1}). Substantial bodies of salt in a basin therefore raise the average conductivity of the sediment and provide low resistance conduits for heat flow.

The increase in thermal conductivity is amplified if the salt rises to form vertical columns. If a sedimentary succession contains 10% salt lying conformably with other sediments, the average conductivity can be estimated from the harmonic mean:

$$\lambda_{av} = 1/[0.9/2.3 + 0.1/5.5] = 2.44 \text{ W m}^{-1} \text{ K}^{-1} \tag{6.26}$$

If the salt rises to form a vertical pillar, the average conductivity is best estimated using an arithmetic mean:

$$\lambda_{av} = 0.9 \times 2.3 + 0.1 \times 5.5 = 2.62 \text{ W m}^{-1} \text{ K}^{-1} \tag{6.27}$$

This represents a 7% increase in average conductivity.

Salt has a low density (2.0–2.2 g cm^{-3}) compared with most other sedimentary minerals. When a thick layer of salt is buried beneath denser material, positive buoyancy forces act to drive it upwards. This is what causes the familiar salt domes, or diapirs, as columns of salt rise through other sediments to a gravitationally stable level. Such columns form pipes of high thermal conductivity that channel heat away from surrounding rocks (Figure 6.6).

A rising column of salt carries heat through advection. The rate at which heat is transported is described using an equation similar to fluid flow (Equation 6.16):

$$(\partial \mathbf{Q}_s/\partial z) = \phi \rho_s c_s w_z (\partial T/\partial z) \tag{6.28}$$

where $\mathbf{Q}_s$ is advective heat transport by the rising salt plume, ρ_s, c_s are density and specific heat of salt ($\sim$2.1 g cm^{-3} and $\sim$ 920 J kg^{-1} K^{-1}, respectively), w_z is vertical velocity of the salt and $\partial T/\partial z$ is vertical thermal gradient.

Question: Given the thermal properties of salt above, at what rate must a salt dome rise against a background thermal gradient of 25°C km^{-1} (as for Figure 6.6) before the effect on heat flow becomes significant?

Answer: We can call the effect 'significant' if it is equivalent to heat generation on the order of 0.5 μW m^{-3}. That is:

$\partial \mathbf{Q}_s/\partial z > 0.5 \times 10^{-6}$ W m^{-2} m^{-1}

We are told $\rho_s = 2.1 \times 10^3$ kg m^{-3}, $c_s = 9.2 \times 10^2$ J kg^{-1} K^{-1}, $\partial T/\partial z = 2.5 \times 10^{-2}$ K m^{-1}

So, from Equation (6.28):

$\partial \mathbf{Q}_s/\partial z = 2.1 \times 10^3 \times 9.2 \times 10^2 \times w_z \times 2.5 \times 10^{-2} > 0.5 \times 10^{-6}$

$4.83 \times 10^4 w_z > 0.5 \times 10^{-6}$

$w_z > 0.5 \times 10^{-6}/4.83 \times 10^4$

$w_z > 1.035 \times 10^{-11}$ m s^{-1} (3.27 $\times 10^{-4}$ m yr^{-1})

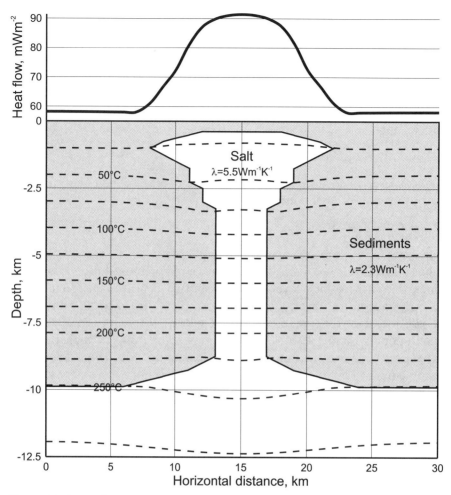

Figure 6.6. A two-dimensional section through a steady-state three-dimensional thermal model of a salt diapir. The top curve indicates the surface heat flow profile over the structure given a basal heat input of 60 mW m^{-2}.

Salt rising at a rate greater than about 0.33 mm yr^{-1} transports an appreciable amount of heat by advection. The effect is the same as if heat is being generated within the salt and conducting through the flanks of the column into the surrounding sediments, affecting heat flow in the vicinity of the salt.

Vertical growth rates of salt domes vary widely according to specific conditions. Published figures range from at least 0.005–0.035 mm yr^{-1} (Remmelts, 1993) to 0.8 mm yr^{-1} (Walters, 1993). The advective heat transfer in the latter instance is equivalent to heat generation of about 2.4 μW m^{-3} in the salt.

6.5. Heat Flow Correlations

Discussion up to this point has concentrated on deep boreholes and direct measurements for an accurate determination of heat flow. We have stressed that core samples are required for thermal conductivity determination, and full casing of the borehole ensures against collapse prior to observations of a stable geothermal gradient. Given the strict conditions required, accurate heat flow data are comparatively rare and there are few opportunities to conduct detailed heat flow mapping comparable to gravity or magnetic surveys. Alternative means are required for regional heat flow mapping.

Heat flow data can be readily correlated with other geophysical and geological parameters. In particular, there are clear associations between heat flow and surface heat production, geologic age, seismic velocity and electrical conductivity. Expected correlations between heat flow and crustal thickness, however, are difficult to establish. There are significant uncertainties in most crustal thickness estimates, and the thermal effects of deep features are obscured at the surface by an effective spatial filtering due to thermal diffusion through the crust.

In view of the many complications associated with local variation in surface geology, only major regional trends can be identified through correlations such as these.

6.5.1. Heat Flow and Heat Production

Roy et al. (1968) demonstrated an empirical linear relationship between surface heat flow and heat generation in local basement rocks (Figure 6.7). As already described in Section 2.1.1, regional trends have since been established for numerous regions or terranes around the globe. Simple statistical measures can be applied to local data to define a heat flow province according to the expression:

$$\mathbf{Q} = \mathbf{q} + A_0 D \tag{6.29}$$

where $\mathbf{Q}$ is surface heat flow (mW m^{-2}), $\mathbf{q}$ is residual heat flowing into the crust from mantle sources (mW m^{-2}), A_0 is surface heat generation (μW m^{-3}) and D is a depth (km) characteristic of the province.

Values for the residual, $\mathbf{q}$, (generally 25–35 mW m^{-2}) indicate the minimum contribution to crustal heat due to global accretion, segregation, and mantle heat generation, as described in Chapter 1. Trends of this type have been directly observed in limited areas with sufficient geological exposure (e.g. Lachenbruch and Bunker, 1971; Hawkesworth, 1974). However, simple mathematical expressions are rarely sufficient in regions of complex geology typical of continental crust (e.g. Smithson and Decker, 1974).

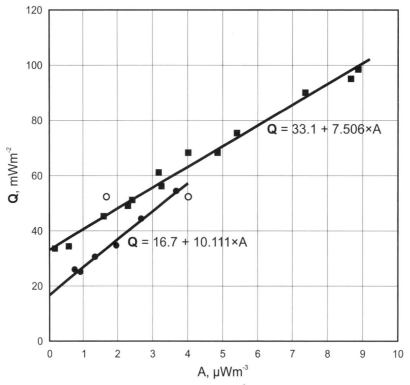

Figure 6.7. Plot of surface heat flow, **Q** (mW m^{-2}), versus heat generation, A (μW m^{-3}), for the Sierra Nevada (circles) and the New England/Central Stable Region of the United States (squares). Linear relationships are apparent. Data from Roy et al. (1968).

6.5.2. Heat Flow and Continental Age

Variations in the concentration of uranium, thorium and potassium within the crust dominate trends in heat flow observed at the surface of the Earth. Consequently, heat flow anomalies may be expected from any geological process impacting on the normal distribution of these elements. An obvious mechanism is melting in active geological terranes, but major variations in radiogenic heating can also be expected from the removal of near-surface layers through erosion and weathering. Models suggesting an exponentially decreasing distribution of heat-generating elements with depth also predict clear correlations between heat generation and crustal age and thickness.

Major correlations between heat flow and geological age are readily estab-lished for Australia. Sass and Lachenbruch (1979) defined three major heat flow provinces (Figure 6.8) based on linear relationships between heat flow and surface heat production. These predominantly coincide with established boundaries for Precambrian, Palaeozoic and Cainozoic basement. Representative values for surface heat flow in each province are consistent with more elaborate determinations provided by Sclater and Francheteau (1970). Mean surface heat flow progressively increases towards younger ter-

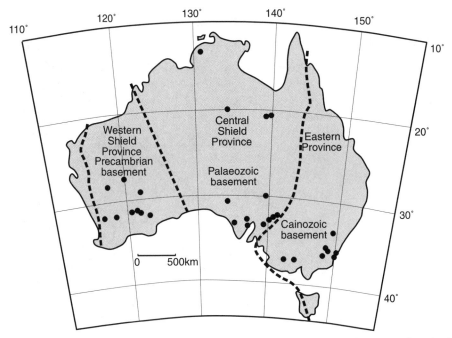

Figure 6.8. The three major Australian heat flow provinces, as defined by Sass and Lachenbruch (1979). Circles indicate locations of heat flow–heat production pairs.

ranes, undisturbed by erosion and depletion, with higher concentrations of U, K, and Th.

Age may affect heat flow in ways other than depleting radiogenic sources. Hamza (1979) identified a correlation between mantle heat flow and the age of the last thermal event (Figure 6.9). For example, there is an apparent increase in heat flow with latitude within the Cainozoic regions of eastern Australia. The trend is assumed to relate to hot-spot activity associated with the northward drift of the Australian plate. Sass and Lachenbruch (1979) presented crustal evolution models for the region based on mantle deformation and high-level magma emplacement. Their models are consistent with the dates obtained for volcanic activity in the same area (Wellman and McDougall, 1974). However, the extent of melting and dyke emplacement is not well constrained and existing data are satisfied with several different models (Cull, O'Reilly and Griffin, 1991).

6.5.3. Heat Flow and Oceanic Age

There are clear correlations between oceanic age and surface heat flow related to the cooling of oceanic lithosphere. Specifically, heat flow increases with the inverse square root of the age of oceanic crust (Figure 6.9). The model is discussed fully in Section 7.1.1 and will not be further examined at this point.

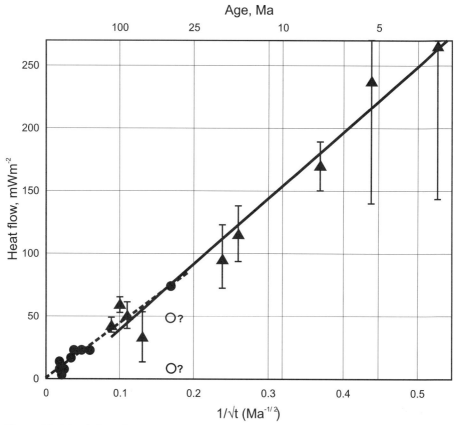

Figure 6.9. Mantle heat flow (corrected for heat generation in the lithosphere beneath the crust) as a function of the age of the last thermal event, t. Data from both continental (circles) and oceanic (triangles) regions. Open circles represent data from the Sierra Nevada and Basin and Range not used in linear regression. Modified after Hamza (1979).

6.5.4. Heat Flow and Seismic Data

Seismic velocity varies with temperature (e.g. Anderson et al. 1968), and the temperature of the upper mantle varies with crustal heat flow. Consequently, we may expect a correlation between crustal heat flow, Q ($\mathrm{mW\,m^{-2}}$), and upper mantle P-wave velocity, V_p ($\mathrm{km\,s^{-1}}$). Cull and Denham (1979) demonstrated such a correlation for Australia, described by the linear expression:

$$Q = 1150 - 135 \times V_p \tag{6.30}$$

Question: The P-wave velocity in the upper mantle beneath a particular locality in Australia is 8.10 ± 0.02 $\mathrm{km\,s^{-1}}$. What is the crustal heat flow suggested by this velocity?

Answer: Simply substitute $V_p = 8.10 \pm 0.02$ into Equation (6.30):
$$\mathbf{Q} = (1150 - (135 \times 8.10)) \pm (135 \times 0.02) = 56.5 \pm 2.7 \ \mathrm{mW\,m^{-2}}$$

Correlations between heat flow and seismic travel time can be made in a more direct way using tomography concepts. Drummond et al. (1989) presented detailed contours of travel time residuals for Australia (Figure 6.10). Systematic trends are evident with early arrivals in the relatively cold Precambrian crust and later arrivals along the warmer east coast, consistent with the heat flow domains in Figure 6.8. Cull and Denham (1979) suggested a linear relationship:

$$\mathbf{Q} = 21.45 \times t_r + 65.3 \tag{6.31}$$

where $\mathbf{Q}$ (mW m^{-2}) is the surface heat flow and t_r (s) is the travel time residual defined by Drummond et al. (1989).

Note: Cleary, Simpson and Muirhead (1972) attributed observed travel time delays to variations in mantle composition, but temperature variations across the Precambrian and Palaeozoic crust are sufficient to explain the data.

Houbolt and Wells (1980) developed an empirical relationship between heat flow and sonic travel times, based on the observation that for most sedimentary rocks the ratio of sound velocity to thermal conductivity increases linearly with

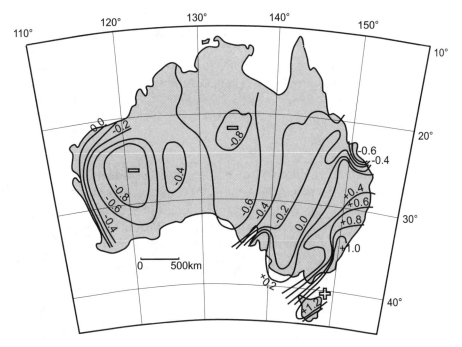

Figure 6.10. Teleseismic travel time residuals for Australia. Contour interval is 0.2 s. Modified after Drummond et al. (1989).

temperature. They used one-way travel time logs in combination with bottom-hole temperatures to derive a relationship for heat flow:

$$Q^* = \ln \frac{T_L + c}{T_U + c} \times \frac{1}{a(t_L - t_U)} \tag{6.32}$$

where $\mathbf{Q}^*$ is heat flow in relative Bolderij units, BU (1 BU $\approx$ 77 mW m^{-2}), T_L and T_U are the temperatures ($^\circ$C) at the bottom and top of a sub-surface interval, t_L and t_U are the one-way sonic travel times (s) to the bottom and top of the interval, $a = 1.039$ and $c = 80.031$.

Question: A particular well in the Browse Basin, Western Australia, has the following properties at two depths:

> At 3224 m, $T = 84.0^\circ$C, $t = 1.12$ s
> At 4421 m, $T = 146.1^\circ$C, $t = 1.47$ s

What is the suggested heat flow in the well?

Answer: In terms of Equation (6.32):

> $T_U = 84.0$, $t_U = 1.12$
> $T_L = 146.1$, $t_L = 1.47$

Substituting into Equation (6.32):

> $\mathbf{Q}^* = \ln[(146.1 + 80.031)/(84.0 + 80.031)] \times 1/[1.039 \times (1.47 - 1.12)] = 0.321 \times 2.750 = 0.883$ BU ≈ 68.0 mW m^{-2}

Houbolt and Wells (1980) concluded that their method works with acceptable accuracy for siliciclastic and carbonate rocks. However, their result for the well Chapman 1 in Texas ($\mathbf{Q}^* = 0.621$ BU, or 47.8 mW m^{-2}) compares unfavourably with another estimate for the same well, 60 mW m^{-2}, obtained by Blackwell et al. (1997) using precision temperature logs and typical values of thermal conductivity.

6.5.5. *Heat Flow and Electrical Conductivity*

Ádám (1978) correlated heat flow with the depths of electrically conducting layers in the crust and upper mantle. Ádám's 'first conducting layer' (FCL) is assumed to coincide with the onset of granitisation and melting in the crust, and the boundary between amphibolite and granulite facies. The 'intermediate conducting layer' (ICL) is apparently related to partial melting at the top of the asthenosphere. Ádám suggested that the depth, h, of each layer be related exponentially to regional heat flow, $\mathbf{q}$ (mW m^{-2}):

$$h = h_0 \times \mathbf{q}^{-a} \tag{6.33}$$

where

> $h_0 = 4493$ km, $a = 1.30$ for the FCL
> $h_0 = 36,167$ km, $a = 1.46$ for the ICL

Magnetotelluric sounding techniques may reveal both h_{FCL} and h_{ICL} at a single locality, giving two constraints for heat flow where no other data may be available.

Question: An electromagnetic sounding at a particular location provided evidence of a significant electrical conductor at a depth of 21 ± 1 km in the crust, and another at a depth of 85 ± 2 km. What regional heat flow do these conductive layers suggest?

Answer: We assume that the conductive layers are Ádám's (1978) FCL and ICL, respectively, in which case $h_{FCL} = 21 \pm 1$ km and $h_{ICL} = 85 \pm 2$ km. We can use Equation (6.33) to estimate $\mathbf{q}$ for each layer:

$$h_{FCL} = 21 \pm 1 = 4493 \times \mathbf{q}^{-1.30}$$

So:

$$\mathbf{q}_{FCL} = \exp[\ln((21 \pm 1)/4493)/-1.30] = 62.0 \pm 2.3 \text{ mW m}^{-2}$$
$$h_{ICL} = 85 \pm 2 = 36{,}167 \times \mathbf{q}^{-1.46}$$

So:

$$\mathbf{q}_{ICL} = \exp[\ln((85 \pm 2)/36167)/-1.46] = 63.2 \pm 1.0 \text{ mW m}^{-2}$$

In this example the two values agree closely with each other, but this may not always be the case.

6.6. Summary

The average heat flow over an interval is the product of the average thermal gradient × average thermal conductivity over the same interval. The average thermal gradient is calculated using only the temperatures at the top and bottom of the interval. The average thermal conductivity is usually the harmonic mean of the conductivities of all layers within the depth interval. High-precision temperature logs allow a best estimate of heat flow to be obtained using only those formations for which accurate conductivity data are available. The *in situ* thermal conductivity of shale layers can be determined by inversion of the heat flow estimate.

A graph of temperature versus thermal resistance is known as a Bullard plot. In a steady-state, conductive heat regime with negligible heat production, all points on a Bullard plot should fall along a straight line, the gradient of which is equal to surface heat flow. Deviations from a straight line may arise from systematic errors in estimating thermal conductivity, anomalously high heat generation, fluid migration, the presence of hydrocarbons, or historical changes in surface conditions.

Non-horizontal surface boundaries, surface boundaries with laterally varying temperature, non-horizontal contacts between formations and formations with laterally varying thermal conductivity all distort heat flow away from vertical. In particular, heat is refracted away from regions of thick sediment cover and preferentially channelled through areas of elevated basement. Also, in mountainous terrain, heat preferentially flows into valleys and surface heat flow varies with elevation.

Diapiric salt columns act to increase surface heat flow in at least three ways. Firstly, salt has a higher thermal conductivity than most other lithologies. Secondly, salt bodies generally resemble vertical dykes more than horizontal layers. Thirdly, salt may transport heat via advection if its rate of ascent is significant.

There are clear associations between heat flow and surface heat production, geologic age, seismic velocity and electrical conductivity. Empirical linear relationships between surface heat flow and heat generation in local basement rocks have been established for numerous regions around the globe. Mean surface heat flow progressively decreases with the age of basement rocks in both continental and oceanic settings. Linear relationships have been identified between crustal heat flow and upper mantle P-wave velocity, and between crustal heat flow and seismic travel time residuals. The depth to electrically conducting layers in the crust and upper mantle is related exponentially to crustal heat flow.

CHAPTER SEVEN

Lithospheric Models

The true law of increase of temperature is inextricably mixed up with the question of the solidity or otherwise of the interior. And if the law of increase of temperature be so, it is evident that the law of cooling, upon which it depends, is also mixed up with the question of the condition of the interior; and the contraction depends on the law of cooling, and the compression on the contraction; so that all these questions are interdependent.

Physics of the Earth's Crust – Rev. Osmond Fisher, 1881, p. 58.

Individual measurements of heat flow, or even a number of measurements made over a region, only have relevance to the immediate area. To understand the significance of individual heat flow measurements in a global sense, we need to investigate the geothermal signatures of various tectonic features. Only in this way can anomalous locations be identified and understood.

Crustal tectonism is the surface expression of phenomena that operate on the lithospheric scale. The *lithosphere* is that section of the upper mantle that exhibits gross plastic behaviour under stress. It can be thought of as a raft of mantle material upon which the crust rides. As such, it is the dominant control on surface heat flow in oceanic crust, and it contributes approximately half of the surface heat flow in continental areas (the other half arises from radiogenic sources within the crust). Tectonic features such as hot spots, subduction zones, and regions of crustal extension can be modelled on the lithospheric scale. Such models provide insights into the occurrence of volcanism and the distribution of heat in sedimentary basins. An understanding of the process of magmatic underplating of the crust is also necessary to fully appreciate the thermal evolution of many localities.

Sherby (1962) showed that the lithosphere lies above the $0.75T_m$ isotherm, where T_m is the absolute pyrolite melting temperature ($T_m \approx 1775$ K; Ringwood, 1969). As temperatures approach T_m, silicates undergo a transition from plastic to viscous behaviour, and at temperatures greater than T_m, mantle material deforms readily and flows under small pressure gradients (Oxburgh and Turcotte, 1976). Sclater and Francheteau (1970) predicted that temperatures rise at a rate to reach the critical $0.75T_m$ level at a depth near 90 km. This depth is consistent with the boundary layer theory of Turcotte and Oxburgh (1969). Mass transfer (convection), rather than steady-state conduction, probably dominates heat transfer at points deeper than the T_m isotherm.

7.1. Stable Lithosphere

Construction of typical lithospheric thermal models must progress separately for oceanic and continental crust. Heat flow through oceanic crust arises from different sources, and is affected by different factors, than those of continental crust. These factors include crustal thickness, age, composition, and the nature of the underlying lithosphere and mantle.

7.1.1. *Oceanic Lithosphere*

The oldest sections of oceanic lithosphere comprise 6–8 km of oceanic crust overlying ultramafic rocks. The uppermost kilometre of crust is sedimentary, overlying basaltic extrusive rocks, which in turn rest upon a mafic and ultramafic complex (Gass, Smith and Vine, 1975). The youngest sections of oceanic lithosphere are at mid-ocean ridge spreading centres, where progressive development of new lithosphere displaces older material away from the point of origin. Over time, the new lithosphere cools, increases in density, contracts and subsides. Three trends should therefore be apparent as the age of oceanic lithosphere increases. There should be an increase in the thickness of the lithosphere, a decrease in surface heat flow, and a deepening of the oceans with thermal subsidence. In most cases, age is directly proportional to distance from a spreading centre, so these trends should be related to distance from a mid-ocean ridge.

The cooling of oceanic lithosphere can be investigated using a simple half-space model. The model approximates a vertical column of magma penetrating to the surface and beginning to cool by conduction once emplaced (Figure 7.1). Basalt and mafic minerals are largely free from heat-producing elements so heat production within oceanic lithosphere may be neglected. The sea floor is defined to be at depth $z = 0$, and at constant temperature, $T = T_0$. All points below the sea floor represent the mantle. The model assumes constant thermal diffusivity, κ, with depth, and an initial constant temperature, $T = T_1$ at time $t = 0$.

This is a one-dimensional problem in space, and the solution for temperature, T, at depth, z, and time, t, is

$$T = (T_1 - T_0) \times \mathrm{erf}[z/(2\sqrt{(\kappa t)})] + T_0 \qquad (7.1)$$

where erf( ) is the error function [see Equation (3.34)].

Equation (7.1) implies that the temperature (T) is the same at all points with equal ($z/\sqrt{t}$). An important conclusion from this observation is that the depth to any specific isotherm is proportional to the square root of lithospheric age.

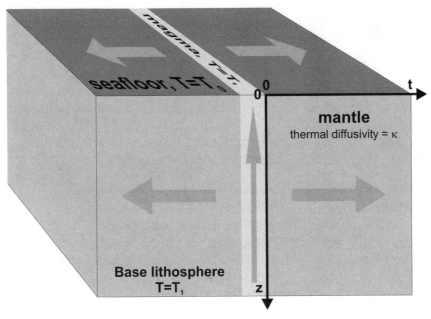

Figure 7.1. Half-space model of cooling oceanic lithosphere.

Question: If 16-Ma-old lithosphere is 35 km thick, how thick will it
be when it is 64 Ma old?

Answer: Lithospheric thickness is defined as the depth to a particular isotherm. Isotherms lie along lines of equal $z/\sqrt{t}$. So a four-
fold increase in t (from 16 Ma to 64 Ma) implies that z must
double. The lithosphere will be 70 km thick at 64 Ma.

If the base of the lithosphere, $L(t)$, is defined by the $0.75T_1$ isotherm, then
by transposing Equation (7.1):

$$L(t) = 2\sqrt{\kappa t} \times \text{erf}^{-1}\left(\frac{0.75T_1 - T_0}{T_1 - T_0}\right) \tag{7.2}$$

The thickness of the oceanic lithosphere derived from Rayleigh wave dispersion data beneath the Pacific Ocean (Leeds, Knopoff and Kausel, 1974)
corresponds closely to the theoretical thickness predicted for a melting temperature $T_1 = 1700$ K. If this is the correct figure, and we assume $T_0 = 275$ K,
then $\text{erf}^{-1}((0.75T_1 - T_0)/(T_1 - T_0))$ is a constant approximately equal to
0.7355, and

$$L(t) = 1.471\sqrt{(\kappa t)} \tag{7.3}$$

Oceanic lithosphere typically has thermal diffusivity $\kappa = 1.5 \times 10^{-6}$ m^2 s^{-1},
implying

$$L(t) = 10.12\sqrt{t} \tag{7.4}$$

where $L(t)$ is in kilometres and t is in millions of years (Ma). This relation predicts the thickness of 80-Ma-old lithosphere to be about 90 km.

Figure 7.2 illustrates the family of theoretical geotherms within oceanic lithosphere, assuming $T_1 = 1700$ K and $T_0 = 275$ K. Also shown are the $T_{0.75}$ solidus curves suggested by Ringwood (1969) and Oxburgh and Turcotte (1976). The base of the lithosphere is defined as where the geotherms cross the $T_{0.75}$ line, so the thickness of the lithosphere depends on which of the two melting models is adopted. In the absence of independent data, it is best to view the two melting models as upper and lower limits of $T_{0.75}$.

Question: What is the theoretical thickness of 9-Ma-old lithosphere?
Answer: Equation (7.4) suggests that the thickness of the lithosphere
  should be about 30 km. Figure 7.2 suggests a value between 23 km
  [the depth at which the 9 Ma geotherm crosses the Ringwood
  (1969) solidus] and 34 km [the depth at which the 9 Ma geotherm
  crosses the Oxburgh and Turcotte (1976) solidus].

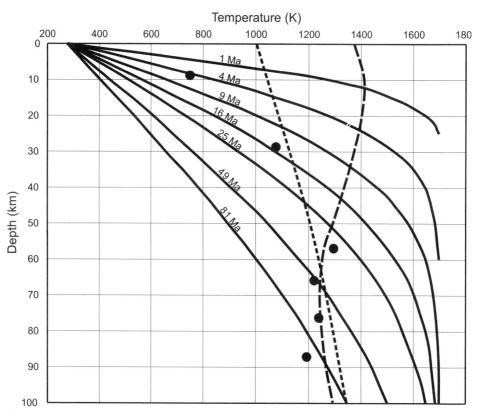

Figure 7.2. Theoretical isotherms calculated for the half-space cooling model. The $T_{0.75}$ solidus curves are as suggested by Ringwood (1969; short-dashed line) and Oxburgh and Turcotte (1976; long-dashed line). The thickness of the oceanic lithosphere under the Pacific Ocean is from studies of Rayleigh wave dispersion data (Leeds et al., 1974; circles).

The thermal gradient is found by differentiating Equation (7.1):

$$\frac{\partial}{\partial z}\left[(T_1 - T_0) \times \text{erf}\left(\frac{z}{2\sqrt{\kappa t}}\right)\right] = (T_1 - T_0) \times \frac{1}{2\sqrt{\kappa t}} \times \frac{2}{\sqrt{\pi}}\exp\left(\frac{-z}{4\kappa t}\right) \quad (7.5)$$

At the surface, $z = 0$. It follows that surface heat flow, $\mathbf{Q}_0$, can be written

$$\mathbf{Q}_0 = \frac{\lambda(T_1 - T_0)}{\sqrt{\pi \kappa t}} \tag{7.6}$$

where λ is the thermal conductivity of the oceanic lithosphere.

Equation (7.6) indicates that surface heat flow is *inversely* proportional to the square root of age. Heat flow data from the Pacific Ocean (Sclater, Crowe and Anderson, 1976; Parsons and Sclater, 1977) suggest the following empirical relation:

$$Q(t) = 470/\sqrt{t} \tag{7.7}$$

where $\mathbf{Q}$ is in milliwatts per square metre (mW m^{-2}), and t is in millions of years (Ma). This, in turn, suggests that oceanic lithosphere has an average thermal conductivity of about $4.0\ \text{W m}^{-1}\text{K}^{-1}$.

Isostatic balance requires that younger, hotter and less dense lithosphere must float higher on the underlying asthenosphere. It follows that water depth should gradually increase with the age of the lithosphere. The half-space cooling model predicts that water depth, h_w, should approximately conform to

$$h_w = h_0 + L\left(\frac{\rho_L - \rho_a}{\rho_a - \rho_w}\right) \tag{7.8}$$

where h_0 is water depth above a spreading centre (m); L is thickness of the lithosphere (m); and ρ_L, ρ_a, ρ_w are density of lithosphere, asthenosphere and water, respectively (kg m^{-3}).

We know from Equation (7.2) that L is proportional to the square root of age, so Equation (7.8) should follow a similar trend:

$$h_w = h_0 + A\sqrt{t} \tag{7.9}$$

where A is a constant.

Observed ocean depths (Parsons and Sclater, 1977) suggest the following empirical relationship:

$$h_w = 2500 + 350\sqrt{t} \tag{7.10}$$

where t is in millions of years and h_w is in metres.

Although simple in concept and practice, empirical evidence suggests that the half-space model breaks down for ages greater than about 70–100 Ma. In particular, there is an apparent flattening of bathymetry at about 6 km below sea level (Figure 7.3), regardless of increasing age. This observation implies that the oceanic lithosphere stops cooling after about 100 Ma. Parsons and

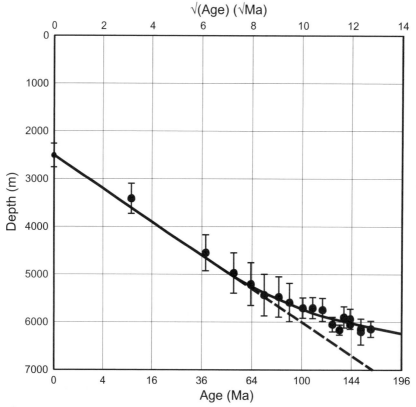

Figure 7.3. Mean ocean depth and standard deviation versus square root of age for the northern Pacific and Atlantic oceans. Dashed line shows prediction for half-space model, solid line is prediction for plate model. Modified after Parsons and Sclater (1977).

Sclater (1977) proposed the *plate model*, which allows a lower temperature boundary to be defined.

In the plate model, a specified lower boundary (at depth, $z = L$) is kept at a constant temperature ($T = T_1$; the half-space model is a special case of the plate model where $L = \infty$). The initial condition (at $t = 0$) is the same as for the half-space model. That is, $T(z) = T_1$ for all z values. The lower boundary condition, however, prevents indefinite cooling, and the plate model converges towards a steady-state solution. If we assume surface temperature is $T_0 = 0°C$, the steady-state solution is

$$T(z) = T_1 \times (z/L) \tag{7.11}$$

where $T(z)$ and T_1 are in degrees Celsius and $0 \leq z \leq L$.

This steady-state solution satisfies both the energy equation and the boundary conditions, but is only valid at $t = \infty$. At real times, the temperature distribution within the slab may be thought of as the sum of the steady-state solution and a transient term that decays with time. The solution for the temperature profile at time t is found using Fourier analysis:

$$T(z, t) = T_1\left(\frac{z}{L}\right) + T_1 \sum_{n=1}^{\infty}\left[a_n \sin\left(\frac{n\pi z}{L}\right)\exp\left(-\kappa\left(\frac{n\pi}{L}\right)^2 t\right)\right] \tag{7.12}$$

where $a_n = 2(-1)^{n+1}/n\pi$.

Each term in this series has a decay constant proportional to n^2, so the higher harmonic terms ($n > 1$) rapidly diminish. A solution to a close approximation at times greater than $t \approx L^2/(\pi^2\kappa)$ is

$$T(z, t) = T_1\left(\frac{z}{L}\right) + T_1 \times \frac{2}{\pi}\sin\left(\frac{\pi z}{L}\right)\exp\left(-\kappa\left(\frac{\pi}{L}\right)^2 t\right) \tag{7.13}$$

where all values of $n > 1$ have been dropped.

For young oceanic lithosphere, the plate model gives the same solutions as the half-space model for bathymetry, lithospheric thickness and heat flow. Differences become apparent at ages older than about 70 Ma. Parsons and Sclater (1977) found that topographic data from the world's oceans supported a plate-type model with a thickness of around $L = 125$ km and a basal temperature of $T_1 = 1350°$C (Figure 7.3). Such a model predicts bathymetry, h_w, to exponentially approach a constant background level of about 6400 m:

$$h_w = 6400 - 3200\exp(-t/62.8) \tag{7.14}$$

for $t > 20$ Ma.

Stein and Stein (1992) re-examined the problem with the benefit of many new oceanic heat flow and bathymetry measurements. They calculated a better fit to the collated heat flow and bathymetry using figures of $L = 95$ km and $T_1 = 1450°$C. Their model predicted a flattening of bathymetry and surface heat flow to background levels of about 5600 m and 48 mW m^{-2}:

$$h_w = 2600 + 365\sqrt{t}\,(\text{m}), \quad \text{for } t < 20 \text{ Ma}$$
$$h_w = 5651 - 2473\exp(-0.0278t)\,(\text{m}), \quad \text{for } t > 20 \text{ Ma} \tag{7.15}$$

$$Q(t) = 510/\sqrt{t}\,(\text{mW m}^{-2}), \quad \text{for } t < 55 \text{ Ma}$$
$$Q(t) = 48 + 96\exp(-0.0278t)\,(\text{mW m}^{-2}), \quad \text{for } t > 55 \text{ Ma} \tag{7.16}$$

Note: Some authors (e.g. Heestand and Crough, 1981) have suggested that data for ages greater than about 100 Ma are anomalous, implying that there may be no physical justification for using a plate model over a half-space model.

Equation (7.16) predicts extremely high values of conductive heat flow for young, hot lithosphere close to a spreading centre. However, such values are not observed in the global data set (Pollack et al., 1993). The discrepancy is due to hydrothermal circulation of seawater, which makes up approximately 34% of the total oceanic heat flux (Stein and Stein, 1994). Hydrothermal convection is thought to be present in oceanic lithosphere extending from spreading centres out to ages of 65 ± 10 Ma, although the magnitude of the effect diminishes rapidly away from the ridge. Approximately 30% of the hydrothermal heat flux is concentrated in lithosphere less than one million years old (Figure 7.4; Stein and Stein, 1994).

The existence of hot-spot swells (see Section 7.2) has consequences for oceanic lithosphere cooling models. Considering that there are over fifty active hot spots with associated swells on the sea floor, and each swell has a diameter averaging 1200 km, over 15% of the sea floor may, in fact, be elevated relative to simple cooling model predictions. Heestand and Crough (1981) pointed out that data used by Parsons and Sclater (1977) to support the hypothesis of the plate model of lithospheric cooling include bathymetric measurements from regions of lithosphere older than eighty million years but affected by hot-spot swells. Heestand and Crough maintained that the apparent flattening of bathymetry at large ages is due entirely to the effect of hot-spot swells, and that data from regions away from the influence of swells fit a half-space model better. They suggest an age–bathymetry relationship:

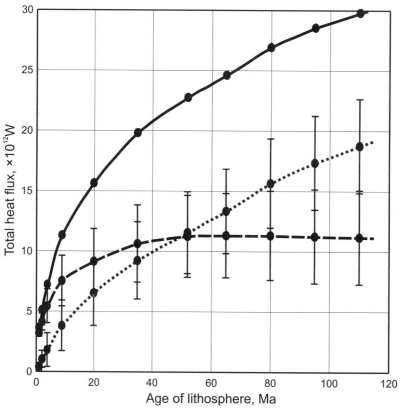

Figure 7.4. Cumulative total heat loss from oceanic lithosphere. Solid line shows amount predicted by thermal models, dotted line is observed conductive heat loss, dashed line is implied hydrothermal heat loss. Modified after Stein and Stein (1994).

$$h_w = 2700 + 295\sqrt{t} \tag{7.17}$$

where h_w is in metres and t in millions of years.

The relationship gives depths very close to those of Stein and Stein's (1992) plate model for ages of 100 Ma [compare with Equation (7.15)], and does not break down at large ages.

7.1.2. Continental Lithosphere

The continental crust is fundamentally different from oceanic crust. Some of the important differences that directly impact on thermal history include the distribution of heat-producing elements, age, thickness and composition. It is not possible to produce a typical geotherm for continental crust because of the large global variation in each of these factors. However, the effect of each factor on the geotherm can be examined.

Various authors have suggested heat generation models for continental crust over the years (e.g. Birch et al., 1968; Lachenbruch, 1968, 1970; Roy et al., 1968; Allis, 1979; Jaupart et al., 1981; Percival and Card, 1983; Arshavskava et al., 1987; Drury, 1989; Cermák et al., 1991; Vigneresse and Cuney, 1991). These were discussed in detail in Chapter 2, so only a short summary is presented here.

Observed linear relationships between surface heat flow and surface heat production on continental crust can be explained by either a step-wise or exponentially decreasing distribution of heat-generating elements with depth. Real data from very deep boreholes and exposed lower crustal sections suggest that a layered model is more realistic (see Section 2.1.1). The chosen model and values of heat generation determine the shape of the calculated geotherm (Figure 7.5). Regardless of the particular model, however, the total contribution of radiogenic heat to surface heat flow, $\mathbf{Q}_R$, is always

$$\mathbf{Q}_R = \int A(z)dz \tag{7.18}$$

where $A(z)$ is heat generation as a function of depth.

The age and thickness of continental crust affect surface heat flow mainly through their effect on crustal heat generation. In general, radiogenic elements in older crust are depleted through natural decay, and as a direct consequence older crust typically exhibits lower surface heat flow (e.g. Sass and Lachenbruch, 1979; Teichmüller and Teichmüller, 1986; Figure 7.6) and thermal gradients. Likewise, the volume of heat-producing material is directly proportional to the thickness of the crust. Therefore, thicker crust contributes more heat than thinner crust of the same age.

Crustal thickness affects thermal gradient in other ways, depending on the conditions at the crust–mantle boundary. If the base of the crust is a constant temperature boundary, then thicker crust results in lower thermal gradient and heat flow. If the base of the crust is a constant heat flux boundary, with a constant flow of heat from the mantle, then crustal thickness has little effect on

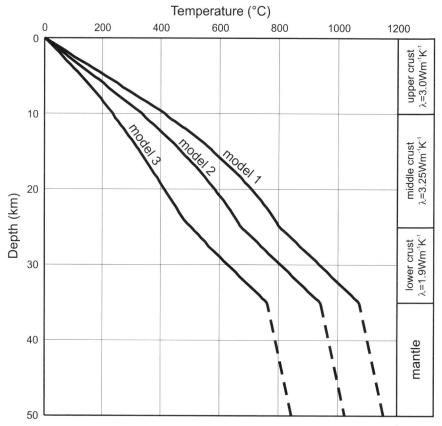

Figure 7.5. Continental geotherms calculated for a basal heat flow of 50 mW m^{-2} using different models of heat generation. Model 1 – upper crust, 1 μW m^{-3}; middle crust, 4.5 μW m^{-3}; lower crust, 0.1 μW m^{-3}. Model 2 – upper crust, 0.75 μW m^{-3}; middle crust, 3.0 μW m^{-3}; lower crust, 0.1 μW m^{-3}. Model 3 – exponential decay of heat generation with depth down to 35 km, $A = 3.5 \times \exp(-0.1016z)$, where z is in kilometres.

thermal gradient beyond that of heat production. In reality, the crust–mantle boundary condition probably lies somewhere between these two end-members and the effect of crustal thickness on thermal gradient varies from region to region.

The effect of crustal composition on the geotherm is the last factor to be examined. The upper continental crust is inhomogeneous and exhibits large variations in conductivity. Excluding sedimentary sections, however, it generally has high thermal conductivity due to the high proportion of quartz-rich rocks. In contrast, the middle and lower crust (deeper than about 5 km) can be modelled with an average composition of pyroxene granulite. The thermal conductivity is described by (Cull, 1975)

$$\lambda = (0.3788 + 0.000283T)^{-1} + 5.86 \times 10^{-10}P \ \text{W m}^{-1}\text{K}^{-1} \qquad (7.19)$$

where T is temperature (K) and P is pressure (Pa).

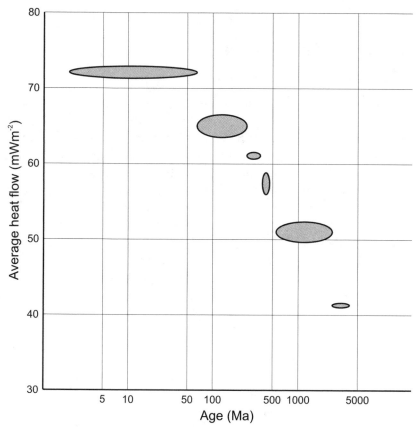

Figure 7.6. Average heat flow as a function of tectonic age. Height of ellipse reflects standard error. Data from Jessop (1990).

Beneath the crust, the upper mantle is considered to be dunite, and to have a total thermal conductivity (accounting for both phonon conduction and radiation components as described in Section 4.1) expressed by (Cull, 1975)

$$\lambda = (0.1433 + 0.000193T)^{-1} + 6.70 \times 10^{-10}P \ \mathrm{W \, m^{-1} \, K^{-1}} \qquad (7.20)$$

Equation (7.20) is valid up to about 1000 K and 2.5×10^{9} Pa (25 kbar, ~90 km), beyond which experimental data are insufficient to justify extrapolation. At any rate, convection is probably the dominant heat transport mechanism beyond those limits, in which case conduction plays only a minor role.

Question: What is the approximate thermal conductivity of the middle crust at 10 km depth and 250°C, and the upper mantle at 75 km and 750°C?

Answer: From Equation (7.19), with $T = 523$ K and $P = 2.8 \times 10^{8}$ Pa:

$$\lambda = (0.3788 + 0.000283 \times 523)^{-1} + 5.86 \times 10^{-10} \times 2.8 \times 10^8 =$$
$$2.062 \text{ W m}^{-1}\text{K}^{-1}$$

From Equation (7.20), with $T = 1023$ K and $P = 2.1 \times 10^9$ Pa:

$$\lambda = (0.1433 + 0.000193 \times 1023)^{-1} + 6.70 \times 10^{-10} \times 2.1 \times 10^9 =$$
$$4.342 \text{ W m}^{-1} \text{ K}^{-1}$$

Thermal conductivity is temperature dependent so an iterative procedure is required to determine an equilibrium thermal profile for continental crust. Schatz and Simmons (1972) and Cull (1975) both attempted such iterations. Cull included the effect of the pressure derivative, which was ignored by Schatz and Simmons. Cull's model better fits xenolith data published by Boyd (1973) and Mercier and Carter (1975), and is thus regarded as a better basis for thermal modelling (Figure 7.7).

7.2. Hot Spots

By far the great majority of volcanic activity on the Earth's surface is associated with tectonic plate margins. In fact, more than 99% of volcanoes are found along spreading centres and subduction zones (Burke and Wilson, 1976). The remaining few are found scattered over isolated parts of the oceans and stable cratons. These few anomalous volcanoes defied credible explanation until the development of plate tectonic theory in the 1960s.

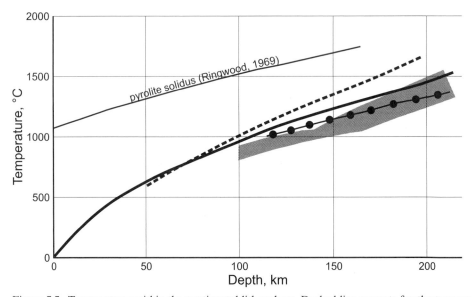

Figure 7.7. Temperatures within the continental lithosphere. Dashed line corrects for the temperature dependence of thermal conductivity (Schatz and Simmons, 1972). Solid line corrects for both temperature and pressure dependence (Cull, 1975). Xenolith data from Boyd (1973; shaded area) and Mercier and Carter (1975, circles) support the latter approach.

7.2.1. Cause of Hot Spots

The presently accepted theory explaining the origin of isolated volcanoes was first suggested by Wilson (1963), who noted that the Hawaiian chain of islands is roughly straight, increases in age towards the north-west and is actively volcanic at the eastern extremity. He rejected the solid-Earth notion that the islands lie along a linear, slowly extending fault zone and proposed that they are the surface expression of a relatively stationary plume of hot mantle material, or *hot spot*, over which the Pacific Ocean lithosphere is drifting in a north-westerly direction.

Wilson's hypothesis has been generally accepted, although the origin of hot-spot magma remains contentious. Hot-spot basalt is enriched in elements (e.g. potassium) incompatible with typical mid-ocean ridge and island arc basalts, implying a different source. Vink, Morgan and Vogt (1985) pointed out that a relatively small increase in temperature (∼100°C) dramatically decreases the viscosity of mantle material, which might accumulate and rise within a narrow conduit to the surface. Regions of anomalously high levels of radioactive material (e.g. potassium) can cause such a temperature increase, but it is unclear whether the bases of such plumes are in the upper or lower mantle, in the centre of convection cells, or in some other position.

At least 122 hot spots active within the past 10 Ma have been identified to date (Burke and Wilson, 1976; Figure 7.8). Only 53 of these are located

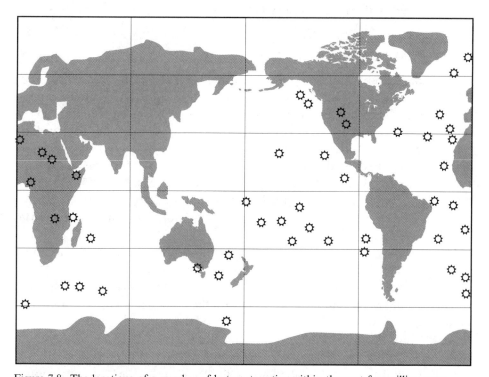

Figure 7.8. The locations of a number of hot spots active within the past few million years.

beneath oceanic basement, while 69 lie beneath continental crust. Considering that the continents account for only about one third of the Earth's surface, this distribution does not appear random. Even amongst the continents themselves, distribution is skewed. Africa holds the lion's share with no fewer than 25 hot spots identified beneath the continental mass, and a further 18 in nearby seas.

The distribution of hot spots provides a clue that absolute plate motions may influence their formation. Sea-floor magnetic anomalies, coastline correlation and plate margin maps make it reasonably simple to deduce the relative motion of one plate with respect to its neighbours. However, the absolute motion of the plates with respect to the Earth's interior can not be deduced from these observations. Hot spots, on the other hand, are assumed to remain relatively stationary with respect to the interior, and thus provide a measurement of absolute plate velocities. The available data suggest that the African continent is virtually stationary, and may be acting somewhat like a blanket. Heat trapped within the mantle results in a temperature increase and a lowering of viscosity, triggering the greater number of observed mantle plumes.

Occasionally a hot spot coincides with a spreading centre. The composition of the basalt at such locations resembles hot-spot basalt more than mid-ocean ridge basalt, and volumes of lava are much greater than those observed at normal spreading centres. Iceland is the most dramatic example of this phenomenon. Iceland lies along the line of the Mid-Atlantic Ridge, but sits several kilometres higher than the average ridge elevation.

7.2.2. *Thermal Effect of Hot Spots*

When passing over a hot spot, both continental and oceanic crust first swell and then gradually sink back to their former altitude. The region immediately surrounding the hot spot can be elevated by as much as 1600 m relative to nearby crust (e.g. Dietz and Menard, 1953), and the swelling typically covers an area on the order of 1200 km diameter. The swelling was first thought to be due to the dynamic support of the upwelling plume of mantle material (e.g. Anderson, McKenzie and Sclater, 1973). Detrick and Crough (1978), however, suggested that the uplift is an isostatic correction after cold, dense lithosphere is replaced by hot, buoyant material from the asthenosphere.

The isostatic theory was tested with a lithospheric heating and thinning model constructed to fit observed data (Detrick et al., 1981; Von Herzen et al., 1982, 1983). Ninety-five heat flow and bathymetry measurements from around Hawaii were compared with the stable oceanic lithosphere model of Parsons and Sclater (1977). The resulting model suggested a thinning of the lithosphere to about 45 km, representing approximately 50% reduction in thickness for 90 Ma lithosphere. The model further predicted a heat flow anomaly exhibiting a time lag of around five million years between uplift and associated heat flow increase. Once beyond the influence of the hot spot, the lithosphere cools, thickens and subsides to its original elevation (Figure 7.9).

Stein and Stein (1992) subsequently reassessed the magnitude of the heat flow anomaly around Hawaii and concluded that it is insufficient to support the thinned lithosphere theory, once again bringing the dynamic support theory back into favour. Seismic data from across the Ninetyeast Ridge (a 5000-km long-hot-spot track in the Indian Ocean) suggest that the passage of a hot spot is associated with a thickening of the oceanic crust from about 7 km in adjacent ocean basin to around 25 km above the ridge (Flueh and Grevemeyer, 1999). A reflector approximately 10 km below the Moho was interpreted as the base of hot-spot-related crustal underplating.

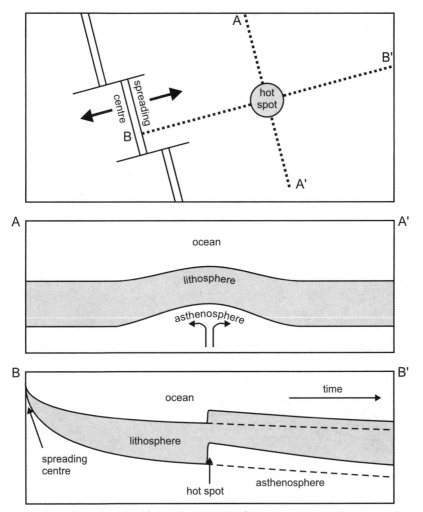

Figure 7.9. Transverse (A–A′) and parallel (B–B′) cross sections of the effect of a hot-spot heat source beneath a moving lithospheric plate. Modified after Detrick and Crough (1978).

> ***Note:*** Although hot spots are responsible for only a small propor-
> tion of the total heat loss from the Earth, they nevertheless have a
> significant influence on topography and the distribution of surface
> heat flow. This influence must be recognised and allowed for when
> real data are used to constrain thermal models of the lithosphere.

7.3. Subduction Zones

Until this point we have restricted our discussion to the thermal state of tectonic plates in their passive phase. We will now continue the discussion with an
examination of the thermal structure of convergent plate boundaries, known
collectively as subduction zones.

7.3.1. Driving Forces

Buoyancy related to the distribution of temperature is most likely the dominant driving force for tectonic plate motions. The effect of mantle convection is
little understood. It may act to help drive the plates, but is equally likely to act
at 90° to the plate motion.

The two main buoyancy forces are *ridge-push* forces at spreading centres
and *slab-pull* forces at subduction zones. Hot, light asthenosphere upwelling
beneath a spreading centre pushes the cooling lithosphere apart (Figure 7.10),
resulting in a ridge-push force acting on all parts of the cooling oceanic lithosphere. Meanwhile, cold subducting slabs have large negative thermal buoyancy relative to their surroundings, and thus pull the lithosphere towards the
subduction zones. In addition, there is a phase change from olivine to spinel at
a depth of around 380 km, with an associated increase in density. The temperature of the olivine–spinel transition increases with pressure (Figure 7.11),
causing the phase change to take place at a shallower depth in a cool subducting slab (Figure 7.12) relative to surrounding lithosphere. This results in extra

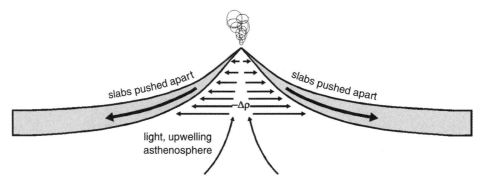

Figure 7.10. Diagrammatic illustration of the ridge-push force on lithospheric plates. Hot, light,
upwelling asthenosphere is less dense than the cooling lithospheric plates and pushes them apart.

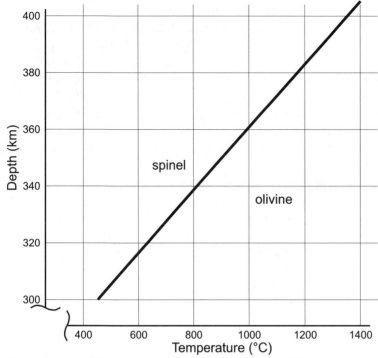

Figure 7.11. Pressure–temperature stability fields for the olivine–spinel system.

downward force acting on the plate. General observations suggest that the speed of plates is proportional to the length of the subducted slabs, implying that the slab-pull force plays a significant role in plate motion.

Note: No evidence of acceleration has been observed on any tectonic plate, so frictional forces on the plate boundaries must exactly equal the driving forces.

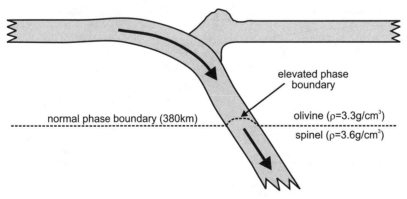

Figure 7.12. Diagrammatic illustration of the slab-pull force on a subducting plate. Phase change to denser spinel occurs at an elevated depth, creating negative buoyancy within the slab.

Uyeda and Kanamori (1979) identified fundamental differences between the two types of subduction zones involving oceanic lithosphere. The first, oceanic–continental collisions, they classified as Chilean-type. These subduction zones have a trench very close to a continental mass, a large accumulation of intensely folded sediment on the leading edge of the continental slab, a shallow angle of subduction, a gravity high on the seaward side of the trench, and large, compressional earthquakes. In contrast, oceanic–oceanic (Mariana-type) collisions occur far from a continental mass, have minimal sediment accumulation, a steep subduction angle and exhibit no seaward gravity high or large earthquakes.

The two types of subduction are explained in terms of absolute plate motions. South America is drifting to the west and overriding the Pacific Plate, which is being buckled and forced downward under considerable compressive stress. Sediments are being scoured from the surface of the Pacific Plate. The Asiatic Plate, on the other hand, is relatively motionless, and slab-pull forces dominantly drive subduction of the Pacific Plate. Elsasser (1971) pointed out that the net force acting on the hinge of the subducting plate is downward and seaward, so the subduction zone should migrate seaward over time.

The generalised heat flow profile across a subduction zone is shown in Figure 7.13. The profile can be divided into three sections, each with its own thermal signature. From the seaward side of the zone, these sections are the trench, the island arc and the back-arc basin. Each of these is examined in more detail below.

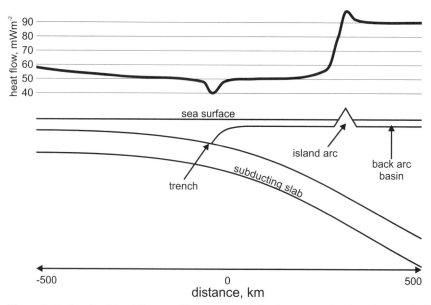

Figure 7.13. Idealised heat flow profile across a subduction zone, showing the trench, island arc volcanism and back-arc basin.

7.3.2. Trenches

A trench is the surface expression of the line at which a subducting slab bends and begins its downward movement. These trenches include the deepest parts of the ocean floor. The thermal conditions above and immediately landward of a trench are related to the reheating of old, cold lithosphere as it sinks into the hot mantle. Heat absorbed by the subducting slab does not reach the Earth's surface. The downward advection of heat with the moving slab further decreases surface heat flow. It follows that surface heat flow should be significantly lower than for most other tectonic regions. Data collected from the vicinity of subduction trenches strongly support this conclusion, with a global average around 45 mW m^{-2} (Jessop, 1990), lower than the average for any other tectonic setting.

The temperature within the subducting slab can be modelled as the inverse of the lithospheric cooling problem. In the subducting slab, the initial temperature profile is of the form $T = T_1 \times (z/L)$, and the boundary conditions are $T = T_1$ at $z = 0$ and $z = L$, the two surfaces of the plate. The solution is

$$T(z, t) = T_1 + T_1 \sum_{n=1}^{\infty}\left[b_n \, \sin\!\left(\frac{n\pi z}{L}\right) \exp\!\left(-\kappa\left(\frac{n\pi}{L}\right)^2 t\right)\right] \qquad (7.21)$$

where $b_n = [2/(LT_1)] \times \int_0^L [T(z, t = 0) - T(z, t = \infty)] \times \sin(n\pi z/L)\partial z$.

Only the equilibrium solution, $T(t = \infty) = T_1$, and the Fourier coefficients b_n, differ from Equation (7.12). Note that the origin for the time-scale ($t = 0$) is the moment of subduction.

7.3.3. Island Arcs

Island arcs are the products of volcanic activity associated with subduction zones. The cold subducting slab somehow promotes volcanism at an intermediate distance from the trench. For example, the distance from the Japan Trench to the Japanese line of volcanoes is in the range 200–300 km. The exact mechanism of this volcanism remains contentious and is irrelevant to this discussion. However, the fact that there is volcanism at all implies that heat flow is elevated.

Heat flow generally rises steeply at the onset of the volcanic arc, as illustrated in Figure 7.13. Typical values exceed 100 mW m^{-2} (Jessop, 1990), but that does not include the advective component related to rising magma.

7.3.4. Back-Arc Basins

At a considerable distance from a subduction trench, beyond the volcanic island arc, there is often a back-arc basin, which also exhibits high heat flow. Most research into back-arc basins has concentrated on the Japan Sea, which lies between the islands of Japan and the Asian continental landmass. Although associated with the subduction of the Pacific Plate beneath the Asian continental plate, the average heat flow in the Japan Sea is on the order of 100 mW m^{-2}. This is high by any standards and difficult to rationalise with the underlying, cold, subducting slab.

Elsasser (1971) suggested that the net force pushing the subduction zone seaward imposes tensional stresses on the back-arc area, initiating high heat flow and sea-floor spreading. Bodri and Bodri (1978), however, modelled asymmetric mantle flow in two wedge-shaped regions located either side of the subducting slab. They calculated two-dimensional time- and temperature-dependent flow patterns, and found that results are very sensitive to the rheological properties of the material. They were able to model island-arc volcanism and high heat flow in the back-arc basin if, but only if, they included non-linear rheological material properties such as yielding and plasticity. In other words, they concluded that the high heat flow may be the result of upper mantle convection.

Sato (1992) approached the problem from another direction. He investigated the thermal structure of the upper wedge of mantle material from its seismic properties and laboratory measurements on dry peridotite. Sato found that over the entire wedge, temperatures were below the melting point for dry peridotite; but beneath the Japanese chain of islands, temperatures were between the dry and wet solidus. Thus, partial melting is made possible by an injection of water from the cold, subducting slab. The uprising molten material produces magmatism and high heat flow in the back-arc basin. The subducting slab itself does not melt.

Note: The thermal structure and mechanisms of island-arc magmatism and high back-arc heat flow are not yet clearly understood. Much work must still be done to understand fully the observed heat flow distribution.

7.3.5. *Continental Collision*

It is generally held that continental crust is too light to be subducted ($\rho_c = 2.9 \text{ g cm}^{-3}$; compared with the mantle, $\rho_m = 3.3 \text{ g cm}^{-3}$). In the event of two sections of continental crust colliding, therefore, the only possible outcome is that the total volume of crust is maintained. This requires considerable crustal thickening and shortening. It is not surprising that the most dramatic topographic features on Earth are related to continental collision.

The Himalayan mountains and Tibetan Plateau, for example, are the result of isostatic uplift due to crustal thickening processes related to the collision of the Indian and Asian plates. Two-dimensional numerical models of deformation (Willett and Pope, 1996) suggest that the Himalayan crust passed through three distinct stages of growth. First was the development of a localised triangular section of uplift. As the lower crust heated up due to increased depth of burial, its viscosity decreased to the point where it could not support extreme topographic highs. The second stage of development, therefore, involved the formation of a topographic plateau with crustal shortening at all depths. The third stage began once the plateau reached a sufficiently large size, thickness and temperature. The lower crust then began outward flow from the centre of the plateau. The model predicts further outward growth of the Tibetan Plateau as a result of lower crustal flow, with no further net uplift or subsidence.

While the crust experiences shortening and thickening, the underlying lithosphere must also thicken, shorten and be pushed deeper. Such a situation has obvious thermal ramifications. The thermal time constant for lithospheric cooling is proportional to the square of the thickness, so if the thickness doubles it follows that cooling must take four times as long. This conclusion assumes that the lithosphere remains cohesive throughout the cooling process and is not broken up by buoyancy forces. Houseman, McKenzie and Molnar (1981) examined the problem numerically to determine the stability of a thick, cold lithosphere submerged in the asthenosphere. They found that for almost any combination of realistic parameters, the unstable lower lithosphere detaches rapidly (faster than crustal deformation, 30–50 Ma, in some zones) and sinks, to be replaced by hot, buoyant asthenospheric material. They concluded that in some cases the entire mantle lithosphere might detach from the lower crust during crustal shortening, exposing the crust to asthenospheric temperature. The overlying plate must warm rapidly, leading to regional metamorphism and the possible development of post-tectonic granites.

The above predictions are necessarily qualitative because many of the key parameters remain poorly constrained. However, the general conclusions remain valid. The lithosphere beneath continents is hotter, and therefore thinner and weaker, than oceanic lithosphere. It is only the cohesive strength of oceanic lithosphere that stops it from sinking into the asthenosphere, so it is not unreasonable to believe that the continental lithosphere does sink once the delicate gravitational balance is upset through collision and thickening.

7.4. Extension

Virtually every economic petroleum deposit is located within an active or past zone of tectonic extension. Not surprisingly, therefore, much effort has been expended to characterise the role that the thermal history of such zones plays in the generation of hydrocarbons. The variations in heat flow through time within specific extensional zones, however, are unique, and only the broadest of generalisations have so far been identified.

7.4.1. Instantaneous Pure Shear

The simplest model for extension is uniform pure shear, where a section of continental lithosphere extends by a factor, β, in one direction and thins to $1/\beta$ of original thickness (Figure 7.14). The parameter β is often referred to as the

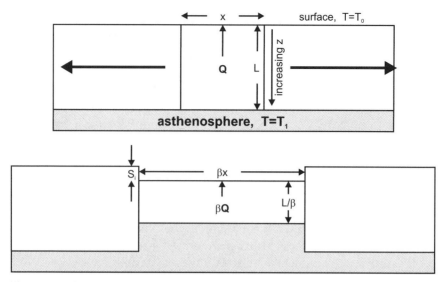

Figure 7.14. Pure shear extension by a factor β. Lithospheric volume is retained, and heat flow is increased due to increased thermal gradient.

stretching factor. This model is obviously a gross simplification of the physical mechanisms involved in extension, but is useful to give us an idea of the resulting thermal disturbance.

Note: Typical values of β are between 1 and 2 for intracontinental basins, while in continental margin basins we find a gradation in β from around 1 onshore to 4–5 offshore, before thinned continental crust gives way to oceanic crust and sea-floor spreading.

If extension is assumed to be instantaneous, the temperature profile changes from $T = T_0 + (T_1 - T_0) \times (z/L)$ to $T = T_0 + (T_1 - T_0) \times (\beta z/L)$, where T_0 is the surface temperature and T_1 is the temperature of the asthenosphere. The equations simplify if we measure T in degrees Celsius and assume $T_0 = 0°C$. Then, the thermal profile immediately following extension is $T = T_1 \times (\beta z/L)$. Thermal gradient and, thus, surface heat flow are increased by a factor of β. The temperature profile subsequently relaxes back to equilibrium, liberating a total amount of heat, **H**:

$$\mathbf{H} = 0.5 \times L \times \rho c \times T_1(1 - 1/\beta) \, \mathrm{J\,m^{-2}} \tag{7.22}$$

where ρ and c are the density and thermal capacity of the lithosphere, respectively, and T_1 is measured in degrees Celsius.

McKenzie (1978) gave the solution to the temperature profile at time t after extension. He showed that if $\beta = \infty$ the problem is identical to oceanic lithospheric cooling, and is closely analogous for other values of β:

$$T(z, t) = T_1\left(\frac{z}{L}\right) + T_1 \sum_{n=1}^{\infty}\left[a_n \sin\left(\frac{n\pi z}{L}\right)\exp\left(-\kappa\left(\frac{n\pi}{L}\right)^2 t\right)\right] \tag{7.23}$$

where

$$a_n = \frac{2\beta}{(n\pi)^2}\sin\left(\frac{n\pi}{\beta}\right)$$

Only the Fourier coefficients, a_n, differ from the lithospheric cooling model [Equation (7.12)]. In the case of pure-shear extension, they are a function of β. The surface heat flow is defined at time t by (Figure 7.15)

$$\mathbf{Q} = \lambda\frac{dT}{dz}\bigg|_{z=0} = \lambda\frac{T_1}{L}\left\{1 + 2\sum_{n=1}^{\infty}\left[\frac{\beta}{n\pi}\sin\frac{n\pi}{\beta}\right]\exp\left(-\kappa\left(\frac{n\pi}{L}\right)^2 t\right)\right\} \tag{7.24}$$

Question: What is the approximate surface heat flow 15 Ma after a 100 km thick section of continental lithosphere ($T_1 = 1425°C$, $\lambda = 4.0 \text{ W m}^{-1}\text{K}^{-1}$, $\kappa = 8 \times 10^{-7} \text{ m}^2 \text{ s}^{-1}$) is instantaneously stretched by a factor $\beta = 2.5$?

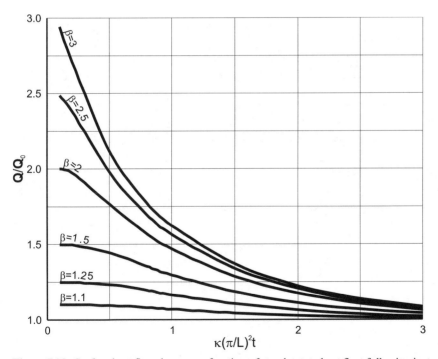

Figure 7.15. Surface heat flow decay as a fraction of steady-state heat flow following instantaneous extension by a factor of β. $\mathbf{Q}$ = heat flow, $\mathbf{Q}_0$ = steady-state heat flow, κ = thermal diffusivity of the continental lithosphere (m^2s^{-1}), L = thickness of the continental lithosphere prior to extension (m), t = time since extension (s).

Answer: For an approximate answer we can use Equation (7.24) and ignore all values of $n > 1$. Then:

$$Q = 4.0 \times (1425/10^5) \times [1 + 2 \times (2.5/\pi)\sin(\pi/2.5) \times \exp(-8 \times$$
$$10^{-7} \times (\pi/10^5)^2 \times 15 \times 3.1557 \times 10^{13})]$$
$$= 0.057 \times [1 + 2 \times (0.7568) \times \exp(-0.3737)]$$
$$= 0.057 \times 2.042$$
$$= 0.116 \text{ W m}^{-2} = 116 \text{ mW m}^{-2}$$

Extensional thinning results in an initial *isostatic subsidence* of the lithosphere, S_i, and associated upwelling of the asthenosphere. The magnitude of S_i depends on β, the initial thickness of the crust (t_c) and lithosphere (L), and the average density of the crust (ρ_c), mantle (ρ_m) and asthenosphere (ρ_a):

$$S_i = \left(1 - \frac{1}{\beta}\right)\frac{[t_c\rho_c + (L - t_c)\rho_m - L\rho_a]}{(\rho_w - \rho_a)} \tag{7.25}$$

where

$\rho_a = \rho_{m0}[1 - \alpha T_1]$ = density of the asthenosphere
$\rho_m = \rho_{m0}[1 - \alpha(T_0 + 0.5 \times (T_1 - T_0)(1 + t_c/L))]$ = density of the mantle
$\rho_c = \rho_{c0}[1 - \alpha(T_0 + 0.5 \times (T_1 - T_0)(t_c/L))]$ = density of the crust
ρ_w = density of the infilling surface load (e.g. seawater= 1050 kg m^{-3})
T_0 = surface temperature (°C)
α = thermal expansion coefficient of the lithosphere (K^{-1})
ρ_{m0} and ρ_{c0} = density of the mantle and crust, respectively, at 0°C

Question: Given that the crust is 30 km thick and has a density at 0°C of $\rho_{c0} = 2800$ kg m^{-3}, the mantle density at 0°C is $\rho_{m0} = 3330$ kg m^{-3}, and $\alpha = 3.28 \times 10^{-5}$, what is the initial subsidence expected in the previous example (assume $T_0 = 0$°C and the load is seawater)?

Answer: From the previous example, $T_1 = 1425$°C, $L = 100,000$ m and $\beta = 2.5$. So:
$\rho_a = 3174.36$ kg m^{-3}, $\rho_m = 3228.83$ kg m^{-3}, $\rho_c = 2780.37$ kg m^{-3} and $\rho_w = 1050$ kg m^{-3}
Thus, from Equation (7.25):
$S_i = (1 - 0.4)(30,000 \times \rho_c + 70,000 \times \rho_m - 100,000\rho_a)/(\rho_w - \rho_a) = 2261$ m

---

Immediately after rifting, the lithosphere begins to cool, becomes denser and contracts. The shrinking leads to further subsidence at the surface and may continue for up to 100 Ma. This second phase of subsidence is usually referred to as *thermal subsidence*, S_t. Where the load is only water, total vertical subsidence is on the order of 1–3 km. However, basins tend to form a focus for sediment deposition. A sediment load alters the isostatic balance and amplifies the vertical scale of subsidence, in some cases to depths of 12–16 km.

Note: The rate of subsidence and sedimentation during the thinning phase can be up to 1 km Ma^{-1} (e.g. Kinsman, 1975). This rate exceeds that at which sedimentation begins to affect surface thermal gradient and heat flow (see Section 6.3.3), which should be recognised if studies are conducted in zones of active subsidence.

7.4.2. Constant Pure Shear

The assumption of an instantaneous stretching event becomes invalid if extension continues for a significant period of time. Jarvis and McKenzie (1980) examined the thermal effects of an alternative model of lithospheric extension. In their model, extension is achieved by stretching the lithosphere at a constant rate, while asthenosphere material flows upward from below to replace the outflowing lithosphere. If the initial thickness of the lithosphere is L, and the initial velocity at which the base of the lithosphere rises with respect to the surface is $\mathbf{V}_0$, then we can define a vertical velocity gradient:

$$\mathbf{s} = \mathbf{V}_0/L \tag{7.26}$$

which is also the extensional strain rate, so

$$\mathbf{u} = \mathbf{s} \times x \tag{7.27}$$

where $\mathbf{u}$ is the horizontal velocity of the lithosphere a distance, x, from the midpoint of the extensional zone.

The conventional stretching factor, β, is related to $\mathbf{s}$ and the duration of stretching, Δt:

$$\beta = \exp(\mathbf{s}\,\Delta t) \tag{7.28}$$

Question: A region of continental crust is undergoing active extension, and has been for the past 5 Ma. The basin is presently 150 km wide and has been stretched by a factor of $\beta = 1.5$. At what rate are the two flanks of the basin moving apart?

Answer: From Equation (7.28):

$$1.5 = \exp(\mathbf{s} \times 5 \times 3.1557 \times 10^{13})$$
$$\mathbf{s} = \ln(1.5)/1.5778 \times 10^{14} = 2.570 \times 10^{-15} \text{ s}^{-1}$$

Then from Equation (7.27), where $x = -75$ km and 75 km, respectively:

$$\mathbf{u} = \mathbf{s} \times x = 2.570 \times 10^{-15} \times \pm 75,000 = \pm 1.927 \times 10^{-10} \text{ m s}^{-1}$$

The two halves of the basin are therefore diverging at a rate of

$$2 \times 1.927 \times 10^{-10} \text{ m s}^{-1} = 12.2 \text{ mm yr}^{-1}.$$

If stretching progresses indefinitely, the thermal structure of the lithosphere approaches a steady-state condition where the temperature varies with depth according to

$$T(z) = \frac{T_1 \text{erf}[(z/L)\sqrt{M}]}{\text{erf}\sqrt{M}} \tag{7.29}$$

$$M = L^2 s/2\kappa \tag{7.30}$$

where T_1 is the temperature at depth, L, and all other symbols have their usual meaning. The steady-state condition is approached if stretching persists for large times relative to the time constant, $L^2/\pi^2\kappa \approx 60$ Ma (Lachenbruch, Sass and Morgan, 1994).

Note: M is a dimensionless parameter relating the velocities associated with advection and diffusion. It is another form of the Peclet number mentioned in Section 6.3.5.

Surface heat flow, $\mathbf{Q}_S$, during the steady-state phase of extension is given by Lachenbruch et al. (1994) as

$$\mathbf{Q}_S = \lambda \frac{T_1}{L} \frac{2}{\pi^{1/2}} \frac{\sqrt{M}}{\text{erf}\sqrt{M}} \tag{7.31}$$

Question: The basin in the previous example continues to stretch until a steady-state condition is achieved. Given that the lithosphere in that region is 100 km thick, with diffusivity $\kappa = 8 \times 10^{-7}$ m^2 s^{-1}, conductivity $\lambda = 4.0$ W m^{-1}K^{-1}, and basal temperature $T_1 = 1425°$C, what is the surface heat flow in the basin?

Answer: From Equation (7.30):

$\quad M = L^2 s/2\kappa = 100,000^2 \times 2.570 \times 10^{-15}/1.6 \times 10^{-6} = 16.06$

Then from Equation (7.31):

$\quad \mathbf{Q}_S = 4.0 \times (1425/100,000) \times (2/\pi^{1/2}) \times (4.0076/1.00000) =$
$\quad 0.26$ W m^{-2} = 260 mW m^{-2}

When extension subsequently ceases, the lithosphere cools at a greater rate than for the case of instantaneous extension because cooling has effectively progressed throughout the extension process. Lachenbruch et al. (1994)

described the temperature distribution as a function of depth, z, and time since stretching ceased, t:

$$T(z, t) = T_1 \frac{z}{L} + \frac{2T_1}{\pi} \sum_{n=1}^{\infty} \frac{1}{n} \left\{ \cos(n\pi) + \frac{1}{\mathrm{erf}\sqrt{M}} \int_0^{n\pi} \mathrm{erf}\left(\frac{\sqrt{M}u}{n\pi}\right) \sin u \, du \right\}$$
$$\times \sin\left(\frac{n\pi z}{L}\right) \exp\left(\frac{-n^2\pi^2\kappa t}{L^2}\right)$$

(7.32)

Heat flow is determined from the depth derivative of Equation (7.32), and for $z = 0$:

$$\mathbf{Q} = \lambda \left[\frac{T_1}{L} + \frac{2T_1}{\pi} \sum_{n=1}^{\infty} \frac{1}{n} \left\{ \cos(n\pi) + \frac{1}{\mathrm{erf}\sqrt{M}} \int_0^{n\pi} \mathrm{erf}\left(\frac{\sqrt{M}u}{n\pi}\right) \sin u \, du \right\} \right.$$
$$\left. \times \frac{n\pi}{L} \cdot \exp\left(\frac{-n^2\pi^2\kappa t}{L^2}\right) \right]$$

(7.33)

Jarvis and McKenzie (1980) concluded that the constant strain model gives results for surface heat flow and thermal subsidence that differ little from the instantaneous stretching model if the time taken to extend by a factor of β, Δt, is less than about $(60/\beta^2)$ Ma. If only heat flow is of concern, the instantaneous stretching model gives sufficient results for Δt less than 60 Ma for $\beta < 2$, or less than $60(2/\beta)^2$ Ma for $\beta > 2$. The instantaneous stretching model predicts thermal subsidence with acceptable accuracy if Δt is less than $60/\beta^2$ Ma for $\beta < 2$ or less than $60(1 - 1/\beta^2)$ Ma for $\beta > 2$.

Note: Equations (7.24) and (7.33) give the upper and lower limits, respectively, to surface heat flow decay given that extension is achieved by pure strain. The solution must lie somewhere between instantaneous extension and a constant strain rate.

7.4.3. Simple Shear

Pure-shear models treat both the crust and mantle as ductile, homogeneous layers. However, it is clear from structural analyses of extensional zones that a pure-shear extension model is inappropriate for the upper crust. In the upper layers, above about 10 km, brittle failure plays a much greater role in deformation than ductile stretching. Another way to model extension is via slip along a continuous, low-angle (10°–20°) detachment surface passing through much of the lithosphere (Figure 7.16; Wernicke, 1985). Such a

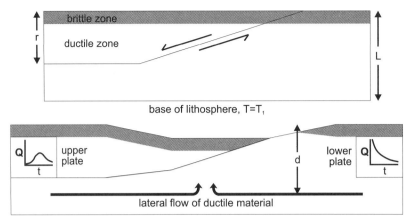

Figure 7.16. Lithospheric extension and thinning via a detachment surface. Lower plate is pulled from under upper plate. Surface heat flow, **Q**, through time, *t*, will be different for the two plates.

detachment appears as a fault in the upper crust and a ductile shear zone at deeper levels.

During extension, the section of lithosphere beneath the detachment is effectively pulled out from under the upper plate. The accommodation space created during the extension is filled by lateral flow of ductile material from either the lower crust or mantle, as opposed to an upwelling of hot asthenosphere as proposed by the pure-shear model. The resultant thinning and subsidence is similar to that of the pure-shear model. That is, an initial isostatic subsidence followed by a longer, sustained subsidence related to cooling. An important difference in this model, however, is that it predicts asymmetry in the heat flow distribution within the resulting basin.

On the upper plate, surface temperature remains constant, while the temperature at the detachment surface initially increases, before relaxing back to equilibrium. In response, surface heat flow initially increases as the thermal pulse conducts through the overlying rock, then gradually decays back to equilibrium. In contrast, on the lower plate, the hot detachment plane is exposed to the surface, resulting in an initial rapid increase in surface heat flow followed by subsequent decay to equilibrium. Lachenbruch et al. (1994) gave the decay in surface heat flow for the lower plate:

$$\mathbf{Q} = \lambda \frac{T_1}{L} + \lambda \frac{T_1}{L} \frac{r}{L} \sum_{n=1}^{\infty} 2 \left\{ 1 - \left(\frac{L}{n\pi r} \right) \left[\sin\left(\frac{n\pi d}{L} \right) - \sin\left(\frac{n\pi(d-r)}{L} \right) \right] \right\}$$
$$\times \exp\left[\frac{-n^2\pi^2\kappa t}{L^2} \right]$$

(7.34)

where T_1 is in degrees Celsius and r, d and L are as given on Figure 7.16.

Question: Consider a section of crust undergoing extension where most of the deep deformation is concentrated along a detachment fault extending to a depth of 12 km. The loss in crustal thickness is compensated by lateral flow within the lower crust at a depth of 25 km. The underlying lithosphere extends to a depth of 100 km. Assuming typical values for λ (3.3 W m^{-1} K^{-1}), κ (8 × 10^{-7} m^2 s^{-1}) and T_1 (1425°C), what is the expected surface heat flow on the lower plate 5, 10 and 20 Ma after the extension event?

Answer: Substituting all the known variables into Equation (7.34):

$$Q(t) = 3.3 \times (1425/10^5) + 3.3 \times (1425/10^5) \times (12 \times 10^3/10^5) \times$$
$$\Sigma_n 2\{1 - (10^5/(n \times \pi \times 12 \times 10^3)) \times [\sin(n \times \pi \times 25 \times$$
$$10^3/10^5) - \sin(n \times \pi \times (25 \times 10^3 - 12 \times 10^3)/10^5)]\} \times$$
$$\exp[-n^2 \times \pi^2 \times 8 \times 10^{-7} \times t/10^{10}]$$

$$Q(t) = 0.04703 + 0.04703 \times 0.12000 \times 2\Sigma_n\{1 - (2.65258n) \times$$
$$[\sin(0.78540n) - \sin(0.40841n)]\} \times \exp[-0.02492n^2 t]$$

where t is in millions of years.

Only the first few terms need be calculated in each series before the exponential term reduces to insignificance:

$Q(5$ Ma$) = 0.04703 + 0.01129 \times \{3.733\} = 0.08917 = 89.2$ mW m^{-2}

$Q(10$ Ma$) = 0.04703 + 0.01129 \times \{0.5429\} = 0.05316 = 53.2$ mW m^{-2}

$Q(20$ Ma$) = 0.04703 + 0.01129 \times \{0.0847\} = 0.04799 = 48.0$ mW m^{-2}

After 20 Ma surface heat flow has almost decayed back to the background level of 47.0 mW m^{-2}.

Models of this kind imply that the thermal history of a passive margin basin (basins lying along the trailing edges of continental masses, formed during a rifting event) depends on whether the basin originally lay on the upper or lower plate. It is even possible for different sections of the same basin to have different tectonic origins. Detailed structural analysis may be required to resolve such anomalies.

7.4.4. A Pure-Shear/Simple-Shear Coupled Model

The Wernicke (1985) simple-shear model of extension is just as much an over-simplification as the pure-shear model. While deformation in the upper crust is undoubtedly by simple shear, there is no evidence of a detachment surface extending into the upper mantle (Kusznir and Matthews, 1988). Pure-shear and simple-shear extension models are end-members of what appears to be the actual mechanism of extension – coupled pure-shear/simple-shear. In the upper 10–15 km of the crust, brittle deformation by simple shear dominates, whereas in the ductile region of the deep crust and upper mantle pure shear accommodates the deformation. Kusznir and Ziegler (1992) presented a model that addresses this duality.

In the upper crust, the basic building block of extension is a half graben formed by movement on a normal fault. The movement alters the isostatic balance of the crust in the vicinity of the fault (Figure 7.17). Unloading of the footwall causes isostatic uplift, while beneath the hanging wall the crust is thinned and net movement is downward. At a moderate distance from the normal fault (>about 40 km for 10 km of extension; Kusznir and Ziegler, 1992), the isostatic balance of the crust remains unaltered, and there is no net vertical movement. This causes the intervening sections of crust to flex, giving rise to the term 'flexural cantilever model', which is used to describe this kind of mechanism. Two normal faults of the same polarity in close proximity (<about 50 km apart) cause the central block to rotate; as one side subsides and the other is uplifted. This results in the characteristic tilt-block structure of extensional basins.

An important difference between the flexural cantilever and earlier models is the method used to calculate the isostatic movements of the crust. The earlier models assumed simple Airy isostasy – balancing the altered load in the vertical direction only, based on the density contrast between crust and mantle. This assumed that the lithosphere has no strength and that each section can move freely in the vertical direction relative to other sections. In reality, continental lithosphere is able to support and distribute a load across a finite lateral distance, and has an effective elastic thickness, $T_e = 3$–5 km (Kusznir and Ziegler, 1992). The vertical deflection, **w**, caused by a distributed horizontal load, $r(x)$, is

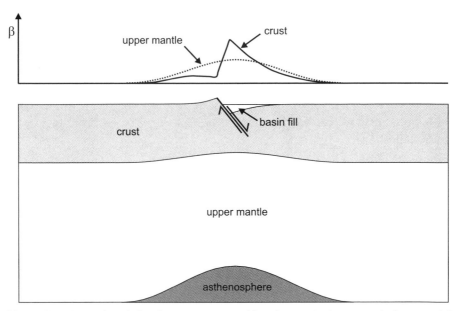

Figure 7.17. Lateral variation in apparent stretching factor, β, due to a single normal fault. Stretching factor varies between the brittle crust (solid line) and ductile mantle (dotted line).

$$\mathbf{w}(x) = \frac{1}{2\pi} \int_{-\infty}^{\infty} R(k) \left[\int_{-\infty}^{\infty} r(x)e^{-ikx}dx \right] \times e^{ikx}dk \qquad (7.35)$$

where

$R(k) = 1/[(\rho_m - \rho_i)g + Dk^4]$

ρ_m = the density of the mantle at the base of the lithosphere

ρ_i = the density of the material filling the resultant depression

k = the wavenumber = 2π/wavelength

$D = YT_e^3/[12(1 - \nu^2)]$ = the flexural rigidity of the lithosphere

Y = Young's modulus

ν = Poisson's ratio

The most important implication of the flexural cantilever model is that there is no simple relationship between surface subsidence (S_i and S_t) and stretching factor (β) – at least, not in a one-dimensional sense. On a major footwall, syn-rift deposits may be very thin or non-existent, whereas they may be very thick a short distance away, above the sediment-loaded hanging wall. Thus, if the thickness of the syn-rift deposit at a single location is used to infer isostatic subsidence, large errors may be expected.

Note: Extension and subsidence are essentially two- or three-dimensional processes, and looking at data from a one-dimensional profile (i.e. a borehole) does not necessarily yield accurate information about the total subsidence of the basin.

There are two ways to estimate β. Firstly, the ratio of the thickness of the ductile lower crust before and after extension can be used. This is fraught with uncertainty, and only a very rough estimate can be expected from any attempt to quantify either of these parameters. Secondly, extension in the ductile lower zones must equal extension by simple-shear faulting in the upper crust. Specifically, Kusznir and Ziegler (1992) suggested that the distribution of extension beneath a single normal fault is bell-shaped:

$$\beta(x) = 1 + C\sin^2(\pi x/W) \qquad (7.36)$$

where

$C = (2E)/(W - E)$

E = extension along the fault (km)

W = the pre-extension width of the pure-shear region (km)

Question: The horizontal movement on a particular normal fault is found to be 10 km. Given that $W = 100$ km, what is the maximum value of β beneath the extension zone? Also, what is the value of β 25 km either side of the maximum?

Answer: Given that $E = 10$ km and $W = 100$ km, Equation (7.36) tells us:

$C = (2 \times 10)/(100 - 10) = 0.222$
So the maximum thinning factor, at $x = W/2 = 50$ km, is
$\beta_{max} = 1 + C \times \sin^2(\pi/2) = 1 + 0.222 = 1.222$
At $x = 25$ km and 75 km:
$\beta(x) = 1 + 0.222 \times \sin^2(25\pi/100) = 1 + (0.222 \times 0.5) = 1.111$

A separate stretching function, $\beta_i(x)$, can be calculated for each normal fault. The composite stretching function for the entire rift zone, $\beta_t(x)$, is then the product of each individual function:

$$\beta_t(x) = \beta_1(x) \times \beta_2(x) \times \beta_3(x) \times \ldots \times \beta_n(x) \tag{7.37}$$

Each fault is offset from the others, so the composite effect is generally a broad zone of thinning, tapering at the edges (Figure 7.18). The stretching function can be used to determine the distribution of elevated heat flow immediately following the rifting event.

Note: Heat flow values calculated in this way should be viewed as the heat flow at the base of the brittle zone, or the top of the ductile zone. Distribution of heat within the upper crust is controlled by basin architecture and conductivity contrasts, and is best determined by numerical modelling.

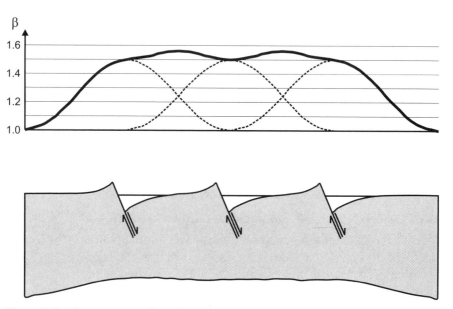

Figure 7.18. The cumulative effect of more than one normal fault. The stretching functions associated with each normal fault multiply to produce the overall stretching function.

7.4.5. The Blanketing Effect of Sediments

All predictions of thermal subsidence, regardless of which models are used, become complicated in the presence of a significant load of sediment. Simple, two-layer models (mantle and crust) assume the thermal properties of the crust to be relatively homogeneous, regardless of the extension mechanism. A load of sedimentary material in the surface depression caused by subsidence affects both cooling rate and further subsidence.

Zhang (1993) showed that the effect of a blanket of relatively low-conductivity sediments should not be ignored when modelling lithospheric cooling and subsidence after extension. Most surface heat flow and subsidence models assume that the surface boundary is at a constant temperature, and that sediment deposited on the subsiding surface has an effective infinite conductivity. Zhang investigated the effect on such models when sediment has a finite conductivity lower than the underlying lithosphere.

Post-rift cooling is much slower when blanketed by sediment. Even 200 Ma after extension, there may still be significant amounts of residual heat within the lithosphere, where simple models predict almost total equilibrium by that time. Assuming a moderate conductivity for the sediment $(2.0 \, \mathrm{W\,m^{-1}K^{-1}})$, Zhang calculated subsidence and compared results with those predicted by McKenzie's (1978) instantaneous pure shear model described in Section 7.4.1. Thermal subsidence was much less for the blanketed model, with a maximum discrepancy between the two models of 2.0 km, 2.6 km and 3.2 km for thinning factors of $\beta = 1.5$, 2 and 3, respectively. Sediment cover can reduce the expected amount of thermal subsidence (as opposed to isostatic subsidence associated with initial extension) by more than half at any given time. Likewise, heat flow may remain high for much longer periods than predicted by McKenzie. Although still an overly simple model, Zhang's results indicate that serious discrepancies can be introduced into a thermal history model if the insulating effect of the sediment burden is ignored.

EXAMPLE

Zhang (1993) modelled subsidence in the Central Graben of the North Sea. With a stretching factor of $\beta = 2.07$ and sediment thermal conductivity of $\lambda_S = 1.67 \, \mathrm{W\,m^{-1}\,K^{-1}}$, Zhang's model predicted present heat flow to be $Q = 58.2 \, \mathrm{mW\,m^{-2}}$, initial tectonic subsidence to be $S_i = 4.566$ km, and total subsidence after 65 Ma, $S_{65} = 6.822$ km. These figures compare favourably with values estimated from geological studies: $\mathbf{Q} = 55 \, \mathrm{mW\,m^{-2}}$ (Andrews-Speed, Oxburgh and Cooper, 1984), $S_i = 4.212$ km, $S_{65} = 6.800$ km. McKenzie's (1978) instantaneous extension model predicts heat flow of about $\mathbf{Q} = 50 \, \mathrm{mW\,m^{-2}}$ after the same time period.

7.4.6. *Underplating*

Lithospheric extension is associated with a thinning of the crust and upper mantle and a simultaneous increase in thermal gradient within those layers. These effects have been discussed in the above sections. What has not yet been discussed is the effect of lithospheric thinning on the underlying asthenosphere, which must rise to fill the vacated space. If stretching and upwelling are rapid, the pressure within the upper asthenosphere and lower lithosphere may drop below the melt solidus, resulting in partial melting (McKenzie and Bickle, 1988). Melt generated in this way may rise through the lithosphere to intrude the crust as dykes, extrude to the surface, or cake and solidify at the moho (e.g. Kusznir and Ziegler, 1992). Basaltic magmas generated beneath continental areas are frequently emplaced at the moho because of the relatively low density of crustal rocks. This process (referred to as *magmatic underplating*) adds heat and mass to the continental crust and can cause, among other effects, regional scale metamorphism, anatexis and surface uplift (Henk et al., 1997).

Here, we are mainly concerned with the effect underplating has on the flow of heat through a section of extended crust. As the magma cools (loses specific heat) and solidifies (loses latent heat), the lost heat is injected into the base of the crust. The magma essentially acts as a deep heat source over a finite time. The magnitude of the effect is proportional to the rate at which magma is accreted to the crust, which is in turn related to the volume of melt generated and the time-frame over which the process takes place.

The rate at which heat is liberated during underplating is fairly simple to calculate. Specific heat is equal to

$$\mathbf{Q}_s = \rho c w \, \Delta T \ \mathrm{W\,m^{-2}} \tag{7.38}$$

where ρ, c are density ($\mathrm{kg\,m^{-3}}$) and specific heat ($\mathrm{J\,kg^{-1}\,K^{-1}}$) of basalt, respectively; w is rate of accretion ($\mathrm{m\,s^{-1}}$) and ΔT is total amount of cooling (K).

Latent heat liberated during crystallisation is equal to

$$\mathbf{Q}_l = l\rho w \ \mathrm{W\,m^{-2}} \tag{7.39}$$

where $l = $ latent heat ($\mathrm{J\,kg^{-1}}$) of basalt.

The total liberated heat is the sum of $\mathbf{Q}_s$ and $\mathbf{Q}_l$. While the duration of elevated heat flow due to underplating may be relatively short (a few million years), the early thermal history of a basin may be significantly affected and the length of time required for lithospheric cooling extended.

Question: Cull (1990) concluded from xenolith data that geothermal gradients within the crust of eastern Australia were once much higher than could be explained by realistic values of crustal heat generation. An excellent fit to the data was obtained, however, by including a 4-km-thick basalt layer within the crustal model at a depth of 28 km, producing heat at a rate of 13 $\mu W\,m^{-3}$. Given that the density of basalt is 2800 $kg\,m^{-3}$, specific heat is 1100 $J\,kg^{-1}\,K^{-1}$, latent heat is 4.18×10^5 $J\,kg^{-1}$, and the basalt cooled 200°C after emplacement, what rate of accretion was required to produce heat at the calculated rate?

Answer: Heat was added to the crust at the rate

$$13 \ \mu W\,m^{-3} \times 4000 \ m = 52 \ mW\,m^{-2}$$

The total heat liberated during underplating is found from Equations (7.38) and (7.39):

$$\mathbf{Q}_{total} = \mathbf{Q}_s + \mathbf{Q}_l = \rho w (c\,\Delta T + l)$$
$$0.052 = 2800 \times w \times (1100 \times 200 + 4.18 \times 10^5)$$
$$0.052 = 1.7864 \times 10^9 w$$
$$w = 2.911 \times 10^{-11} \ m\,s^{-1} = 920 \ m\,Ma^{-1}$$

This implies that underplating continued for approximately 4.3 Ma to accrete 4 km of basalt.

Stretching factor and the temperature of the asthenosphere prior to extension control the volume of melt, with greater stretching and higher temperatures both leading to greater volume (McKenzie and Bickle, 1988). Kusznir and Ziegler (1992) suggested an ambient temperature for the asthenosphere on the order of 1280°C, possibly increased by several hundred degrees within about 1500 km of a mantle plume or upwelling limb of a mantle convection cell. Figure 7.19 shows the thickness of melt expected for different stretching factors, asthenosphere temperature and lithospheric thickness.

Note: McKenzie and Bickle (1988) concluded that significant volumes of melt are only produced if $\beta > 2$ and $T_{0.75} > 1380$°C. The melts are generally alkali-basalts but become tholeiitic as the degree of melting increases.

Although simple in principle, underplating further complicates the task of unravelling the thermal history of a region. The thickness, timing and rate of accretion of the underplated material are critical parameters in determining the overall thermal effect. Of these, only the thickness can be approximated with any degree of accuracy, and then only if appropriate deep seismic data indicate the presence of an intermediate velocity layer at the base of the crust. The timing of emplacement might be estimated if there are associated extruded basalts in the region that can be reliably dated. The rate of emplacement, however, is very difficult to constrain, and only independent data such as xenoliths may yield clues.

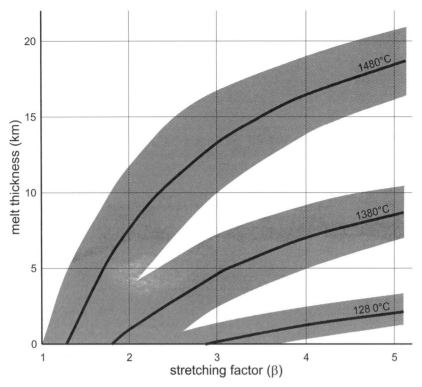

Figure 7.19. Average thickness of melt produced during an extensional event as a function of asthenosphere potential temperature and stretching factor. The solid lines are the predictions for 100-km-thick lithosphere, while the grey zones represent the range 70–130 km (thinner lithosphere produces *more* melt). Modified after McKenzie and Bickle (1988).

7.4.7. Passive Margins

Occasionally extension does not cease. The lithosphere continues to stretch, thin and weaken until it is no longer strong enough to withstand the applied extensional stress, at which point it ruptures. At the moment of rupture, sea-floor spreading commences and the region ceases to be a rift or extensional zone, becoming instead a *passive margin*, sometimes called a *trailing edge*.

Passive margins differ in a number of fundamental ways from intracontinental rift basins. Most important is the distribution of stretching and thinning across the basin. Whereas the stretching factor across an intracontinental basin is generally fairly constant (Figure 7.18), on passive margins it is typical to find a gradation in β from around 1 onshore to 4–5 offshore, before thinned continental crust gives way to oceanic crust and sea-floor spreading. This has ramifications for heat flow history and both isostatic and thermal subsidence, which all vary with β across the basin. Those parts of the basin with $\beta > 2$ may also be affected by magmatic underplating, as explained above.

Following rupture, the edge of a continental plate is no longer supported by its own internal rigidity and thus sinks into the hot asthenosphere. This second period of isostatic subsidence is superimposed on the thermal subsidence and it is important to distinguish the relative contribution of each before using the data to derive heat flow history. The isostatic subsidence is not constant across the width of the basin, but rather increases in magnitude seaward to a maximum at the rupture point.

Note: In general, the tectonic and subsidence history of passive margins must be well understood before an accurate reconstruction of thermal history can be attempted.

7.5. Summary

The lithosphere is composed of the crust and that part of the upper mantle that exhibits gross plastic behaviour under stress. The base of the lithosphere is effectively defined by the $0.75T_m$ isotherm, where T_m is the melting temperature of pyrolite ($T_m \approx 1775$ K). Heat transport below the lithosphere, is dominated by convection, while heat transport in the lithosphere is mainly by conduction.

Oceanic lithosphere is formed at mid-ocean ridges, and the subsequent cooling controls surface heat flow, bathymetry and lithospheric thickness beneath oceans. Evidence is currently inconclusive as to whether oceanic lithosphere cools according to a half-space model or a plate model.

Continental lithosphere is generally older and thicker than oceanic lithosphere. Surface heat flow on continental lithosphere is controlled by a number of factors, including age, thickness and crustal composition.

Hot spots are assumed to be plumes of hot mantle material that remain stationary in respect to the mantle. The tectonic plates move above these plumes, causing characteristic chains of volcanic activity at the surface. The thermal effects of hot spots include increased surface heat flow and elevation, and are observed to a radius of several hundred kilometres around the axis of the plume. These effects slowly dissipate after the hot spot moves on.

Where two tectonic plates collide, one necessarily descends beneath the other, and the result is a subduction zone. Heat flow across a subduction zone follows a characteristic profile. Below the subduction trench, heat flow is low. It increases slowly towards the island arc, where it rises quickly to a maximum. Beyond the island arc, heat flow often remains high through the back-arc basin. The reasons for the heat flow profile relate to downward advection of heat with the subducting plate, loss of heat due to reheating of the subducted plate, and probably convective effects due to the addition of fluid to the upper mantle of the overlying plate.

Most of the economic petroleum accumulations in the world are hosted within zones of continental extension. Petroleum generation is controlled by the thermal history of source beds within the extensional zone. The thermal history is controlled predominantly by the amount of stretching undergone by the lithosphere. Secondary controls on heat flow include the rate of rifting and the degree of basaltic underplating.

The upper crust deforms by brittle failure, while the lower crust and mantle undergo ductile deformation. Subsidence is due to isostatic balance and then thermal decay, with the effect significantly amplified by the addition of a sediment load.

CHAPTER EIGHT

Numerical Modelling

When I model I pretty much blank. You can't think too much or it doesn't work.

Paulina Porizkova, actress and model, 1965–   .

General lithospheric models are unable to resolve the details of temperature distribution in the upper crust. Yet it is in the upper crust that we are most often interested when we are constructing thermal models. Much of this book, for example, has focussed on petroleum exploration in the upper few kilometres of sedimentary basins. Heat entering the crust from below flows through to the surface along paths controlled by thermal conductivity contrasts and the structural geometry of the intervening layers, resulting in a unique temperature distribution. This chapter introduces the reader to a powerful tool for resolving the temperature distribution in this critical region – finite difference numerical modelling.

8.1. Finite Difference Approximations

Heat flow is controlled by conservation of energy laws that can be written as differential equations. These equations can be solved using finite difference approximations, assuming appropriate boundary conditions are enforced. The precision of solutions is limited by time constraints and the memory capacity of the computer at the scientist's disposal, but the evolution of digital computers has improved the efficiency of solving finite difference equations to the point where detailed, three-dimensional thermal models of entire basins can now be solved in a matter of hours.

In finite difference numerical modelling, spatial and time derivatives within governing equations are replaced by polynomial approximations. This is justified because the definition of a derivative is

$$\frac{dF}{dx} = \lim_{\Delta x \to 0} \frac{F(x + \Delta x) - F(x)}{\Delta x} \tag{8.1}$$

The right-hand side of Equation (8.1) can be solved using a finite value of Δx in order to approximate the derivative on the left. In general, the smaller we make Δx, the more accurate will be the approximation. If we wish to approximate a continuous solution to the derivative over a particular domain, the domain can be divided into discrete elements, each Δx in size. The domain is thus $N \Delta x$ wide, where N is the total number of elements in the domain. To find a solution we need to know $\{x, F(x)\}$ at each of $N + 1$ nodes, where each node is separated from the next by Δx. To distinguish these nodes from each other, they are labelled x_0 to x_N in the direction of increasing x.

EXAMPLE

Consider the function $F = 2x^2 + 4x - 3$. We wish to approximate the derivative, dF/dx, between $x = 0$ and $x = 5$.

We begin by dividing the domain into ten discrete elements, each of width $\Delta x = 0.5$. The edges of the elements define eleven nodes. We can calculate the pairs $\{x, F(x)\}$ at each node:

Node	x	$F(x)$	Node	x	$F(x)$
0	0.0	−3.0	6	3.0	27.0
1	0.5	−0.5	7	3.5	35.5
2	1.0	3.0	8	4.0	45.0
3	1.5	7.5	9	4.5	55.5
4	2.0	13.0	10	5.0	67.0
5	2.5	19.5			

With Equation (8.1) we can approximate the dF/dx at any node.
For example, at node 1:

$$F(x_1) = -0.5$$
$$F(x_1 + \Delta x) = F(x_2) = 3.0$$
$$\Delta x = 0.5$$
$$dF/dx = [F(x_2) - F(x_1)]/\Delta x = (3.0 - -0.5)/0.5 = 7.0$$

At node 6:

$$F(x_6) = 27.0$$
$$F(x_6 + \Delta x) = F(x_7) = 35.5$$
$$\Delta x = 0.5$$
$$dF/dx = [F(x_7) - F(x_6)]/\Delta x = (35.5 - 27.0)/0.5 = 17.0$$

Note that the derivative can be solved at nodes 0 to 9, but not at node 10. Note also that the solution is not exact. In this example, the derivative is overestimated by $2 \Delta x$, or 1.0.

The vertical thermal gradient within the crust is defined as a simple differential equation, dT/dz, where $T(z)$ is the temperature distribution function with depth. In reality, the temperature versus depth function is continuous, but the derivative can be numerically approximated using a finite difference. The depth domain is divided into discrete elements of height Δz, the edges of which define nodes (Figure 8.1). The distance between adjacent nodes is Δz, and

$$\frac{dT}{dz} \approx \frac{\Delta T}{\Delta z} = \frac{T_{k+1} - T_k}{\Delta z} \tag{8.2}$$

where T_k is the temperature at node k.

This kind of approximation is called a *forward* difference. It is asymmetric, in that it actually approximates the gradient at the depth, $z_{k+\frac{1}{2}}$, mid-way between nodes k and $k + 1$. Similarly:

$$\frac{dT}{dz} \approx \frac{\Delta T}{\Delta z} = \frac{T_k - T_{k-1}}{\Delta z} \tag{8.3}$$

is called a *backward* difference, is also asymmetric, and approximates the gradient at depth, $z_{k-\frac{1}{2}}$.

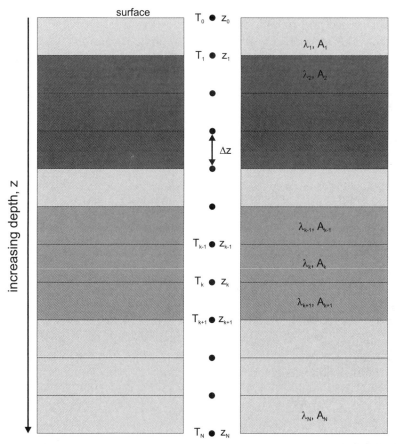

Figure 8.1. The discretisation of a vertical section of crust. The section is divided into segments of height Δz, each with homogeneous thermal properties. The edges of the segments define $N + 1$ nodes, and each node is associated with a temperature (T_k) and depth (z_k).

A gradient estimate centred on z_k can be found by averaging the forward and backward differences:

$$\frac{dT}{dz}\bigg|_{z_k} \approx \frac{T_{k+1} - T_{k-1}}{2\Delta z} \tag{8.4}$$

Question: Use a centred difference approximation to estimate the derivative at node 6 in the previous example.

Answer: From the previous example, and using Equation (8.4):

    $F_{k-1} = F(x_5) = 19.5$

    $F_{k+1} = F(x_7) = 35.5$

    $\Delta x = 0.5$

    $(F_{k+1} - F_{k-1})/2\,\Delta x = (35.5 - 19.5)/2 \times 0.5 = 16.0$

In this case the finite difference approximation gives the correct answer.

The finite difference approximation approaches the true derivative as Δz approaches zero. The smaller Δz, the better the approximation. The error inherent in a forward or backward difference can be shown to be proportional to Δz, while the error in the centred difference is proportional to $(\Delta z)^2$. In other words, if we halve Δz, the accuracy of the centred difference improves by a factor of four.

Second, and subsequent, derivatives can also be written as centred finite difference approximations. The second derivative of temperature with depth, for example, is

$$\frac{d^2T}{dz^2}\bigg|_{z_k} \approx \frac{1}{\Delta z}\left[\frac{dT}{dz}\bigg|_{z_{k+\frac{1}{2}}} - \frac{dT}{dz}\bigg|_{z_{k-\frac{1}{2}}}\right] \approx \frac{1}{\Delta z^2}\left[T_{k+1} + T_{k-1} - 2T_k\right] \tag{8.5}$$

Similar expressions can be found for partial derivatives such as $\partial[T(z, t)]/\partial t$ and $\partial[T(z, t)]/\partial z$.

In practice, temperature (T) is usually the unknown quantity that we wish to resolve. The data we generally have at our disposal are the surface temperature distribution, the basal heat flow or temperature distribution, and the distribution of thermal properties (thermal conductivity and heat generation) within the volume under investigation. The governing equation for steady-state heat flow in one dimension was given back in Section 1.3, and can be written as

$$\frac{d}{dz}\left(\lambda(z)\frac{dT}{dz}\right) = -A(z) \tag{8.6}$$

where $\lambda(z)$ is thermal conductivity function with depth and $A(z)$ is heat generation function with depth.

In three dimensions, Equation (8.6) becomes

$$\nabla(\lambda(x, y, z)\nabla T) = -A(x, y, z) \tag{8.7}$$

where

$$\nabla F = (\partial F/\partial x)\mathbf{i} + (\partial F/\partial y)\mathbf{j} + (\partial F/\partial z)\mathbf{k}$$

and $\mathbf{i}, \mathbf{j}$ and $\mathbf{k}$ are the unit vectors in the x, y and z directions, respectively.

It is Equation (8.7) that must be solved over our model space in order to discover the equilibrium temperature distribution consistent with our structural model and boundary conditions. There are many techniques of varying sophistication for solving three-dimensional finite difference problems, but we will restrict our discussion to a class that is relatively simple to understand and put into practice. Appropriately, the class of techniques is known as *relaxation*.

Although slower than direct matrix-based methods, relaxation techniques have several advantages over the others. For example, round-off errors (Section 8.3.2) do not significantly affect relaxation methods because of the constant checking of equilibrium between points. Relaxation is also generally easier to implement and more flexible. For example, thermal conductivity can be made a function of temperature.

Note: Care should be taken when modelling temperature dependence. Some temperature dependence functions may render the system unstable, in which case no numerical technique will converge to a solution.

8.2. Relaxation

The essence of the relaxation technique is to ascertain whether each node in a lattice is in thermal equilibrium with its nearest neighbours. If not, then a new temperature is assigned to the node to bring it into equilibrium. In this way, each node becomes relaxed with respect to its nearest neighbours.

8.2.1. Discretising the Model

Modern three-dimensional seismic techniques and visualisation software allow the development of detailed models of the structure of the crust. In order to estimate numerically the temperature distribution within such models, however, we must first define a three-dimensional network of cells and nodes to represent the observed geological conditions. This involves dividing the known geology into a finite number of cells, where each cell has homogeneous thermal properties. The smaller we are able to make the cells, the more accurately they will represent the actual conditions (Figure 8.2). It is our aim to keep the procedure simple for the purposes of this discussion, so we will deal only with rectangular prismatic cells, where each cell is the same size as every other, and the edges lie along orthogonal axes. Translating geological models

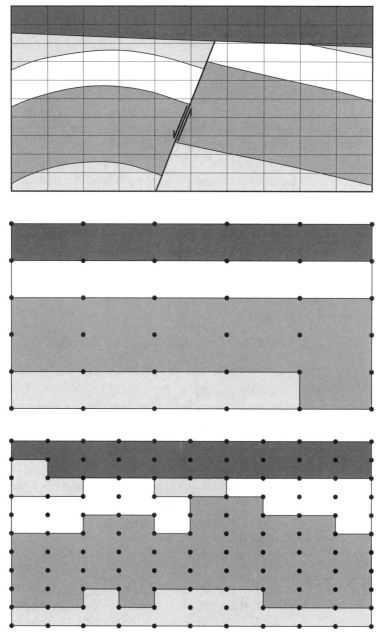

Figure 8.2. Dividing a cross section of known geology (top figure) into 25 (middle figure) and 100 (bottom figure) discrete homogeneous cells. The corners of the cells define a network of nodes at which temperature may be calculated. Note that the greater the number of cells, the more detail is preserved in the model.

into cells and nodes for numerical analysis requires several decisions to be made.

The first decision is how big to make the cells. The cell size is necessarily a trade-off between accurately reproducing the observed geological structure, and restrictions imposed by available computing power. On the side of accuracy, we need to consider the total volume to be modelled, the thickness of individual formations, and the horizontal wavelength of structures. Ideally, the cell size should be such that the smallest significant feature is adequately resolved.

Question: We wish to model a section of crust 20 km × 20 km × 10 km deep, deformed by regional folds with a wavelength of 5 km and amplitude of 2 km. The folds deform a sequence of 2 km of fluvial sandstone overlain by 3 km of shale and 1 km of limestone, and plunge at 15° to the north (Figure 8.3). What is the coarsest cell size appropriate in this situation, and how many cells are required for the full model?

Answer: Vertically, the thinnest unit we need to model is the 1-km-thick limestone formation, but to resolve the shape of the folds requires a vertical resolution of about 250 m. This means a mini-

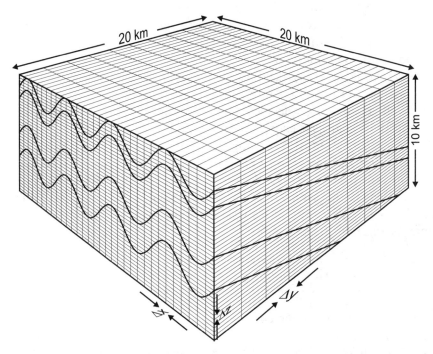

Figure 8.3. Example of a geological structure divided into discrete cells for numerical finite difference modelling. Each cell is $\Delta x \times \Delta y \times \Delta z$ in size, and is bounded by eight nodes.

mum of forty intervals to cover the 10 km vertical section of the model.

In the east–west direction, the model must resolve folds with a wavelength of 5 km. To do this requires about eight cells per wavelength, or cells of length 625 m, making a total of thirty two intervals across the 20 km extent of the model.

The folds plunge to the north, and increase in depth by approximately 5 km over the 20 km extent of the model. The cell width should be a maximum of about 2 km. Any coarser than this will negate the vertical precision of the model. This means a minimum of ten intervals in the north–south direction.

The minimum number of cells required to resolve the observed geology adequately is therefore:

$40 \times 32 \times 10 = 12,800$

Resolution of geological structures cannot be improved beyond a certain point because the number of cells involved quickly becomes very large. A model dividing a volume of crust into just ten segments in the east–west direction, ten in the north–south direction and ten vertically, is very simplistic, but contains 1000 cells $(10 \times 10 \times 10)$. Moreover, it includes 1331 nodes $(11 \times 11 \times 11)$ at which we must calculate temperature. If we double the number of segments on each side of the model, the number of cells increases by a factor of eight, and the number of nodes by almost as much. The large numbers of cells and nodes involved places a severe limitation on the possible resolution of the model. An increase in resolution requires increases in both computing power and computing time, both of which are generally limited. In general, models should be constructed to the highest resolution allowed by the available computer and time constraints.

Once the cell size has been decided, each cell must be assigned values of thermal conductivity and heat generation appropriate for the piece of crust it represents. Cell properties must be homogeneous, in that the entire cell has the same properties, but they may also be anisotropic, having different values in orthogonal directions. For example, the vertical thermal conductivity assigned to a particular cell is the same over the entire volume of the cell, and the horizontal thermal conductivity is likewise constant over the entire cell, but the two may be different from each other.

Note: Heat generation cannot be anisotropic.

Each node and cell is a unique entity and must be distinguishable from all other nodes and cells. For this reason, a self-consistent labelling convention is required so that individual elements of the model can be easily identified. The following conventions will be used in all subsequent discussions.

The three orthogonal axes of the grid correspond to east–west, north–south and vertical orientations. The x-coordinate increases to the east, the y-coordinate increases to the south and the z-coordinate increases downward, which establishes a right-handed coordinate system. Nodes are labelled $n_{i,j,k}$, with $n_{0,0,0}$ in the top north–west corner. Subscripts i, j and k increase in the x, y and z directions, respectively. Cells are labelled after the node in their south-east bottom corner. Thus, $cell_{1,1,1}$ is the north-west top cell of the model, and $cell_{i,j,k}$ is defined by the nodes as illustrated in Figure 8.4. The total number of cells in the east–west direction is NX, in the north–south direction is NY, and in the vertical direction is NZ, so the south-east bottom cell is $cell_{NX,NY,NZ}$. The thermal conductivity in $cell_{i,j,k} = \lambda_{i,j,k}$, and the heat generation equals $A_{i,j,k}$.

Relaxation iterations proceed more efficiently if the average heat generation and directional thermal conductivities are calculated at each node prior to beginning the relaxation process. Recall that each internal node in the model is at the junction of eight different cells (edge nodes connect with fewer cells). The heat generation represented by a particular node is the sum of the heat generation from one eighth of each of the surrounding cells:

$$A = \frac{\begin{aligned}(A_{i,j,k} + A_{i+1,j,k} + A_{i,j+1,k} + A_{i+1,j+1,k} + A_{i,j,k+1} + A_{i+1,j,k+1} + \\ A_{i,j+1,k+1} + A_{i+1,j+1,k+1})\end{aligned}}{8} \qquad (8.8)$$

The thermal conductivity at a node will generally be anisotropic. The conductivity in the upward direction, for example, will be the arithmetic mean of the conductivities of the four cells lying above the node (see Section 4.2.2). If we use the subscripts N, S, E, W, U, D (north, south, east, west, up, down) to

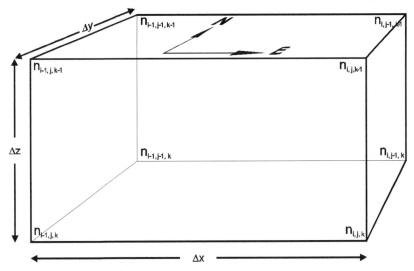

Figure 8.4. Nomenclature for $cell_{i,j,k}$. The cell has dimensions $\Delta x \times \Delta y \times \Delta z$, properties $\lambda_{i,j,k}$ and $A_{i,j,k}$, and its corners define eight nodes.

represent the thermal conductivity in each of the six cardinal directions from a particular node, then

$$\lambda_U = \frac{\lambda_{i,j,k} + \lambda_{i,j+1,k} + \lambda_{i+1,j,k} + \lambda_{i+1,j+1,k}}{4} \tag{8.9a}$$

$$\lambda_D = \frac{\lambda_{i,j,k+1} + \lambda_{i,j+1,k+1} + \lambda_{i+1,j,k+1} + \lambda_{i+1,j+1,k+1}}{4} \tag{8.9b}$$

$$\lambda_W = \frac{\lambda_{i,j,k} + \lambda_{i,j+1,k} + \lambda_{i,j,k+1} + \lambda_{i,j+1,k+1}}{4} \tag{8.9c}$$

$$\lambda_E = \frac{\lambda_{i+1,j,k} + \lambda_{i+1,j+1,k} + \lambda_{i+1,j,k+1} + \lambda_{i+1,j+1,k+1}}{4} \tag{8.9d}$$

$$\lambda_N = \frac{\lambda_{i,j,k} + \lambda_{i+1,j,k} + \lambda_{i,j,k+1} + \lambda_{i+1,j,k+1}}{4} \tag{8.9e}$$

$$\lambda_S = \frac{\lambda_{i,j+1,k} + \lambda_{i+1,j+1,k} + \lambda_{i,j+1,k+1} + \lambda_{i+1,j+1,k+1}}{4} \tag{8.9f}$$

Note: If each cell is assigned an anisotropic thermal conductivity, then λ_U and λ_D must be calculated using the vertical conductivity and λ_N, λ_S, λ_E and λ_W using the horizontal conductivity.

8.2.2. *Boundary Conditions*

The number of boundary conditions needed to define a problem fully is equal to the order of the governing differential equation. In the case of the three-dimensional steady-state problem, the defining equation is of second order (the highest term is a second-derivative) along each axis, so it requires two boundary conditions along each axis.

There are two main kinds of boundary condition that are commonly used in thermal modelling. A *Dirichlet* boundary condition has the temperature of the edge node fixed at a constant value, whereas a *Neumann* boundary condition has the heat flow (or thermal gradient) fixed at a constant value. In general, at least one boundary of the model must be of Dirichlet type in order to define a problem completely. The other boundary conditions can be of either type.

A constant temperature boundary is fairly self-explanatory. All nodes at the boundary in question are maintained at a constant temperature throughout the equilibration process. This is generally the condition assumed at the surface of the model, where each node is assigned a temperature equal to the average surface temperature at that point. The value is maintained throughout the relaxation process. A constant heat flow boundary is a little more involved. We know that vertical heat flow is defined as

$$\mathbf{Q} = \lambda \times (dT/dz) \tag{8.10}$$

Using a backward finite difference approximation, this can be written for the bottom boundary:

$$\mathbf{Q} = \lambda \times (T_{i,j,ZN} - T_{i,j,ZN-1})/\Delta z \tag{8.11}$$

If we know $\mathbf{Q}$, λ and Δz, we can calculate $T_{i,j,ZN}$ from $T_{i,j,ZN-1}$:

$$T_{i,j,ZN} = T_{i,j,ZN-1} + \mathbf{Q}\,\Delta z/\lambda \tag{8.12}$$

The four vertical sides of the model are generally assumed to be insulated. That is, no heat flows into or out of the model through the sides. Zero lateral heat flow can be enforced by making $\lambda_N = 0$ at all nodes on the north side, $\lambda_S = 0$ at all nodes on the south side, $\lambda_E = 0$ at all east-side nodes and $\lambda_W = 0$ for all nodes on the west side. Nodes on the four vertical edges will have two of these conductivities equal to zero.

Heat generation, A, for each of the side and edge nodes is found by adding the heat generation of all adjacent cells and dividing by eight. Nodes on the side of the model have only four adjacent cells, and will therefore have heat generation values approximately half that of the interior nodes. Likewise, nodes on the vertical edges will have heat generation values of about a quarter of the interior nodes.

Note: The sum of the heat generation assigned to all nodes should equal the sum of the heat generation of all cells.

The final model usually has boundary conditions of zero heat flow across the four vertical sides of the model, constant temperature at the surface, and either constant temperature or constant heat flow at the base.

8.2.3. Equilibrium Nodal Temperature

The relaxation technique requires that an initial temperature be assigned to each node. The initial guess at temperature distribution need not be accurate, but the closer it is to the equilibrium condition, the fewer iterations will be required to converge to an acceptable solution. The initial conditions might be set to a constant temperature or thermal gradient throughout the model, or more sophisticated algorithms may be employed to solve the one-dimensional steady-state equation for each vertical column of nodes.

The relaxation process requires that we calculate, for each node, the temperature that is in equilibrium with the current temperatures of the six surrounding nodes. In order to do this, we must first construct a finite difference equation. For a node to be in thermal equilibrium, the total heat generated at the node must equal the net heat flowing away from the node.

The total amount of heat generated at a node is

$$A_T = \Delta x \times \Delta y \times \Delta z \times A \tag{8.13}$$

Question: Consider our earlier example with the lattice of cells of dimensions 250 m × 625 m × 2 km. Assume that the average heat generation at a particular node (the average of the heat generation in the eight surrounding cells) is 1.25 μW m^{-3}. What is the total rate of heat generation represented at the node?

Answer: Total heat generation is found using Equation (8.13):
$A_T = \Delta x \times \Delta y \times \Delta z \times A = 250 \times 625 \times 2000 \times 1.25 \times 10^{-6} =$
390.625 W

The total heat flowing upward from the node is the product of the vertical heat flow and the top surface area of a single cell. The finite difference approximation is

$$\mathbf{H}_U = \Delta x \times \Delta y \times \lambda_U \times (T - T_U)/\Delta z \tag{8.14a}$$

where T_U is the temperature of the adjacent node in the upward direction.

Likewise, the total heat flowing to the east from the node is

$$\mathbf{H}_E = \Delta y \times \Delta z \times \lambda_E \times (T - T_E)/\Delta x \tag{8.14b}$$

where T_E is the temperature of the adjacent node in the eastward direction.

The equations for the total heat flowing in the remaining four orthogonal directions are

$$\mathbf{H}_D = \Delta x \times \Delta y \times \lambda_D \times (T - T_D)/\Delta z \tag{8.14c}$$

$$\mathbf{H}_W = \Delta y \times \Delta z \times \lambda_W \times (T - T_W)/\Delta x \tag{8.14d}$$

$$\mathbf{H}_N = \Delta x \times \Delta z \times \lambda_N \times (T - T_N)/\Delta y \tag{8.14e}$$

$$\mathbf{H}_S = \Delta x \times \Delta z \times \lambda_S \times (T - T_S)/\Delta y \tag{8.14f}$$

For equilibrium:

$$A_T = \mathbf{H}_U + \mathbf{H}_D + \mathbf{H}_W + \mathbf{H}_E + \mathbf{H}_N + \mathbf{H}_S \tag{8.15}$$

Substituting all the appropriate equations and rearranging the formula, the temperature that brings a particular node into thermal equilibrium with all six surrounding nodes is

$$T = \frac{\left[A + \dfrac{T_U \lambda_U + T_D \lambda_D}{\Delta z^2} + \dfrac{T_W \lambda_W + T_E \lambda_E}{\Delta x^2} + \dfrac{T_N \lambda_N + T_S \lambda_S}{\Delta y^2}\right]}{\left[\dfrac{\lambda_U + \lambda_D}{\Delta z^2} + \dfrac{\lambda_W + \lambda_E}{\Delta x^2} + \dfrac{\lambda_N + \lambda_S}{\Delta y^2}\right]} \tag{8.16}$$

Note: The heat generation parameter in Equation (8.16) is A, not A_T.

Question: The initial temperature distribution in the model illustrated in Figure 8.3 is set to a constant vertical thermal gradient of $30°\text{C km}^{-1}$. Surface temperature is $0°\text{C}$. At node $n_{16,5,10}$:
$$\lambda_E = \lambda_W = \lambda_N = \lambda_S = \lambda_U = 2.3 \text{ W m}^{-1}\text{K}^{-1},$$
$$\lambda_D = 3.1 \text{ W m}^{-1}\text{ K}^{-1}, \qquad A = 0.5 \ \mu\text{ W m}^{-3}$$
To what temperature does this node relax according to Equation (8.16)?

Answer: Node $n_{16,5,10}$ is at a depth of $10 \Delta z$, or $10 \times 250 = 2500$ m. The constant thermal gradient implies that the initial temperature at this node is $2.5 \times 30 + 0 = 75°\text{C}$. The gradient is purely vertical, so all nodes on the same horizontal layer are at the same temperature. That is:
$$T_E = T_W = T_N = T_S = 75°\text{C}$$
Due to the imposed thermal gradient:
$$T_U = 75 - (30 \times 0.25) = 67.5°\text{C}$$
$$T_D = 75 + (30 \times 0.25) = 82.5°\text{C}$$
We know from the earlier example that $\Delta x = 625$ m, $\Delta y = 2000$ m and $\Delta z = 250$ m, so from Equation (8.16):
$$T = [0.5 \times 10^{-6} + ((67.5 \times 2.3 + 82.5 \times 3.1)/250^2) + ((75 \times 2.3 + 75 \times 2.3)/625^2) + ((75 \times 2.3 + 75 \times 2.3)/2000^2)]/[((2.3 + 3.1)/250^2) + ((2.3 + 2.3)/625^2) + ((2.3 + 2.3)/2000^2)]$$
$$T = [0.5 \times 10^{-6} + (411/62500) + (345/390625) + (345/4000000)]/[(5.4/62500) + (4.6/390625) + (4.6/4000000)]$$
$$T = 0.00754595/0.000099326 = 75.9715°\text{C}$$
The temperature at the node increases by almost $1°$ to bring it into equilibrium.

Equation (8.16) is generally valid for all nodes except those on the top and bottom layers of the model. The temperature at those nodes must be calculated using the appropriate boundary condition. If the basal boundary condition is one of constant heat flow, **Q**, then the equilibrium temperature of nodes at the base of the model is found by adapting Equation (8.12):

$$T = T_U + \mathbf{Q} \, \Delta z / \lambda_U \tag{8.17}$$

Question: If in the previous question we impose a constant heat flow boundary condition through the base of the model, $\mathbf{Q} = 60$ mW m^{-2}, what will be the initial equilibrium temperature of node $n_{16,5,40}$ if the conductivity of the bottom row of cells is 2.3 W m^{-1} K^{-1}?

Answer: T_U is found from the initial constant thermal gradient, 30°C km. In this case, T_U is the temperature of the node immediately above the bottom row, or at a depth of 10,000 $-250 = 9750$ m. So:

$T_U = 0 + 0.030 \times 9750 = 292.5$°C

We know $\mathbf{Q} = 0.060$ W m^{-2}, $\Delta z = 250$ m and $\lambda_U = 2.3$ W m^{-1} K^{-1}, so from Equation (8.17):

$T = 292.5 + (0.060 \times 250/2.3) = 299.0217$°C

Note that on the sides of the model, Equation (8.16) refers to nodes that are outside the domain of the problem. For example, for nodes on the eastern side of the lattice, T_E is undefined. However, recall that the usual boundary condition on the east side of a model (as for the other three sides, too) is one of zero heat loss, and to enforce this, $\lambda_E = 0$. This effectively removes the term in Equation (8.16) that contains T_E. The same goes for nodes on the other three sides.

Equation (8.16) can be solved sequentially for every node in the lattice (except where boundary conditions override). However, the lattice is unlikely to be in equilibrium after just one complete pass because each node was equilibrated with nodes that subsequently changed temperature. But repeated iterations bring the lattice progressively closer to equilibrium, and the process may be stopped when a predetermined precision is attained.

8.2.4. Precision

Prior to implementing the relaxation technique, we must decide at what point we want our iterations to cease. Relaxation converges on an equilibrium solution, but an exact solution is never reached (except to the precision of the computer being used). As the technique relies on convergence *towards* a steady-state solution, and not an *exact* solution, we need to know when to stop the procedure. That is, we must define a necessary precision for our solution.

The relaxation process can be halted when the change in nodal temperatures with each iteration no longer significantly affects the parameter in which we are interested. We should continue the relaxation process until successive iterations alter the calculated parameter by no more than about 1% of the required precision. For example, if we are interested in calculating the surface thermal gradient to a precision of $0.1°C\,km^{-1}$ or better, then we should continue the relaxation until successive iterations alter the surface gradient by no more than $0.001°C\,km^{-1}$.

In general, the change in temperature from one iteration to the next is different for each node. It is the node that shows the *greatest* change in temperature that indicates how close the relaxation process is to completion. If the maximum change in nodal temperature from one iteration to the next is represented by ΔT, then the relaxation process can be halted when ΔT falls below a predetermined cut-off value related to the parameter in which we are interested.

Note: We are interested in the absolute value of ΔT, ignoring sign.

Question: For the example previously described, we are interested in the surface thermal gradient and wish to continue the relaxation process until the gradient is determined to a precision better than $0.1°C\,km^{-1}$. What should be our cut-off value for ΔT?

Answer: We should continue the relaxation process until ΔT alters the vertical thermal gradient by less than 1% of the required precision. That is, less than 1% of $0.1°C\,km$, or less than $0.001°C\,km^{-1}$. The vertical thermal gradient at the surface is approximately

$dT/dz = (T_D - T_0)/\Delta z$

where T_0 is constant (usually) and T_D is the temperature at the node immediately below the surface node. After one iteration, the limit of the surface thermal gradient is

$dT/dz = (T_D + \Delta T - T_0)/\Delta z$

The maximum change in thermal gradient between iterations is therefore

$[(T_D + \Delta T - T_0)/\Delta z] - [(T_D - T_0)/\Delta z] = \Delta T/\Delta z$

We need to continue the relaxation process until this value falls below $0.001°C\,km^{-1}$, or $10^{-6}\,°C\,m^{-1}$:

$\Delta T/\Delta z < 10^{-6}$

$\Delta T < 10^{-6} \times \Delta z$

$\Delta T < 10^{-6}\,°C\,m^{-1} \times 250\ m$

$\Delta T < 2.5 \times 10^{-4}\,°C$

The relaxation process can be stopped when successive iterations alter the nodal temperature by no more than $2.5 \times 10^{-4}\,°C$.

8.3. Errors

All finite difference methods solved using a computer contain inherent errors which must be understood prior to utilising the above technique. The two most important types of error are discretisation error and round-off error.

8.3.1. Discretisation Error

Approximating a derivative by a finite difference inherently introduces an error into our calculations. The error is due to the discretisation of a continuous function, and is thus called *discretisation error*.

For a given function, $F(x)$, differentiation theory states:

$$\frac{dF}{dx} = \lim_{\Delta x \to 0} \frac{F(x + \Delta x) - F(x)}{\Delta x} \tag{8.18}$$

In practice, Δx cannot be made infinitely small, so the approximation will generally not be exact. The centred finite difference approximations of the first and second derivatives are

$$F'(x) = 0.5 \, \Delta x^{-1} [F(x + \Delta x) - F(x - \Delta x)] \tag{8.19a}$$

$$F''(x) = \Delta x^{-2} [F(x + \Delta x) + F(x - \Delta x) - 2F(x)] \tag{8.19b}$$

The Taylor series expansions of $F(x + \Delta x)$ and $F(x - \Delta x)$ are

$$F(x + \Delta x) = F(x) + \Delta x F'(x) + (\Delta x^2/2)F''(x) + (\Delta x^3/6)F'''(x) + \ldots + (\Delta x^n/n!)F^n(x) + \ldots$$

$$F(x - \Delta x) = F(x) - \Delta x F'(x) + (\Delta x^2/2)F''(x) - (\Delta x^3/6)F'''(x) + \ldots + (-1)^n(\Delta x^n/n!)F^n(x) + \ldots$$

When the Taylor series are substituted into the right-hand side of Equations (8.19a, b):

$$F'(x) = 0.5 \, \Delta x^{-1} [2 \, \Delta x F'(x) + (2 \, \Delta x^3/6)F'''(x) \ldots] = F'(x)$$
$$+ (\Delta x^2/6)F'''(x) \ldots \tag{8.20a}$$

$$F''(x) = \Delta x^{-2} [\Delta x^2 F''(x) + (2 \, \Delta x^4/4!)F''''(x) \ldots] = F''(x)$$
$$+ (\Delta x^2/12)F''''(x) \ldots \tag{8.20b}$$

The finite difference approximation of the first derivative effectively discards terms of third and higher order while the approximation of the second derivative discards terms of fourth and higher order. The error in the finite difference approximation is proportional to Δx^2 in both cases. In general, the $F'''(x)$ and higher terms are unknown and variable, but we can be confident that halving Δx will reduced the discretisation error by a factor of four.

The discretisation errors for the forward and backward finite difference approximations of the first derivative are easily found using a procedure similar to that shown above. They can be shown to be proportional to Δx and $F''(x)$, so halving Δx only reduces the discretisation error by a factor of two.

8.3.2. Round-Off Error

Finite difference numerical methods are practically impossible to implement without the aid of digital computers. However, a major limitation of digital computers is that they are only able to store numbers of finite precision. That is, any real number must be rounded off in the computer's memory to a finite number of significant figures (the actual number of significant figures depends on the system being used). This rounding-off causes an error in any number that cannot be precisely represented as a decimal (for example, $\frac{1}{3}$). The loss of precision may be insignificant at first, but in a recursive process the round-off error is carried from one step to the next, and may actually be amplified in the process. After a relatively small number of steps, the results may become totally meaningless. Luckily, relaxation methods are not sensitive to round-off errors and they do not amplify from one iteration to the next. However, round-off errors should be addressed when creating computer applications to solve heat flow problems using other numerical methods.

8.4. Summary

Temperature distribution in the upper crust is controlled by the amount of heat entering from below, the amount of heat generated within the section, the structural geometry and the thermal conductivity contrasts of the rocks. Lithospheric scale models are unable to resolve the temperature distribution in this region. Finite difference numerical models, however, are able to do so.

Heat flow is controlled by conservation of energy laws that can be written as differential equations. The differential equations can be replaced by polynomial approximations. In finite difference numerical modelling, the volume to be modelled is divided into discrete cells of uniform thermal properties, defined by a lattice of nodes over which temperature is estimated. Six boundary conditions are required to define a problem fully in three dimensions. Generally, these boundary conditions will consist of constant temperature at the surface, zero heat loss through the four sides of the model, and either constant temperature or constant heat flow through the base of the model.

Relaxation is a numerical technique that is relatively simple to understand and put into practice. The temperature of each node in the lattice is successively altered to bring the node into thermal equilibrium with its nearest neighbours. Although slower than some matrix-based methods, the relaxation technique has the advantages of being flexible, easy to implement, and relatively insensitive to round-off errors.

The relaxation process should cease when successive iterations alter the parameter in which we are interested by no more than about 1% of the required precision.

Errors inherent in all numerical methods should be kept in mind when implementing relaxation techniques. This is especially true of the discretisation error.

CHAPTER NINE

Unravelling the Thermal History of Sedimentary Basins

We cannot explore the sub-oceanic crust, and do not know from direct evidence whether it has been compressed, as we know has happened to the continental areas. Many geologists feel assured that the oceanic areas have always been oceanic, and that they have never interchanged places with the continents. If that be the case they have never been compressed and elevated.

Physics of the Earth's Crust – Rev. Osmond Fisher, 1881, p. 282.

By this point, the task of recreating the thermal history of a prospective petroleum exploration province probably seems fraught with uncertainty and complicated mathematics. Not only do w e require estimates of such tenuous parameters as the thickness, temperature and thermal properties of the lithosphere, crust and sediments, but we must also interpret often-ambiguous geological evidence for stretching factor, extension rate, age, subsidence and magmatic underplating. Even then, all of these parameters intelligently and carefully chosen only provide a history of heat flow to be superimposed over a background value, which is, itself, often a matter of some conjecture.

With good-quality data, however, most of these complications can be largely overcome. The discussions in Chapter 7 focussed on forward modelling of heat flow history – estimating what the heat flow history of a basin should have been based on its general tectonic age and style. The alternative is to reverse model the thermal history – use physical data from the basin to deduce what the heat flow history must have been. This chapter details the steps for such a procedure. As with any modelling exercise, good-quality data vastly improve the reliability of the final result, so care should be taken with each and every step to ensure the most accurate outcome.

9.1. Determining Present Heat Flow Distribution

The most important constraint on heat flow history is the present heat flow distribution. As a matter of routine, this should be the first priority of any serious attempt to unravel the thermal history of a basin. Every effort should be made to maximise the accuracy of the study using the methods detailed in Chapters 2, 3, 4 and 6. Heat flow should be determined at as many points

293

across the basin as practical, bearing in mind the bias in drill holes toward areas of shallow basement. The aim should be for a statistically reliable sample of the entire basin.

Of particular importance is the heat flow at locations away from the basin, especially if such locations are representative of the continental crust prior to extension. If such is the case, then the average heat flow in these regions can be regarded as a good approximation of the background heat flow, upon which the elevated heat flow due to extension is superimposed. Unfortunately, drill holes of sufficient depth and quality for heat flow measurements may be difficult to find beyond the basin margin.

Note: Background heat flow estimates should be collected at a distance from the basin sufficient to avoid the effects of heat diffraction. A minimum of 10 km beyond a major basin margin fault or unconformity is suggested.

A simple comparison between the average heat flow within the basin and beyond its margins tells us whether the basin is still undergoing thermal subsidence. If there is no appreciable difference across the basin margin, then thermal subsidence has probably run its course. If, however, heat flow within the basin is significantly higher than the surrounding region, this is strong evidence that there is still residual heat in the lithosphere from the extensional event. If heat flow is higher in the surrounding areas, this may be due to a greater thickness of heat-generating crust.

The possibility of advective redistribution of heat within the basin should also be examined at this point. Evidence might include non-linear Bullard plots or anomalous patterns of heat flow distribution not related to basement topography. Physical processes such as known aquifer systems or surface springs should be assessed for possible thermal effects. If possible, heat flow should be calculated *below* the effects of fluid migration.

9.2. Understanding the Tectonic History of the Basin

The tectonic and thermal evolutions of basins are closely linked. Understanding one gives important insights into the other. One of the most important tools for thermal modelling, therefore, is *palinspastic restoration*, which serves to unravel the tectonic development of the basin.

Working with detailed seismic reflection interpretations and reliable compaction models, it is possible to reconstruct the structure and geometry of a basin prior to the deposition of the youngest stratigraphic layer of the basin sequence. From that point, the process can be repeated to reconstruct the form of the basin prior to the deposition of the next youngest formation. Subsequent restorations provide a picture of the basin at progressively earlier times and the process can ideally be repeated right back to the initial rifting event. Such a procedure is referred to as palinspastic restoration.

When completed, palinspastic restoration provides a series of snapshots of the development of the basin. If undertaken with sufficient accuracy, such restorations provide information on the timing and magnitude of movement along faults, progressive rotation of fault blocks, variations in the rate of subsidence, and periods, locations and magnitudes of uplift and erosion.

9.3. Determining the Stretching Factor, β

Palinspastic restoration yields two independent sets of data that may be used to estimate the stretching factor, β. Firstly, by tracking the development of normal faults throughout the history of the basin, their combined effect gives a direct measurement of the amount of horizontal extension that the upper crust has undergone. The total stretching factor, β_t, is simply:

$$\beta_t = \frac{\text{present basin width}}{\text{initial width of extension zone}} \tag{9.1}$$

The spatial distribution of β across the basin can be determined by applying Equations (7.36) and (7.37) to each individual fault. In general, the distribution should resemble Figure 7.18.

Secondly, palinspastic restoration should reveal the average amount of isostatic subsidence, S_i, experienced by the basin during the rifting event. This value can also be derived from the average thickness of syn-rift sediments penetrated by wells, so long as the number and distribution of wells available is statistically reliable. If the latter method is utilised, then the present thickness of syn-rift formations must first be corrected for the effects of compaction. If porosity, ϕ, is represented as a function of depth, z (as described in Section 4.3.5), it can be shown that

$$S_i = z_b - z_t + \int_0^{S_i} \phi(z)dz - \int_{z_t}^{z_b} \phi(z)dz \tag{9.2}$$

where z_t and z_b are the depths to the top and bottom of the syn-rift layer, respectively. Note that S_i is a function of itself, so an iterative approach may be required to determine a solution.

Question: The syn-rift sequence in a particular well is a 2-km-thick section of sandstone, presently at 4500–6500 m depth. A number of porosity measurements suggest a compaction function for the sandstone of the Sclater and Christie (1980) type:

$$\phi = \phi_0 \exp(-Az)$$

where $\phi_0 = 0.6$ and $A = 0.000734$.
What is a good estimate for S_i?

Answer: From the question, we know that $z_b = 6500$ m and $z_t = 4500$ m.

From Equation (9.2):

$S_i = z_b - z_t + (-\phi_0/A)[\exp(-AS_i) - 1] -$
$(-\phi_0/A)[\exp(-6500A) - \exp(-4500A)]$

$S_i = 6500 - 4500 + (-817.44)[\exp(-AS_i) - 1] -$
$(-817.44)[0.00847 - 0.03677]$

$S_i = 2000 + 817.44 - 817.44 \exp(-AS_i) - 23.134$

$S_i = 2794.3 - 817.44 \exp(-AS_i)$

Solved iteratively, $S_i = 2680$ m.

With S_i and some sensible approximations for a number of other parameters, the stretching factor can be approximated. McKenzie (1978) related β to S_i using an instantaneous pure-shear model, as described earlier by Equation (7.26). Zhang (1993) described an alternative relationship that took into account the thermal blanketing effect of the sediment load. The two methods may be compared using the spreadsheet 'IsoSub.XLS', available from the web site mentioned in the Preface.

A third estimate of β can be obtained from deep seismic data if they are available:

$$\beta = \frac{\text{thickness of unextended crust}}{\text{thickness of extended crust}} \tag{9.3}$$

Note: The sediment column is not included in the calculation of the thickness of extended crust.

If all estimates of β agree closely with each other and $\beta < 2$, then we can be confident of the value. For example, Kusznir, Roberts and Morley (1995) predicted $\beta = 1.2$ for Upper Jurassic extension using subsidence data from the Northern Viking Graben in the North Sea. This was found to be equivalent to the amount of extension on upper crustal faults deduced from palinspastic restorations. However, this will not always be the case, and we should recall some of the limitations imposed by β on certain models.

Recall from Section 7.4.2, for example, that the assumption of instantaneous rifting is valid only if $\Delta t < 60/\beta^2$, where Δt is the actual rifting time in millions of years, or the age difference between the top and bottom of the syn-rift sequence. If this condition is not met, then the instantaneous shear model is inadequate and the initial thermal conditions will lie somewhere between the McKenzie (1978) model and the Jarvis and McKenzie (1980) steady-state strain model.

If $\beta > 2$, McKenzie and Bickle (1988) concluded that there is a significant chance of crustal underplating by volumes of mantle melt (see Figure 7.19). The volume of melt, and thus the extra heat carried to the base of the crust, depends on the temperature at the base of the lithosphere, as explained in Section 7.4.6. If melting and underplating has taken place, it may lead to an overestimate of crustal thickness from seismic data, and a subsequent under-estimate of β.

Question: Consider the basin from the previous question, with 2680 m of syn-rift subsidence (filled with sandstone; $\lambda = 2.05 \ \mathrm{W \, m^{-1} \, K^{-1}}$, $\rho = 2150 \ \mathrm{kg \, m^{-3}}$). The basin in which the subsidence was measured is presently 175 km wide, but after palin-spastic restoration it is estimated that the same region was origin-ally 115 km wide. Nearby crust is 37 km thick and the average surface temperature in the basin is 15°C. Assuming that the nearby crust is representative of pre-rift conditions, what is the best esti-mate of β?

Answer: Using the spreadsheet 'IsoSub.XLS', with realistic values for all undefined parameters, the best estimate for β using Zhang's (1993) method is 1.5. This value is quite sensitive to crus-tal thickness. For example, for 40-km-thick crust, the best estimate drops to $\beta = 1.4$, and for 35 km thickness it rises to $\beta = 1.6$. The structural work suggests that $\beta = 175/115 = 1.52$, in excellent agreement with the subsidence model. We would not expect any underplating following such a small amount of stretching, so the crust beneath the basin should be approximately $37/1.5 = 24.7$ km thick.

9.4. Determining Heat Flow Anomaly and Distribution for Each Time Step

Mareschal (1991) explained that thermal subsidence is very closely related to surface heat flow. Recall that thermal subsidence is the result of the cooling and contracting of the lithosphere after stretching and thinning. The heat dissipated through cooling is added to the heat that flows into the base of the lithosphere to cause the elevated heat flow observed at the surface. The magnitude of the heat flow anomaly with respect to background levels is directly proportional to the rate of subsidence, but the actual ratio depends on the basal boundary condition.

    If the base of the lithosphere is modelled as a constant heat flux boundary, then *all* of the heat lost from the contracting lithosphere dissipates through the surface, and the heat flow anomaly, $\mathbf{Q}(t)$ can be shown to be

$$\mathbf{Q}(t) = \frac{\rho c}{\alpha} \times \frac{\partial h}{\partial t} \qquad (9.4)$$

where ρ is density of the lithosphere ($kg\,m^{-3}$); c is specific heat of the lithosphere ($J\,kg^{-1}\,K^{-1}$), α is coefficient of thermal expansion of the lithosphere (K^{-1}) and $\partial h/\partial t$ is rate of thermal subsidence ($m\,s^{-1}$).

Question: If the base of the lithosphere is modelled as a constant heat flux boundary, and the average thermal subsidence in a particular basin over a particular time period is determined to be $10\,m\,Ma^{-1}$, by how much was heat flow elevated above background levels during that time period?

Answer: We will assume the following parameters:

$$\alpha = 3.28 \times 10^{-5}, \quad \rho = 3228.83\ kg\,m^{-3}, \quad c = 700\ J\,kg^{-1}\,K^{-1}$$

Now it is simply a matter of inserting the relevant values into Equation (9.4), after converting the subsidence rate into the correct units:

$$Q(t) = (3228.83 \times 700)/(3.28 \times 10^{-5}) \times (10/(3.1557 \times 10^{13})) = $$
$$6.89 \times 10^{10} \times 3.17 \times 10^{-13} = 0.0218\ W\,m^{-2}$$

Average surface heat flow during the time interval in question was elevated by $21.8\ mW\,m^{-2}$ above background levels. It should be noted, however, that the extra heat is unlikely to have been evenly distributed over the basin.

The relationship between thermal subsidence and the surface heat flow anomaly becomes more complicated if the basal condition is modelled as a constant temperature boundary. Unfortunately, if we recall that the base of the lithosphere is *defined* as an isotherm, we cannot in good conscience avoid the added complications. The complications arise because, in order to maintain a constant temperature at a given depth, some of the heat being lost through the cooling of the lithosphere must flow downward, which reduces the amount of heat dissipating through the surface. Mareschal (1991) demonstrated that the proportion of heat that exits through the surface approaches 50% at times long after the rifting event:

$$Q(t) \rightarrow \frac{1}{2} \times \frac{\rho c}{\alpha} \times \frac{\partial h}{\partial t}, \qquad \text{as } t \rightarrow \infty \tag{9.5}$$

At times soon after the rifting event, the proportion of heat that dissipates through the surface depends on the stretching factor, β. Immediately following rifting:

$$Q(t) = \frac{\beta - 1}{\beta} \times \frac{\rho c}{\alpha} \times \frac{\partial h}{\partial t} \tag{9.6}$$

As time passes, the amount of heat flowing through the surface approaches that defined by Equation (9.5). At any time, t, the proportion of heat flowing through the surface can be approximated by

$$\mathbf{Q}(t) \approx \left[\frac{1}{2} + \left(\frac{(\beta/2) - 1}{\beta}\right)e^{-0.066t}\right] \times \frac{\rho c}{\alpha} \times \frac{\partial h}{\partial t} \qquad (9.7)$$

where t is in millions of years.

Question: Reassess the heat flow anomaly in the previous question assuming a constant temperature basal condition and a stretching factor, $\beta = 2.5$. The time period in question was 25 Ma after the rifting event.

Answer: As well as the parameters from the previous example, we also have

$$\beta = 2.5, \qquad t = 25 \text{ Ma}$$

The answer to Equation (9.7) will be the product of Equation (9.4) and the additional factor in square brackets, so using the results from the previous example we can say:

$$\mathbf{Q}(t) \approx [0.5 + (((2.5/2) - 1)/2.5) \times \exp(-0.066 \times 25)] \times$$
$$0.0218 = 0.5192 \times 0.0218 = 0.0113 \text{ W m}^{-2}$$

Average surface heat flow during the time interval in question was elevated 11.3 mW m^{-2} above background levels.

The average heat flow anomaly given by Equation (9.7) should be thought of as extra heat entering the base of the crust. It should be stressed that this is an *average* heat flow anomaly, and is unlikely to be evenly distributed across the basin. The distribution of anomalous heat will mimic the distribution of β, with regions of higher β having correspondingly higher heat flow anomalies. For intracontinental basins, this may mean that the heat flow anomaly is reasonably constant across the basin (see Figure 7.18), but for passive margins the anomaly may increase offshore.

Note: The heat flow anomaly is added to the background mantle heat flow to find the total heat flow at the base of the crust at any specific time. Surface heat flow is the sum of mantle heat flow and the radiogenic heat generated within the crust. The background level of mantle heat flow must be determined independently, possibly using the heat flow province concept discussed in Section 2.1.1.

Mareschal (1991) made these calculations assuming instantaneous rifting. As for the determination of β, the subsidence–heat flow calculations are valid only if $\Delta t < 60/\beta^2$, where Δt is the rifting time in millions of years, or the age difference between the top and bottom of the syn-rift sequence. For slow rifting events, the heat flow may be very difficult to derive from the subsidence history.

9.5. Solving for Temperature Distribution in Crust

At this stage, we can reconstruct the heat flow history at the base of the crust by superimposing the heat flow anomaly associated with thermal subsidence onto the background heat flow. We have already determined the structural configuration of the basin at different times through palinspastic restoration, and know the heat flow at the base of the crust at each of the time slices. All that remains is to determine the heat distribution within the crust.

Note: Temperature equilibrates within a sequence on a time-scale proportional to the time constant, $\tau = L^2/\pi^2\kappa$, where L is thickness and κ is the thermal diffusivity. For the lithosphere (e.g. $L = 100$ km, $\kappa = 8 \times 10^{-7}$ m^2 s^{-1}), $\tau = 40$ Ma. For a thick sequence of sediments (e.g. $L = 10$ km, $\kappa = 1.5 \times 10^{-7}$ m^2 s^{-1}), $\tau = 2.1$ Ma. It is apparent that the equilibration time for the sediments is rapid compared with the lithosphere as a whole, and, to a reasonable approximation, can be thought of as instantaneous.

Heat, like many other forms of energy, prefers to flow along paths of least resistance. In the case of extended crust with heat conducting from below, this means that heat is preferentially channelled through regions of elevated basement. Low-conductivity sediments act as a retardant to heat flow. The best way to determine the temperature distribution resulting from heat flow partitioning is through numerical modelling of the basin architecture. A reasonably simple method of doing this was presented in Chapter 8.

Three-dimensional numerical models developed for a number of time steps throughout the history of a basin should accurately represent the physical and thermal conditions prevailing at the time they represent. If constructed accurately, the results allow us to determine the temperature at any point and time within the basin's development. Initially, models should be developed at the crustal scale, and the basal boundary condition should be one of constant heat flow. The distribution of heat flow across the base of the crust at any particular time is found following the procedures detailed in Sections 9.3 and 9.4. The solution to the crustal scale numerical models determines the heat flow history at the base of the basin sequence. A second series of numerical models focussing in detail on the basin itself, and using the heat flow results from the crustal scale models, should fully determine the temperature distribution within the basin at each time step.

The temperature history of specific points, such as the top of a potential source bed penetrated by a well, can be effectively traced using the above procedure. With the temperature history fully determined, organic maturity can be calculated using the methods described in Chapter 5, and the results compared with actual measurements. Any discrepancies must be due to either errors in the maturity measurements or problems with the thermal model. A high level of confidence can be had in the thermal history model if due care was taken at every step in the construction of the model and the modelled maturity matches the observed maturity.

9.6. Summary

The thermal history of a sedimentary basin is intrinsically a three-dimensional problem and can only be adequately addressed following a detailed structural analysis and palinspastic restoration. By carefully following a rigid course of model development, the thermal history of key sequences can be reconstructed with reasonable accuracy and compared with observed organic maturity.

The general sequence of investigation into the thermal history of a sedimentary basin should begin with a detailed investigation of the present heat flow distribution both within and outside the basin. A careful palinspastic restoration of the basin should follow this, in order to determine the structure of the basin at past times and the total amount of stretching undergone by the lithosphere. Subsidence data give an excellent indication of the heat flow history at the base of the crust, and this data can be input into three-dimensional numerical models of crust and basin architecture in order to calculate the history of temperature distribution within the basin.

All these steps, if carefully followed, give the most accurate picture possible of the thermal development of a basin. This is of primary importance in the development of petroleum generation models from potential source beds and a necessary step in any serious hydrocarbon exploration strategy.

References

Abeles, B. (1963). Lattice thermal conductivity of disordered semi-conductor alloys at high temperatures. *Physical Review*, 131(5), 1906–11.

Ádám, A. (1978). Geothermal effects in the formation of electrically conducting zones and temperature distribution in the Earth. *Physics of the Earth and Planetary Interiors*, 17, P21–8.

Adams, J.A.S. and Weaver, C.E. (1958). Thorium to uranium ratios as indicators of sedimentary processes: Example of the concept of geochemical facies. *American Association of Petroleum Geologists Bulletin*, 42, 387–430.

Agrawal, P.K., Thakur, N.K. and Negi, J.G. (1992). MAGSAT data and Curie-depth below Deccan Flood Basalts (India). *Pageoph*, 138(1), 61–75.

Allis, R.G. (1979). A heat production model for stable continental crust. *Tectonophysics*, 57, 151–65.

Amundsen, H.E.F., Griffin, W.L. and O'Reilly, S.Y. (1987). The lower crust and upper mantle beneath northwestern Spitsbergen: Evidence from xenoliths and geophysics. *Tectonophysics*, 139, 169–85.

Anderson, D.L., Schreiber, E., Liebermann, R.C. and Soga, N. (1968). Some elastic constant data on minerals relevant to geophysics. *Reviews of Geophysics*, 6(4), 491–524.

Anderson, R.N., McKenzie, D. and Sclater, J.G. (1973). Gravity, bathymetry, and convection in the Earth. *Earth and Planetary Science Letters*, 18, 391–402.

Andrews-Speed, C.P., Oxburgh, E.R. and Cooper, B.A. (1984). Temperatures and depth-dependent heat flow in the western North Sea. *American Association of Petroleum Geologists Bulletin*, 68, 1764–81.

Arshavskava, N.I., Galdin, N.E., Karus, E.W., Kuznetsov, O.L., Lubimova, E.A., Milanovsky, S.Y., Nartikoev, V.D., Semashko, S.A. and Smirnova, E.V. (1987). Geothermic investigations. In *The Superdeep Well of the Kola Peninsula*, ed. Y.A. Kozlovsky, pp. 387–93. Berlin: Springer-Verlag.

Ashwal, L.D., Morgan, P., Kelley, S.A. and Percival, J.A. (1987). Heat production in an Archean crustal profile and implications for heat flow and mobilization of heat-producing elements. *Earth and Planetary Science Letters*, 85, 439–50.

Asquith, G. (1991). *Log Evaluation of Shaly Sandstones: A Practical Guide*. Continuing Education Course Note Series No. 31. Tulsa, Oklahoma: American Association of Petroleum Geologists.

Asquith, G.B. and Gibson, C.R. (1982). *Basic Well Log Analysis for Geologists*. Tulsa, Oklahoma: American Association for Petroleum Geologists.

Auld, M.J. (1948). Temperature gradients for convection in well models. *Journal of Applied Physics*, 19, 218.

Baldwin, B. and Butler, C.O. (1985). Compaction curves. *American Association of Petroleum Geologists Bulletin*, 69(4), 622–6.

Barin, I. (1993). *Thermochemical Data of Pure Substances*, 2nd ed., vol. 1. Weinheim, Germany: VCH.

Barker, C. (1996). *Thermal Modeling of Petroleum Generation: Theory and Applications.* Amsterdam: Elsevier.

Barker, C.E. and Pawlewicz, M.J. (1986). The correlation of vitrinite reflectance with maximum temperature in humic organic matter. In *Lecture Notes in Earth Sciences*, vol. 5, *Paleogeothermics*, eds. G. Buntebarth and L. Stegena, pp. 79–93. Berlin: Springer-Verlag.

Beach, R.D., Jones, F.W. and Majorowicz, J.A. (1987). Heat flow and heat generation estimates for the Churchill Basement of the Western Canadian Basin in Alberta, Canada. *Geothermics*, 16, 1–16.

Beardsmore, G.R. (1996). The Thermal History of the Browse Basin and its Implications for Petroleum Exploration. Ph.D. dissertation, Monash University, Victoria, Australia.

Beardsmore, G.R. and O'Sullivan, P.B. (1994). *Report on Vitrinite Reflectance and Apatite Fission Track Analysis in Six Browse Basin Oil Exploration Wells.* Destructive analysis report to the Australian Geological Survey Organisation.

Beardsmore, G.R. and O'Sullivan, P.B. (1995). Uplift and erosion on the Ashmore Platform, North West Shelf: Conflicting evidence from maturation indicators. *Australian Petroleum Exploration Association Journal*, 35(1), 333–43.

Beck, A.E. (1977). Climatically perturbed temperature gradients and their effect on regional and continental heat flow means. *Tectonophysics*, 41, 17–39.

Beck, A.E. (1988). Thermal properties. In *Handbook of Terrestrial Heat-Flow Density Determination*, eds. R. Haenel, L. Rybach and L. Stegena, pp. 87–124. Dordrecht: Kluwer Academic Publishers.

Beck, A.E. and Balling, N. (1988). Determination of virgin rock temperatures. In *Handbook of Terrestrial Heat-Flow Density Determination*, eds. R. Haenel, L. Rybach and L. Stegena, pp. 59–85. Dordrecht: Kluwer Academic Publishers.

Beck, A.E., Anglin, F.M and Sass, J.H. (1971). Analysis of heat flow data: *In situ* thermal conductivity measurements. *Canadian Journal of Earth Sciences*, 8, 1–19.

Benfield, A.E. (1939). Terrestrial heat flow in Great Britain. *Proceedings of the Royal Society of London*, 173A, 428–50.

Bennett, R.H., Bryant, W.R. and Keller, G.H. (1981). Clay fabric of selected submarine sediments: Fundamental properties and models. *Journal of Sedimentary Petrology*, 51, 217–32.

Bethke, C.M. and Altaner, S.P. (1986). Layer-by-layer mechanism of smectite illitization and application to a new rate law. *Clays and Clay Minerals*, 34(2), 136–45.

Birch, F. and Clark, H. (1940). The thermal conductivity of rocks and its dependence upon temperature and composition. *American Journal of Science*, 238, 529–58.

Birch, F., Roy, R.F. and Decker, E.R. (1968). Heat flow and thermal history in New England and New York. In *Studies of Appalachian Geology: Northern and Maritime*, ed. E-an Zen, pp. 437–51. New York: Wiley Interscience.

Blackwell, D.D. and Steele, J.L. (1989a). Thermal conductivity of sedimentary rocks: Measurement and significance. In *Thermal History of Sedimentary Basins*, eds. N.D. Naeser and T.H. McCulloch, pp. 45–96. New York: Springer-Verlag.

Blackwell, D.D. and Steele, J.L. (1989b). Heat flow and geothermal potential of Kansas. In *Kansas Geological Survey Bulletin*, vol. 226, *Geophysics in Kansas*, ed. D.W. Steeples, pp. 267–96. Wichita, Kansas: Kansas Geological Survey.

Blackwell, D.D., Wisian, K.W. and Beardsmore, G.R. (1997). Application of temperature logging technology to increasing the accuracy of basin thermal models. In *Applications of Emerging Technologies: Unconventional Methods in Exploration for Petroleum and Natural Gas*, vol. 5, eds. R.J. Kruizenga et al., pp. 41–65. Dallas, Texas: Institute for the Study of Earth and Man, Southern Methodist University.

Blackwell, D.D., Beardsmore, G.R., Nishimori, R.K. and McMullin, R.J. Jr. (1999). High-resolution temperature logs in a petroleum setting: examples and applications. In *Geothermics in Basin Analysis*, eds. A. Förster and D.F. Merriam, pp. 1–34. Dordrecht: Kluwer Academic/Plenum Publishers.

Bodri, L. and Bodri, B. (1978). Numerical investigation of tectonic flow in island-arc areas. *Tectonophysics*, 50, 163–75.

Bodri, L. and Cermák, V. (1993). Heat production in the continental lithosphere, part II: Variational approach. *Tectonophysics*, 225, 29–34.

Boreham, C.J., Crick, I.H. and Powell, T.G. (1988). Alternative calibration of the Methylphenanthrene Index against vitrinite reflectance: Application to maturity measurements on oils and sediments. *Organic Geochemistry*, 12, 289–94.

Boudreau, A.E., Love, C. and Hoatson, D.M. (1993). Variation in the composition of apatite in the Munni Munni Complex and associated intrusions of the West Pilbara Block, Western Australia. *Geochimica et Cosmochimica Acta*, 57(18), 4467–77.

Boville, B.A. and Gent, P.R. (1998). The NCAR climate system model, version one. *Journal of Climate*, 11(6), 1115–30.

Boyd, F.R. (1973). A pyroxene geotherm. *Geochimica et Cosmochimica Acta*, 37(12), 2533–46.

Boyd, F.R. (1984). Siberian geotherm based on lherzolite xenoliths from the Udachnaya kimberlite, USSR. *Geology*, 12, 528–30.

Brennan, R.P. (1997). *Heisenburg Probably Slept Here: The Lives, Times and Ideas of the Great Physicists of the 20th Century*. New York: John Wiley and Sons.

Brown, A.F. (1968). *Statistical Physics*. Edinburgh: Edinburgh University Press.

Brown, G.C. and Mussett, A.E. (1993). *The Inaccessible Earth. An Integrated View of its Structure and Composition*. 2nd ed. London: Chapman and Hall.

Bruce, R.H., Middleton, M.F., Holyland, P., Loewenthal, D. and Bruner, I. (1996). Modelling of petroleum formation associated with heat transfer due to hydrodynamic processes. *Petroleum Exploration Society of Australia Journal*, 24, 6–12.

Brune, J.N., Henyey, T.L. and Roy, R.F. (1969). Heat flow, stress, and rate of slip along the San Andreas Fault, California. *Journal of Geophysical Research*, 74(15), 3821–7.

Bücker, C. and Rybach, L. (1996). A simple method to determine heat production from gamma logs. *Marine and Petroleum Geology*, 13(4), 373–5.

Bullard, E.C. (1939). Heat flow in South Africa. *Proceedings of the Royal Society of London, A*, 173, 428–50.

Bullard, E.C. (1947). The time taken for a bore hole to attain temperature equilibrium. *Monthly Notices of the Royal Astronomical Society, Geophysics Supplement*, 5, 127–30.

Bullard, E.C. (1954). The flow of heat through the floor of the Atlantic Ocean. *Proceedings of the Royal Society of London, A*, 222, 408–29.

Bullard, E.C., Maxwell, A.E. and Revelle, R. (1956). Heat flow through the deep sea floor. *Advanced Geophysics*, 3, 153–81.

Buntebarth, G. and Stegena, L. (eds.) (1986). *Lecture Notes in Earth Sciences*, vol. 5, *Paleogeothermics*. Berlin: Springer-Verlag.

Burke, K.C. and Wilson, J.T. (1976). Hot spots on the Earth's surface. *Scientific American*, 235(2), 46–57.

Burnham, A.K. and Sweeney, J.J. (1989). A chemical kinetic model of vitrinite maturation and reflectance. *Geochimica et Cosmochimica Acta*, 53, 2649–57.

Cameron, A.G.W. and Benz, W. (1991). The origin of the Moon and the single impact hypothesis IV. *Icarus*, 92, 204–16.

Carmichael, R.S. (1989). *Practical Handbook of Physical Properties of Rocks and Minerals*. Boca Raton, Florida: CRC Press.

Carslaw, H.A. and Jaeger, J.C. (1959). *Conduction of Heat in Solids*, 2nd ed. Oxford: Oxford University Press.

Cermák, V. (1979). Heat flow map of Europe. In *Terrestrial Heat Flow in Europe*, eds. V. Cermák and L. Rybach, pp. 3–40. Berlin: Springer-Verlag.

Cermák, V. and Bodri, L. (1986). Two dimensional temperature modelling along five East-European geotraverses. *Journal of Geodynamics*, 5, 133–63.

Cermák, V. and Bodri, L. (1993). Heat production in the continental crust, part I: Data converted from seismic velocities and their attempted interpretation. *Tectonophysics*, 225, 15–28.

Cermák, V. and Rybach, L. (eds.) (1991). *Terrestrial Heat Flow and the Lithosphere Structure*. Berlin: Springer-Verlag.

Cermák, V., Bodri, L., Rybach, L. and Buntebarth, G. (1990). Relationship between seismic velocity and heat production: Comparison of two sets of data and test of validity. *Earth and Planetary Science Letters*, 99, 48–57.

Cermák, V., Bodri, L. and Rybach, L. (1991). Radioactive heat production in the continental crust and its depth dependence. In *Terrestrial Heat Flow and the Lithosphere Structure*, eds. V. Cermák and L. Rybach, pp. 23–69. Berlin: Springer-Verlag.

Cermák, V., Safanda, J., Bodri, L., Pollack, H.N. and Chapman, D.S. (1996). Climate change derived from borehole temperatures: Information from Europe (abs.). *Eos, Transactions, American Geophysical Union*, 77(46 suppl.), F43.

Chamley, H. (1989). *Clay Sedimentology*. Berlin: Springer-Verlag.

Chen, J., Fu, J., Sheng, G., Liu, D. and Zhang, J. (1996). Diamondoid hydrocarbon ratios: Novel maturity indices for highly mature crude oils. *Organic Geochemistry*, 25(3/4), 179–90.

Clark, S.P. (1957). Radiative transfer in the Earth's mantle. *Transcripts of the American Geophysical Union*, 38, 931–8.

Cleary, J.R., Simpson, D.W., and Muirhead, K.J. (1972). Variations in Australian upper mantle structure from observations of the Cannikin explosion. *Nature*, 236, 111–12.

Collins, M. (1996). Apatite Fission Track Study of Tectonics and Provenance in the Sawtooth Range, Sverdrup Basin, Canadian Arctic Archipelago. B.Sc.(Honours) dissertation, Dalhousie University, Nova Scotia, Canada.

Cooper, G.T., O'Sullivan, P.B., Sherwood, N. and Hill, K.C. (1998). Assessing maturity and the effects of diagenetic oxidation through the integration of fluorescence alteration of multiple macerals (FAMM^{TM}) and apatite fission track thermochronology (AFTT). *Petroleum Exploration Society of Australia Journal*, 26, 151–60.

Cooper, L.R. and Jones, C. (1959). The determination of virgin strata temperatures from observations in deep survey boreholes. *Geophysical Journal of the Royal Astronomical Society*, 2, 116–31.

Correia, M. (1967). Relations possibles entre l'état de conservation des éléments figurés de la matière organique et l'existence de gisements d'hydrocarbures. *Revue de L'Institut Français du Pétrole*, 22(9), 1285–306.

Craig, G.Y. and Jones, E.J. (1982). *A Geological Miscellany*. Oxford: Oxford Orbital Press.

Crain, I.K. (1968). The glacial effect and the significance of continental terrestrial heat flow measurements. *Earth and Planetary Science Letters*, 4, 69–72.

Cull, J.P. (1974). Thermal conductivity probes for rapid measurements in rock. *Journal of Physics, E: Scientific Instruments*, 7, 771–4.

Cull, J.P. (1975). The Pressure and Temperature Dependence of Thermal Conductivity within the Earth. Ph.D. dissertation, Oxford University, Oxford, Great Britain.

Cull, J.P. (1979). Climatic corrections to Australian heat flow data. *BMR Journal of Australian Geology and Geophysics*, 4, 303–7.

Cull, J.P. (1980). Geothermal records of climatic change in New South Wales. *Search*, 11(6), 201–3.

Cull, J.P. (1982). An appraisal of Australian heat flow data. *BMR Journal of Australian Geology and Geophysics*, 7, 11–21.

Cull, J.P. (1990). Underplating of the crust and xenolith geotherms in Australia. *Geophysical Research Letters*, 17(8), 1133–6.

Cull, J.P. (1991). Heat flow and regional geophysics in Australia. In *Terrestrial Heat Flow and the Lithosphere Structure*, eds. V. Cermák and L. Rybach, pp. 486–500. Berlin: Springer-Verlag.

Cull, J.P. and Denham, D. (1979). Regional variations in Australian heat flow. *BMR Journal of Australian Geology and Geophysics*, 4, 1–13.

Cull, J.P., O'Reilly, S.Y. and Griffin, W.L. (1991). Xenolith geotherms and crustal models in Eastern Australia. *Tectonophysics*, 192, 359–66.

Davis, E.E. (1988). Oceanic heat flow density. In *Handbook of Terrestrial Heat-Flow Density Determination*, eds. R. Haenel, L. Rybach and L. Stegena, pp. 223–60. Dordrecht: Kluwer Academic Publishers.

Deaton, B.C., Nestell, M. and Balsam, W.L. (1996). Spectral reflectance of conodonts: A step toward quantitative color alteration and thermal maturity indexes. *American Association of Petroleum Geologists Bulletin*, 80(7), 999–1007.

Deer, W.A., Howie, R.A. and Zussman, J. (1990). *An Introduction to the Rock-Forming Minerals*, 2nd ed. London: Longman Scientific and Technical.

Defant, A. (1961). *Physical Oceanography*, vol. 1. London: Pergamon Press.

Deming, D. (1994). Estimation of the thermal conductivity anisotropy of rock with application to the determination of terrestrial heat flow. *Journal of Geophysical Research*, 99(B11), 22087–91.

Demongodin, L., Pinoteau, B., Vasseur, G. and Gable, R. (1991). Thermal conductivity and well logs: A case study in the Paris Basin. *Geophysics Journal International*, 105, 675–91.

Detrick, R.S. and Crough, S.T. (1978). Island subsidence, hot spots, and lithospheric thinning. *Journal of Geophysical Research*, 83(B3), 1236–44.

Detrick, R.S., Von Herzen, R.P., Crough, S.T., Epp, D. and Fehn, U. (1981). Heat flow on the Hawaiian Swell and lithospheric reheating. *Nature*, 292(5819), 142–3.

DeVries, D.A. and Peck, A.J. (1958). On the cylindrical probe method of measuring thermal conductivity with special reference to soils. *Australian Journal of Physics*, 11, 255–71.

Dietz, R.S. and Menard, H.W. (1953). Hawaiian swell, deep, and arch and subsidence of the Hawaiian Islands. *Journal of Geology*, 61, 99–113.

Diment, W.H. (1967). Thermal regime of a large diameter borehole: Instability of the water column and comparison of air- and water-filled conditions. *Geophysics*, 32(4), 720–6.

Diment, W.H. and Pratt, H.R. (1988). *Thermal Conductivity of Some Rock-Forming Minerals: A Tabulation*. USGS open file report no. 88–690. Reston, Virginia: United States Geological Survey.

Dowdle, W.L. and Cobb, W.M. (1975). Static formation temperature from well logs – an empirical method. *Journal of Petroleum Technology*, 27, 1326–30.

Drummond, B.J., Muirhead, K.J., Wright, C., and Wellman, P. (1989). A teleseismic travel time residual map of the Australian continent. *BMR Journal of Australian Geology and Geophysics*, 11, 101–5.

Drury, M.J. (1986). *Thermal Conductivity, Thermal Diffusivity, Density and Porosity of Crystalline Rocks*. Earth Physics Branch open file report no. 86–5. Ottawa: Earth Physics Branch.

Drury, M.J. (1989). The heat flow–heat generation relationship: Implications for the nature of continental crust. In *Tectonophysics*, vol. 164, *Heat Flow and Lithospheric Structure*, eds. V. Cermák, L. Rybach and E.R. Decker, pp. 93–106. Amsterdam: Elsevier.

Durand, B. (1980). Sedimentary organic matter and kerogen. Definition and quantitative importance of kerogen. In *Kerogen: Insoluble Organic Matter from Sedimentary Rocks*, ed. B. Durand, pp. 13–34. Paris: Paris Editons Technip.

Durand, B. and Monin, J.C. (1980). Elemental analysis of kerogens (C, H, O, N, S, Fe). In *Kerogen: Insoluble Organic Matter from Sedimentary Rocks*, ed. B. Durand, pp. 113–42. Paris: Editons Technip.

Dziewonski, A.M. and Anderson, D.L. (1981). Preliminary reference Earth model. *Physics of the Earth and Planetary Interiors*, 25, 297–356.

Elsasser, W.M. (1971). Sea-floor spreading as thermal convection. *Journal of Geophysical Research*, 76, 1101–12.

Emsley, J. (1989). *The Elements*. Oxford: Clarendon Press.

Epp, D., Grim, P.J. and Langseth, M.G. (1970). Heat flow in the Caribbean and Gulf of Mexico. *Journal of Geophysical Research*, 75(29), 5655–69.

Epstein, A.G., Epstein, J.B. and Harris, L.D. (1977). Conodont color alteration – An index to organic metamorphism. *United States Geological Survey, Professional Paper*, 995, 1–27.

Faas, R.W. and Crocket, D.S. (1983). Clay fabric development in deep-sea core: Site 515, Deep Sea Drilling Project leg 72. In *Initial Report of the Deep Sea Drilling Project*, vol. 72, eds. P.F. Backer et al., pp. 519–35. Washington DC: United States Government Printing Office.

Falvey, D. and Middleton, M. (1981). Passive continental margins: Evidence for a pre-breakup, deep crustal metamorphic subsidence mechanism. *Oceanologica Acta*, 4, 103–14.

Fisher, A.T., Von Herzen, R.P., Blum, P., Hoppie, B. and Wang, K. (1999). Evidence may indicate recent warming of shallow slope bottom water off New Jersey shore. *Eos, Transactions, American Geophysical Union*, 80(15), 165.

Fisher, O. (1881). *Physics of the Earth's Crust*. London: MacMillan and Co.

Fleischer, R.L., Price, P.B. and Walker, R.M. (1975). *Nuclear Tracks in Solids: Principles and Applications*. Berkeley: University of California Press.

Flueh, E.R. and Grevemeyer, I. (1999). Ocean site survey reveals anatomy of a hotspot track. *Eos, Transactions, American Geophysical Union*, 80(7), 77.

Foland, S.S. (1997). Comparison of palaeotemperature indicator techniques: AFTA, VR, FAMM and ICA – Considerations for designing sampling programs. *APPEA Journal*, 37(1), 455–71.

Foldvik, A. and Kvinge, T. (1974). Conditional instability of sea water at the freezing point. *Deep-sea Research*, 21, 169–74.

Forbes, J.D. (1849). Account of some experiments on the temperature of the Earth at different depths, and in different soils, near Edinburgh. *Proceedings of the Royal Society of Edinburgh*, 189–222.

Förster, A. and Merriam, D.F. (eds.) (1999). *Geothermics in Basin Analysis*. Dordrecht: Kluwer Academic/Plenum Publishers.

Fournier, R.O. and Rowe, J.J. (1966). Estimation of underground temperatures from the silica content of water from hot springs and wet-steam wells. *American Journal of Science*, 264, 685–97.

Fournier, R.O. and Truesdell, A.H. (1973). An empirical Na–K–Ca geothermometer for natural waters. *Geochimica et Cosmochimica Acta*, 37(5), 1255–75.

Fukao, Y., Mizutani, H. and Uyeda, S. (1968). Optical absorption spectra at high temperatures and radiative thermal conductivity of olivines. *Physics of the Earth and Planetary Interiors*, 1(2), 57–62.

Furlong, K.P, Hanson, R.B. and Bowers, J.R. (1991). Modeling thermal regimes. In *Reviews in Mineralogy*, vol. 26, *Contact Metamorphism*, ed. D.M. Kerrick, pp. 437–505. Washington DC: Mineralogical Society of America.

Galbraith, R.F. (1981). On statistical models for fission track counts. *Mathematical Geology*, 13, 471–88.

Gallagher, K. (1990). Some strategies for estimating present day heat flow from exploration wells, with examples. *Exploration Geophysics*, 21, 145–59.

Gasparini, P., Mantovani, M.S.M., Corrado, G. and Rapolla, A. (1979). Depth of Curie temperature in continental shields: A compositional boundary? *Nature*, 278, 845–6.

Gass, I.G., Smith, A.G. and Vine, F.J. (1975). *Geodynamics Today: Origin and Emplacement of Ophiolites*. London: The Royal Society.

Gates, W.L. (1976). Modelling the ice age climate. *Science*, 191, 1138–44.

George, S.C., Smith, J.W. and Jardine, D.R. (1994). Vitrinite reflectance suppression in coal due to a marine transgression: A case study of the organic geochemistry of the Greta Seam, Sydney Basin. *Australian Petroleum Exploration Association Journal*, 34(1), 241–55.

Gerard, R., Langseth, M.G. and Ewing, M. (1962). Thermal gradient measurements in the bottom water and bottom sediment of the western Atlantic. *Journal of Geophysical Research*, 67(2), 785–803.

Giancoli, D.C. (1984). *General Physics*. London: Prentice/Hall International.

Gleadow, A.J.W., Duddy, I.R. and Lovering, J.F. (1983). Fission track analysis: A new tool for the evaluation of thermal histories and hydrocarbon potential. *Australian Petroleum Exploration Association Journal*, 23, 93–102.

Gleadow, A.J.W., Duddy, I.R., Green, P.F. and Lovering J.F. (1986). Confined fission track lengths in apatite: A diagnostic tool for thermal history analysis. *Contributions to Mineral Petrology*, 94, 405–15.

Gosnold, W.D., Todhunter, P.E. and Schmidt, W. (1997). The borehole temperature record of climate warming in the mid-continent of North America. *Global and Planetary Change*, 15(1–2), 33–45.

Green, P.F. (1986). On the thermo-tectonic evolution of northern England: Evidence from fission track analysis. *Geology*, 5, 493–506.

Green, P.F., Duddy, I.R., Gleadow, A.J.W., Tingate, P.T. and Laslett, G.M. (1985). Fission-track annealing in apatite: Track length measurements and the form of the Arrhenius plot. *Nuclear Tracks*, 10, 323–8.

Green, P.F., Duddy, I.R., Gleadow, A.J.W., Tingate, P.T. and Laslett, G.M. (1986). Thermal annealing of fission tracks in apatite: 1 – a qualitative description. *Isotope Geoscience*, 59, 237–53.

Green, P.F., Duddy, I.R., Gleadow, A.J.W. and Lovering, J.F. (1989a). Apatite fission track analysis as a paleotemperature indicator for hydrocarbon exploration. In *Thermal History of Sedimentary Basins – Methods and Case Histories*, ed. N.D. Naeser, pp. 181–95. New York: Springer-Verlag.

Green, P.F., Duddy, I.R., Laslett, G.M., Hegarty, K.A., Gleadow, A.J.W. and Lovering, J.F. (1989b). Thermal annealing of fission tracks in apatite 4. Qualitative modelling techniques and extensions to geological time scales. *Chemical Geology (Isotope Geoscience)*, 79, 155–82.

Gretener, P.E. (1967). On the thermal instability of large diameter wells – an observational report. *Geophysics*, 32(4), 727–38.

Gretener, P. E. (1981). *Geothermics: Using Temperature in Hydrocarbon Exploration*. Tulsa, Oklahoma: American Association of Petroleum Geologists.

Gretener, P.E. and Curtis, C.D. (1982). Role of temperature and time on organic metamorphism. *American Association of Petroleum Geologists Bulletin*, 66(8), 1124–49.

Grivet, M., Rebetez, M., Ben Ghouma, N., Chambaudet, A., Jonckheere, R. and Mars, M. (1993). Apatite fission-track age correlation and thermal history analysis from projected track length distributions. *Chemical Geology*, 103, 157–69.

Großwig, S., Hurtig, E. and Kühn, K. (1996). Fibre optic temperature sensing: A new tool for temperature measurements in boreholes. *Geophysics*, 61(4), 1065–7.

Gutjahr, C.C.M. (1966). Carbonization measurements of pollen-grains and spores and their application. *Leidse Geologische Mededelingen*, 38, 1–29.

Haenel, R., Rybach, L. and Stegena, L. (eds.) (1988). *Handbook of Terrestrial Heat-Flow Density Determination*. Dordrecht: Kluwer Academic Publishers.

Hales, A.L. (1937). Convection currents in geysers. *Monthly Notices of the Royal Astronomical Society, Geophysics Supplement*, 4, 122–31.

Hamilton, E.L. (1976). Variations of density and porosity with depth in deep-sea sediments. *Journal of Sedimentary Petrology*, 46(2), 280–300.

Hamilton, L.J. (1986). Statistical features of the oceanographic area off south-western Australia, obtained from bathythermograph data. *Australian Journal of Marine and Freshwater Research*, 37, 421–36.

Hamilton, R.M. (1965). Temperature variation at constant pressures of the electrical conductivity of periclase and olivine. *Journal of Geophysical Research*, 70, 5679–92.

Hamza, V.M. (1979). Variation of continental mantle heat flow with age: Possibility of discriminating between thermal models of the lithosphere. *Pure and Applied Geophysics*, 117(1/2), 65–74.

Hamza, V.M. and Beck, A.E. (1972). Terrestrial heat flow, the neutrino problem, and a possible energy source in the core. *Nature*, 240, 343–4.

Harker, A. (1932). *Metamorphism*. London: Methuen and Co.

Hawkesworth, C. J. (1974). Vertical distribution of heat production in the basement of the Eastern Alps. *Nature*, 249, 435–6.

Hays, J.D., Imbrie, J. and Shackleton, N.J. (1976). Variations in the Earth's orbit: Pacemaker of the ice ages. *Science*, 194(4270), 1121–32.

Heestand, R.L. and Crough, S.T. (1981). The effect of hot spots on the oceanic age-depth relation. *Journal of Geophysical Research*, 86(B7), 6107–14.

Helsen, S., David, P. and Fermont, W.W.J. (1995). Calibration of conodont color alteration using color image analysis. *Journal of Geology*, 103(3), 257–67.

Henk, A., Franz, L., Teufel, S. and Oncken, O. (1997). Magmatic underplating, extension, and crustal reequilibration: Insights from a cross-section through the Ivrea Zone and Strona-Ceneri Zone, Northern Italy. *Journal of Geology*, 105, 367–77.

Hermanrud, C., Cao, S. and Lerche, I. (1990). Estimates of virgin rock temperature derived from BHT measurements: Bias and error. *Geophysics*, 55(7), 924–31.

Hood, A., Gutjahr, C.C.M. and Heacock, R.L. (1975). Organic metamorphism and the generation of petroleum. *American Association of Petroleum Geologists Bulletin*, 59(6), 986–96.

Horai, K. and Simmons, G. (1969). Thermal conductivity of rock-forming minerals. *Earth and Planetary Science Letters*, 6, 359–68.

Horner, D.R. (1951). Pressure build-up in wells. *Proceedings of the Third World Petroleum Congress, The Hague*, 2, 924–31.

Horváth, F., Dövényi, P., Szalay, A. and Royden, L.H. (1988). Subsidence, thermal and maturation history of the Great Hungarian Plain. In *Memoir 45: The Pannonian Basin – A Study in Basin Evolution*, eds. L.H. Royden and F. Horváth, pp. 355–72. Tulsa, Oklahoma: American Association of Petroleum Geologists.

Houbolt, J.J.H.C. and Wells, P.R.A. (1980). Estimation of heat flow in oil wells based on a relation between heat conductivity and sound velocity. *Geologie en Mijnbouw*, 59(3), 215–24.

Houseman, G.A., McKenzie, D.P. and Molnar, P. (1981). Convective instability of a thickened boundary layer and its relevance for the thermal evolution of continental convergent belts. *Journal of Geophysical Research*, 86(B7), 6115–32.

Houseman, G.A., Cull, J.P., Muir, P.M. and Paterson, H.L. (1989). Geothermal signatures and uranium ore deposits on the Stuart Shelf of South Australia. *Geophysics*, 54(2), 158–70.

Howard, L.E. and Sass, J.H. (1964). Terrestrial heat flow in Australia. *Journal of Geophysical Research*, 69, 1617–26.

Hunt, J.M., Lewan, M.D. and Hennet, R.J.-C. (1991). Modeling oil generation with time–temperature index graphs based on the Arrhenius equation. *American Association of Petroleum Geologists Bulletin*, 75(4), 795–807.

Hurford, A.J. and Green, P.F. (1982). A users' guide to fission-track dating calibration. *Earth and Planetary Science Letters*, 59, 343–54.

Hurford, A.J. and Green, P.F. (1983). The zeta age calibration of fission-track dating. *Isotopic Geoscience*, 1, 285–317.

Hyndman, R.D. and Everett, J.E. (1968). Heat flow measurements in a low radioactivity area of the Western Australian Precambrian Shield. *Geophysics Journal*, 14, 479–86.

Hyndman, R.D., Rodgers, G.C., Bone, M.N., Lister, C.R.B., Wade, U.S., Barrett, D.L., Davis, E.E., Lewis, T.J., Lynch, S. and Seemann, D. (1978). Geophysical measurements in the region of the Explorer Ridge of Western Canada. *Canadian Journal of Earth Sciences*, 15, 1508–25.

Issler, D.R. (1984). Calculation of organic maturation levels for offshore Eastern Canada; Implications for general application of Lopatin's method. *Canadian Journal of Earth Sciences*, 21(4), 477–88.

Issler, D.R. and Beaumont, C. (1989). A finite element model of the subsidence and thermal evolution of extensional basins: Application to the Labrador continental margin. In *Thermal History of Sedimentary Basins: Methods and Case Histories*, eds. N.D. Naeser and T.H. McCulloh, pp. 239–67. New York: Springer-Verlag.

Jaeger, J.C. (1956). Conduction of heat in an infinite region bounded internally by a circular cylinder of a perfect conductor. *Australian Journal of Physics*, 9(2), 167–79.

Jamieson, J.C. and Lawson, A.W. (1958). High temperature heat conductivity of some metal oxides. *Journal of Applied Physics*, 29, 1313–14.

Jarvis, G.T. and McKenzie, D.P. (1980). Sedimentary basin formation with finite extension rates. *Earth and Planetary Science Letters*, 48, 42–52.

Jaupart, C., Sclater, J.G., and Simmons, G. (1981). Heat flow studies: Constraints on the distribution of uranium, thorium, and potassium in the continental crust. *Earth and Planetary Science Letters*, 52, 328–44.

Jeffreys, H. (1937). Notes on Mr. Hales's paper. *Monthly Notices of the Royal Astronomical Society, Geophysics Supplement*, 4, 131.

Jessop, A.M. (1990). *Developments in Solid Earth Geophysics*, vol. 17, *Thermal Geophysics*. Amsterdam: Elsevier.

Jessop, A.M., Hobart, M.A., and Sclater, J.G. (1976). *Geothermal Series*, vol. 5, *The World Heat Flow Data Collection – 1975*. Ottawa: Earth Physics Branch.

Keen, C.E. and Lewis, T. (1982). Measured radiogenic heat production in sediments from the continental margin of eastern North America: Implications for petroleum generation. *American Association of Petroleum Geologists Bulletin*, 66, 1402–7.

Kern, H. and Siegesmund, S. (1989). A test of the relationship between seismic velocity and heat production for crustal rocks. *Earth and Planetary Science Letters*, 92, 89–94.

Khavari Khorasani, G. and Michelsen, J.K. (1993). The thermal evolution of solid bitumens, bitumen reflectance, and kinetic modeling of reflectance: Application in petroleum and ore prospecting. *Energy Sources*, 15(2), 181–204.

Kinsman, D.J.J. (1975). Rift valley basins and sedimentary history of trailing continental margins. In *Petroleum and Global Tectonics*, eds. G. Fischer and S. Judson, pp. 83–126. Princeton, New Jersey: Princeton University Press.

Kisch, H.J. (1980). Illite crystallinity and coal rank associated with lowest-grade metamorphism of the Taveyanne greywacke in the Helvetic zone of the Swiss Alps. *Eclogae Geologicae Helvetiae*, 73(3), 753–77.

Klemens, P.G. (1958). Thermal conductivity and lattice vibration modes. *Solid State Physics: Advances in Research and Applications*, 7, 1–98.

Kozlovsky, Y.A. (ed.) (1987). *The Superdeep Well of the Kola Peninsula*. Berlin: Springer-Verlag.

Kreyszig, E. (1983). *Advanced Engineering Mathematics*. New York: John Wiley and Sons.

Kübler, B. (1967). La crystallinité de l'illite et les zones tout à fait supérieures du index of illite métamorphisme. In *Etages Tectoniques (Colloque de Neuchâtel, 18–21 Avril 1966)*, pp. 105–22. Neuchâtel: Neuchâtel University Institute Geology.

Kukkonen, I.T. and Peltonen, P. (1999). Xenolith-controlled geotherm for the central Fennoscandian Shield: Implications for lithosphere–asthenosphere relations. *Tectonophysics*, 304, 310–15.

Kusznir, N.J. and Matthews, D.H. (1988). Deep seismic reflections and the deformational mechanics of the continental lithosphere. In *Journal of Petrology*, Special Volume, *Oceanic and Continental Lithosphere: Similarities and Differences*, eds. M.A. Menzies and K.G. Cox, pp. 63–87. Oxford, United Kingdom: Clarendon Press.

Kusznir, N.J. and Ziegler, P.A. (1992). The mechanics of continental extension and sedimentary basin formation: A simple-shear/pure-shear flexural cantilever model. In *Tectonophysics*, vol. 215, *Geodynamics of Rifting, Volume III. Thematic Discussions*, ed. P.A. Ziegler, pp. 117–31. Amsterdam: Elsevier.

Kusznir, N.J., Roberts, A.M. and Morley, C.K. (1995). Forward and reverse modelling of rift basin formation. In *Special Publication*, vol. 80, *Hydrocarbon Habitat in Rift Basins*, ed. J.J. Lambiase, pp. 35–56. London: Geological Society of London.

Kutasov, I.M. (1999). *Developments in Petroleum Science*, vol. 48, *Applied Geothermics for Petroleum Engineers*. Amsterdam: Elsevier.

Lachenbruch, A.H. (1968). Preliminary geothermal model of the Sierra Nevada. *Journal of Geophysical Research*, 73, 6977–89.

Lachenbruch, A.H. (1970). Crustal temperature and heat productivity: Implications of the linear heat flow relation. *Journal of Geophysical Research*, 75, 3291–300.

Lachenbruch, A.H. and Brewer, M.C. (1959). Dissipation of the temperature effect of drilling a well in Arctic Alaska. *United States Geological Survey Bulletin*, 1083-C, 73–109.

Lachenbruch, A.H. and Bunker, C.M. (1971). Vertical gradients of heat production in the continental crust, 2. Some estimates from borehole data. *Journal of Geophysical Research*, 76, 3852–60.

Lachenbruch, A.H. and Sass, J.H. (1980). Heat flow and energetics of the San Andreas Fault Zone. *Journal of Geophysical Research*, 85, 6185–223.

Lachenbruch, A.H., Sass, J.H. and Morgan, P. (1994). Thermal regime of the southern Basin and Range Province: 2. Implications of heat flow for regional extension and metamorphic core complexes. *Journal of Geophysical Research*, 99(B11), 22121–33.

Lal, D., Rajan, R.S. and Tamhane, A.S. (1969). Chemical composition of nuclei of Z > 22 in cosmic rays using meteoric minerals as detectors. *Nature*, 221, 33–7.

LaPlante, R.E. (1974). Hydrocarbon generation in the Gulf Coast Tertiary sediments. *American Association of Petroleum Geologists Bulletin*, 58, 1281–9.

Laslett, G.M., Kendall, W.S., Gleadow, A.J.W. and Duddy, I.R. (1982). Bias in measurement of fission-track length distributions. *Nuclear Tracks*, 6, 79–85.

Laslett, G.M., Green, P.F., Duddy, I.R. and Gleadow, A.J.W. (1987). Thermal modelling of fission tracks in apatite: 2. A quantitative analysis. *Chemical Geology*, 65, 1–13.

Lawson, A.W. (1957). On the high temperature heat conductivity of insulators. *Journal of the Physics and Chemistry of Solids*, 3, 155–6.

Lee, D.-C., Halliday, A.N., Snyder, G.A. and Taylor, L.A. (1997). Age and origin of the moon. *Science*, 278(5340), 1098–103.

Lee, W.H.K. (ed.) (1965). *Monograph 8, Terrestrial Heat Flow*. Washington DC: American Geophysical Union.

Lee, W.H.K. and Uyeda, S. (1965). Review of heat flow data. In *Monograph 8, Terrestrial Heat Flow*, ed. W.H.K. Lee, pp. 87–190. Washington DC: American Geophysical Union.

Leeds, A.R., Knopoff, L. and Kausel, E.G. (1974). Variations of upper mantle structure under the Pacific Ocean. *Science*, 186, 141–3.

Lees, C.H. (1910). On the shapes of the isogeotherms under mountain ranges in radio-active districts. *Proceedings of the Royal Society of London, A*, 83, 339–46.

Liebfried, G. and Schlömann, E. (1954). Wärmeleitung in elektrisch isolierenden Kristallen. *Nachrichten der Akademie der Wissenschaften in Göttingen, IIa. Mathematisch-Physikalische Klasse*, 4, 71–93.

Lilley, F.E.M., Sloane, M.N. and Sass, J.H. (1978). A compilation of Australian heat flow measurements. *Journal of the Geological Society of Australia*, 24, 439–45.

Lisk, M. and Eadington, P. (1994). Oil migration in the Cartier Trough, Vulcan Sub-basin. In *The Sedimentary Basins of Western Australia. (Proceedings of Symposium, Perth, WA)*, eds. P.G. Purcell and R.R. Purcell, pp. 301–12. Perth: Petroleum Exploration Society of Australia.

Lopatin, N.V. (1971). Temperature and geologic time as a factor in coalification. *Izveotiya Akademii. Nauk SSSR, Seriya Geolicheskya*, 3, 95–106.

Lopatin, N.V. (1976). The influence of temperature and geologic time on the catagenetic processes of coalification and petroleum and gas formation. In *Isseledovaniya*

Organicheskogo Veshchestva Sovremennykh i Iskopayemykh Osakdov (Study of Organic Matter in Recent and Old Sediments), pp. 361–6. Moscow: Nauka Press.

Lubimova, E.A. (1958). Thermal history of the Earth with consideration of the variable thermal conductivity in the mantle. *Geophysical Journal*, 1, 115–34.

Lubimova, E.A. (1969). Thermal history of the Earth. In *Geophysical Monograph 13. The Earth's Crust and Upper Mantle*, ed. P.J. Hart, pp. 63–77. Washington DC: American Geophysical Union.

MacDonald, G.J.F. (1959). Calculations on the thermal history of the Earth. *Journal of Geophysical Research*, 64, 1967–2000.

MacKenzie, A.S. and McKenzie, D. (1984). Isomerization and aromatization of hydrocarbons in sedimentary basins formed by extension. *Geological Magazine*, 120(5), 417–70.

MacKenzie, A.S., Patience, R.L., Maxwell, J.R., Vandenbroucke, M. and Durand, B. (1980). Molecular parameters of maturation in the Toarcian shales, Paris Basin, France – I. Changes in the configurations of acyclic isoprenoid alkanes, steranes and triterpanes. *Geochimica et Cosmochimica Acta*, 44, 1709–21.

MacKenzie, A.S., Lewis, C.A. and Maxwell, J.R. (1981). Molecular parameters of maturation in the Toarcian shales, Paris Basin, France – IV. Laboratory thermal alteration studies. *Geochimica et Cosmochimica Acta*, 45, 2369–76.

Majorowicz, J.A. and Jessop, A.M. (1981). Regional heat flow patterns in the Western Canadian sedimentary basin. *Tectonophysics*, 74, 109–238.

Majorowicz, J.A., Gough, D.I. and Lewis, T.J. (1993). Electrical conductivity and temperature in the Canadian Cordilleran crust. *Earth and Planetary Science Letters*, 115, 57–64.

Mareschal, J.-C. (1991). Determination of past heat flow from subsidence data in intracontinental basins and passive margins. In *Terrestrial Heat Flow and the Lithosphere Structure*, eds. V. Cermák and L. Rybach, pp. 70–85. Berlin: Springer-Verlag.

McKenna, T.E. and Sharp J.M. (1998). Radiogenic heat production in sedimentary rocks of the Gulf of Mexico Basin, South Texas. *American Association of Petroleum Geologists Bulletin*, 82(3), 484–96.

McKenna, T.E., Sharp J.M. and Lynch, F.L. (1995). Thermal conductivity and radiogenic heat production from South Texas coast (Gulf of Mexico Basin)(abs.). *Abstracts with Programs, Geological Society of America*, 27(6), 443.

McKenzie, D. (1978). Some remarks on the development of sedimentary basins. *Earth and Planetary Science Letters*, 40, 25–32.

McKenzie, D.P. (1981). The variation of temperature with time and hydrocarbon maturation in sedimentary basins formed by extension. *Earth and Planetary Science Letters*, 55, 87–98.

McKenzie, D. and Bickle, M.J. (1988). The volume and composition of melt generated by extension of the lithosphere. *Journal of Petrology*, 29(3), 625–79.

Mercier, J. and Carter, N.L. (1975). Pyroxene geotherms. *Journal of Geophysical Research*, 80(23), 3349–62.

Middleton, M.F. (1993). A transient method of measuring the thermal properties of rocks. *Geophysics*, 58(3), 357–65.

Milankovic, M. (1969). *Special Publications 132. Canon of Insolation and the Ice-Age Problem.* Jerusalem: Israel Program for Scientific Translations. [English translation of Milankovic, M. (1941). *Kanon der Erdbestrahlung und seine Anwendung auf das Eiszeitenproblem.* Belgrade: Royal Serbian Academy.]

Molnar, P. and England, P. (1990). Temperatures, heat flux, and frictional stress near major thrust faults. *Journal of Geophysical Research*, 95(B4), 4833–56.

Montagner, J.-P. and Anderson, D.L. (1989). Constrained reference mantle model. *Physics of the Earth and Planetary Interiors*, 58, 205–27.

Mooney, D.L. and Steg, R.G. (1969). Pressure dependence of the thermal conductivity and ultrasonic attenuation of non-metallic solids. *High Temperatures – High Pressures*, 1(2), 237–40.

Moore, A., Bradshaw, J. and Edwards, D. (1996). Geohistory modelling of hydrocarbon migration and trap formation in the Arafura Sea. *Petroleum Exploration Association of Australia Journal*, 24, 35–51.

Moore, M.E., Gleadow, A.J.W. and Lovering, J.F. (1986). Thermal evolution of rifted continental margins: New evidence from fission tracks in basement apatites from southeastern Australia. *Earth and Planetary Science Letters*, 78, 255–70.

Morrow, D.W. and Issler, D.R. (1993). Calculation of vitrinite reflectance from thermal histories: A comparison of some methods. *American Association of Petroleum Geologists Bulletin*, 77(4), 610–24.

Mory, A.J., Nicoll, R.S. and Gorter, J.D. (1998). Lower Palaeozoic correlations and thermal maturity, Carnarvon Basin, WA. In *The Sedimentary Basins of Western Australia 2. (Proceedings of Symposium, Perth, WA)*, eds. P.G. Purcell and R.R. Purcell, pp. 599–611. Perth: Petroleum Exploration Society of Australia.

Naeser, C.W. (1967). The use of apatite and sphene for fission track age determinations. *Geological Society of America Bulletin*, 78, 1523–6.

Naeser, N.D. and McCulloh, T.H. (eds.) (1989). *Thermal History of Sedimentary Basins*. New York: Springer-Verlag.

Newstead, G. and Beck, A. (1953). Borehole temperature measuring equipment and the geothermal flux in Tasmania. *Australian Journal of Physics*, 6, 480–9.

Nicoll, R.S. and Foster, C.B. (1994). Late Triassic conodont and palynomorph biostratigraphy and conodont thermal maturation, North West Shelf, Australia. *AGSO Journal of Australian Geology and Geophysics*, 15(1), 101–18.

Nicoll, R.S. and Gorter, J.D. (1984). Conodont colour alteration, thermal maturation and the geothermal history of the Canning Basin, Western Australia. *Australian Petroleum Exploration Association Journal*, 24(1), 414–29.

Oleskevich, D.A., Hyndman, R.D. and Wang, K. (1999). The updip and downdip limits to great subduction earthquakes: Thermal and structural models of Cascadia, south Alaska, SW Japan, and Chile. *Journal of Geophysical Research*, 104(B7), 14965–91.

O'Reilly, S.Y. and Griffin, W.L. (1985). A xenolith-derived geotherm for southeastern Australia and its geophysical implications. *Tectonophysics*, 111, 41–63.

Oxburgh, E.R. and Turcotte, D.L. (1976). The physico-chemical behaviour of the descending lithosphere. *Tectonophysics*, 32(1–2), 107–28.

Parsons, B. and Sclater, J.G. (1977). An analysis of the variation of ocean floor bathymetry and heat flow with age. *Journal of Geophysical Research*, 82(5), 803–27.

Pearson, N.J., O'Reilly, S.Y. and Griffin, W.L. (1995). The crust–mantle boundary beneath cratons and craton margins: A transect across the south-west margin of the Kaapvaal craton. *Lithos*, 36, 257–87.

Percival, J.A. and Card, K.D. (1983). Archean crust as revealed in the Kapuskasing Uplift, Ontario. *Geology*, 11, 323–6.

Peters, K.E. and Moldowan, J.M. (1993). *The Biomarker Guide. Interpreting Molecular Fossils in Petroleum and Ancient Sediments*. New Jersey: Prentice Hall.

Pollack, H.N. and Chapman, D.S. (1977). Mantle heat flow. *Earth and Planetary Science Letters*, 34, 174–84.

Pollack, H.N., Hurter, S.J. and Johnson, J.R. (1993). Heat flow from the Earth's interior: Analysis of the global data set. *Reviews of Geophysics*, 31(3), 267–80.

Press, F. and Siever, R. (1986). *Earth*, 4th ed. New York: W.H. Freeman and Co.

Prestwich, J. (1886). On underground temperatures, with observations on the conductivity of rocks, on the thermal effects of saturation and imbibition, and on a special source of heat in mountain ranges. *Proceedings of the Royal Society*, 41, 1–116.

Price, L.C. and Barker, C.E. (1985). Suppression of vitrinite reflectance in amorphous rich kerogen – a major unrecognised problem. *Journal of Petroleum Geology*, 8, 59–84.

Price, P.B. and Walker, R.M. (1963). Fossil tracks of charged particles in mica and the age of minerals. *Journal of Geophysical Research*, 68, 4847–62.

Price, P.J. (1955). Ambipolar thermodiffusion of electrons and holes in semi-conductors. *Philosophical Magazine*, 46, 1252–60.

Pytte, A.M. and Reynolds, R.C. (1989). The thermal transformation of smectite to illite. In *Thermal History of Sedimentary Basins: Methods and Case Histories*, eds. N.D. Naeser and T.H. McCulloh, pp. 133–40. New York: Springer-Verlag.

Radke, M. and Welte, D.H. (1983). The methylphenanthrene index (MPI): A maturity parameter based on aromatic hydrocarbons. In *Advances in Organic Geochemistry, 1981. (Proceedings of the 10th International Meeting on Organic Geochemistry, University of Bergen, Norway, 14–18 September 1981)*, eds. M. Bjoroy et al., pp. 504–12. Chichester, United Kingdom: Wiley and Sons.

Raznjevic, K. (1976). *Handbook of Thermodynamic Tables and Charts*. Washington DC: Hemisphere Publishing Corp.

Reiter, M.A. and Jessop, A.M. (1985). Estimates of terrestrial heat flow in offshore Eastern Canada. *Canadian Journal of Earth Sciences*, 22, 1503–17.

Reiter, M.A. and Tovar, R.J.C. (1982). Estimates of terrestrial heat flow in Northern Chihuahua, Mexico. *Canadian Journal of Earth Sciences*, 22, 1503–17.

Reiter, M.A., Mansure, A.J. and Peterson, B.K. (1980). Precision continuous temperature logging and comparison with other types of logs. *Geophysics*, 45(12), 1857–68.

Remmelts, G. (1993). Salt tectonics in the Southern North Sea, Netherlands (abs.). *American Association of Petroleum Geologists Bulletin*, 77(9), 1658.

Rider, M.H. (1991). *The Geological Interpretation of Well Logs*, 2nd ed. Caithness, United Kingdom: Whittles Publishing.

Ringwood, A.E. (1969). Composition and evolution of the upper mantle. In *Geophysical Monograph 13. The Earth's Crust and Upper Mantle*, ed. P.J. Hart, pp. 1–17. Washington DC: American Geophysical Union.

Ringwood, A.E. (1979). Composition and origin of the Earth. In *The Earth – Its Origin, Structure and Evolution*, ed. M.W. McElhinny, pp. 1–58. New York: Academic Press.

Robert, P. (1988). *Organic Metamorphism and Geothermal History*. Dordrecht: Elf-Aquitaine and D. Reidel Publishing Co.

Roedder, E. (1981). Origin of fluid inclusions and changes that occur after trapping. In *Fluid Inclusions: Applications to Petrology*, eds. L.S. Hollister and M.L. Crawford, pp. 101–37. Toronto, Ontario: Mineralogical Association of Canada.

Ross, E.W., Vagelatos, N., Dickerson, J.M. and Nguyen, V. (1982). Nuclear logging and geothermal log interpretation: Formation temperature sonde evaluation. In *Geothermal Log Interpretation Handbook*, ed. J.K. Hallenburg, pp. V7–52. Tulsa, Oklahoma: Society of Professional Well Log Analysts.

Roux, B., Sanyal, S.K and Brown, S.L. (1982). An improved approach to estimating true reservoir temperature from transient temperature data. In *Geothermal Log Interpretation Handbook*, ed. J.K. Hallenburg, pp. V53–60. Tulsa, Oklahoma: Society of Professional Well Log Analysts.

Roy, R.F., Blackwell, D.D. and Birch, F. (1968). Heat generation of plutonic rocks and continental heat flow provinces. *Earth and Planetary Science Letters*, 5, 1–12.

Roy, R.F., Beck, A.E. and Touloukian, Y.S. (1981). Thermophysical properties of rocks. In *Physical Properties of Rocks and Minerals*, eds. Y.S. Touloukian, W.R. Judd and R.F. Roy, pp. 409–502. New York: McGraw-Hill.

Rybach, L. (1979). The relationship between seismic velocity and radioactive heat production in crustal rocks: An exponential law. *Pure and Applied Geophysics*, 117(1/2), 75–82.

Rybach, L. (1986). Amount and significance of radioactive heat sources in sediments. In *Collection Colloques et Seminares 44, Thermal Modeling of Sedimentary Basins*, ed. J. Burrus, pp. 311–22. Paris: Paris Editons Technip.

Rybach, L., and Buntebath, G. (1984). The variation of heat generation, density and seismic velocity with rock type in the continental lithosphere. *Tectonophysics*, 103, 335–44.

Sandiford, M. (1999). Mechanics of basin inversion. *Tectonophysics*, 305, 109–20.

Sakaguchi, K. and Matsushima, N. (1995). Temperature profile monitoring in geothermal wells by distributed temperature sensing technique. *Transactions of the Geothermal Resource Council*, 19, 355–8.

Sass, J.H. and Lachenbruch, A.H. (1979). The thermal regime of the Australian continental crust. In *The Earth – Its Origin, Structure and Evolution*, ed. M.W. McElhinny, pp. 301–52. New York: Academic Press.

Sass, J.H., Lachenbruch, A.H. and Jessop, A.M. (1971a). Uniform heat flow in a deep hole in the Canadian Shield and its palaeoclimatic implications. *Journal of Geophysical Research*, 76, 8586–96.

Sass, J.H., Lachenbruch, A.H. and Munroe, R.J. (1971b). Thermal conductivity of rocks from measurements on fragments and its application to heat flow determinations. *Journal of Geophysical Research*, 76(14), 3391–401.

Sass, J.H., Jaeger, J.C. and Munroe, R.J. (1976). *Heat Flow and Near Surface Radioactivity in the Australian Continental Crust*. USGS open file report no. 76–250. Reston, Virginia: United States Geological Survey.

Sass, J.H., Lachenbruch, A.H., Moses, T.H. and Morgan, P. (1992). Heat flow from a scientific research well at Cajon Pass, California. *Journal of Geophysical Research*, 97(4), 5017–30.

Sato, H. (1992). Thermal structure of the mantle wedge beneath northeastern Japan: Magmatism in an island arc from the combined data of seismic anelasticity and velocity and heat flow. *Journal of Volcanology and Geothermal Research*, 51, 237–52.

Sawyer, D.S., Bangs, N.L and Golovchenko, X. (1994). Deconvolving Ocean Drilling Program temperature logging tool data to improve borehole temperature estimates: Chile triple junction. *Journal of Geophysical Research*, 99(B6), 11995–2003.

Schatz, J.F. and Simmons, G. (1972). Thermal conductivity of Earth materials at high temperatures. *Journal of Geophysical Research*, 77(35), 6966–83.

Schenk, H.J., Horsfield, B., Krooss, B., Schaefer, R.G. and Schwochau, K. (1997). Kinetics of petroleum formation and cracking. In *Petroleum and Basin Evolution: Insights from Petroleum Geochemistry, Geology and Basin Modelling*, eds. D.H. Welte, B. Horsfield and D.R. Baker, pp. 231–69. Berlin: Springer-Verlag.

Schopf, J.M. (1948). Variable coalification: The process involved in coal formation. *Economic Geology*, 43(3), 207–25.

Sclater, J.G. and Christie, P.A.F. (1980). Continental stretching: An explanation of the post Mid-Cretaceous subsidence of Central North Sea Basin. *Journal of Geophysical Research*, 85(B7), 3711–39.

Sclater, J.G. and Francheteau, J. (1970). The implication of terrestrial heat flow observations on current tectonic and geochemical models of the crust and upper mantle of the Earth. *Geophysical Journal of the Royal Astronomical Society*, 20, 509–42.

Sclater, J.G., Crowe, J. and Anderson, R.N. (1976). On the reliability of oceanic heat flow averages. *Journal of Geophysical Research*, 81(17), 2997–3006.

Sears, F.W., Zemansky, M.W. and Young, H.D. (1978). *University Physics*, 5th ed. Reading, Massachusetts: Addison-Wesley Publishing Co.

Sekiguchi, K. (1984). A method for determining terrestrial heat flow in oil basinal areas. *Tectonophysics*, 103, 67–79.

Serra, O. (1984). *Fundamentals of Well-Log Interpretation; 1. The Acquisition of Logging Data*. Amsterdam: Elsevier.

Sherby, O.D. (1962). Factors affecting the high temperature strength of polycrystalline solids. *Acta Metallurgica*, 10, 135–47.

Sieber, K.G. (1986). Compositional Variation in Apatites. B.Sc. (Honours) dissertation, University of Melbourne, Victoria, Australia.

Simmons, G. and Horai, K. (1968). Heat flow data 2. *Journal of Geophysical Research*, 73(20), 6608–29.

Singh, R.N. and Negi, J.G. (1980). A variational approach to model of depth dependence of radiogenic heat in the crust. *Geophysical Research Letters*, 7, 209–10.

Slichter, L.B. (1941). Cooling of the Earth. *Bulletin of the Geological Society of America*, 52, 561–600.

Smithson, S.B. and Decker, E.R. (1974). A continental crustal model and its geothermal implications. *Earth and Planetary Science Letters*, 22, 215–25.

Somerton, W.H. and Mossahebi, M. (1967). Ring heat source probe for rapid determination of thermal conductivity of rocks. *Review of Scientific Instruments*, 38, 1368–71.

Spector, A. and Grant, F.S. (1970). Statistical models for interpreting aeromagnetic data. *Geophysics*, 35(2), 293–302.

Staplin, F.L. (1969). Sedimentary organic matter, organic metamorphism, and oil and gas occurrence. *Bulletin of Canadian Petroleum Geology*, 17(1), 47–66.

Staplin, F.L. (1982). Determination of thermal alteration index from color of exinite (pollen, spores). In *Shortcourse Number 7. How to Assess Maturation and Paleotemperatures*, ed F.L.Staplin, pp. 7–9. Tulsa, Oklahoma: Society of Economic Paleontologists and Mineralogists.

Stein, C.A. and Stein, S. (1992). A model for the global variation in oceanic depth and heat flow with lithospheric age. *Nature*, 359, 123–8.

Stein, C.A. and Stein, S. (1994). Constraints on hydrothermal heat flux through the oceanic lithosphere from global heat flow. *Journal of Geophysical Research*, 99(B2), 3081–95.

Suzuki, N., Matsubayashi, H. and Waples, D.W. (1993). A simpler kinetic model of vitrinite reflectance. *American Association of Petroleum Geologists Bulletin*, 77(9), 1502–8.

Swanberg, C.A. and Morgan, P. (1979). The linear relation between temperatures based on the silica content of groundwater and regional heat flow: A new heat flow map of the United States. *Pure and Applied Geophysics*, 117, 227–41.

Sweeney, J.J. and Burnham, A.K. (1990). Evaluation of a simple model of vitrinite reflectance based on chemical kinetics. *American Association of Petroleum Geologists Bulletin*, 74(10), 1559–70.

Taylor, A.E., Judge, A. and Allen, V. (1986). Terrestrial heat flow from Project Cesar, Alpha Ridge, Arctic Ocean. *Journal of Geodynamics*, 6, 137–76.

Teichmüller, R. and Teichmüller, M. (1986). Relations between coalification and palaeo-geothermics in Variscan and Alpidic Foredeeps of Western Europe. In *Lecture Notes in Earth Sciences*, vol. 5, *Paleogeothermics*, eds. G. Buntebarth and L. Stegena, pp. 53–78. Berlin: Springer-Verlag.

Thomas, B.M. (1982). Land-plant source rocks for oil and their significance in Australian basins. *Australian Petroleum Exploration Association Journal*, 22(1), 164–78.

Thomson, W. (1860). On the reduction of observations of underground temperature, with application to Professor Forbes' Edinburgh observations, and the continued Calton Hill series. *Transactions of the Royal Society of Edinburgh*, 22, 405–39.

Thomson, W. (1862). On the secular cooling of the Earth. *Proceedings of the Royal Society of Edinburgh*, 23, 157–69.

Ting, T.F.C. (1978). Petrographic techniques in coal analysis. In *Analytical Methods for Coal and Coal Products*, vol. 1, ed. C. Karr, pp. 3–26. Orlando: Academic Press.

Tissot, B. (1984). Recent advances in petroleum geochemistry applied to hydrocarbon exploration. *American Association of Petroleum Geologists Bulletin*, 68(5), 545–63.

Tissot, B.P. and Welte, D.H. (1984). *Petroleum Formation and Occurrence*, 2nd ed. Berlin: Springer-Verlag.

Tissot, B, Durand, B., Espitalié, J. and Combaz, A. (1974). Influence of nature and diagenesis of organic matter in formation of petroleum. *American Association of Petroleum Geologists Bulletin*, 58(3), 499–506.

Toole, J.M. (1981). Sea ice, winter convection, and temperature minimum layer in the Southern Ocean. *Journal of Geophysical Research*, 86, 8037–47.

Touloukian, Y.S., Liley, P.E. and Saxena, S.C. (1970a). *Thermal Properties of Matter*, vol. 3, *Thermal Conductivity, Non-Metallic Liquids and Gases*. New York and Washington: IFI/Plenum.

Touloukian, Y.S., Powell, R.W., Ho, C.Y. and Klemens, P.G. (1970b). *Thermal Properties of Matter*, vol. 2, *Thermal Conductivity, Non-Metallic Solids*. New York and Washington: IFI/Plenum.

Tupper, N.P. and Burckhardt, D.M. (1990). Use of the methylphenanthrene index to characterise expulsion of Cooper and Eromanga Basin oils. *Australian Petroleum Exploration Association Journal*, 30(1), 373–85.

Turcotte, D.L. (1980). On the thermal evolution of the Earth. *Earth and Planetary Science Letters*, 48, 53–8.

Turcotte, D.L. and Oxburgh, E.R. (1969). Convection in a mantle with variable physical properties. *Journal of Geophysical Research*, 74(6), 1458–74.

Urban, T.C., Diment, W.H. and Nathenson, M. (1978). East Mesa geothermal anomaly, Imperial County, California: Significance of temperatures in a deep drill hole near thermal equilibrium. In *Geothermal Energy: A Novelty Becomes Resource (Transactions of Meeting, Hilo, Hawaii, July 25–27)*, vol. 2(2), ed. J. Combs, pp. 667–70. Davis, California: Geothermal Resource Council.

Uyeda, S. and Kanamori, H. (1979). Back-arc opening and the mode of subduction. *Journal of Geophysical Research*, 84, 1049–60.

Van Gijzel, P. (1982). Characterization and identification of kerogen and bitumen and determination of thermal maturation by means of qualitative and quantitative microscopical techniques. In *Shortcourse Number 7. How to Assess Maturation and Paleotemperatures*, ed. F.L. Staplin, pp. 159–216. Tulsa, Oklahoma: Society of Economic Paleontologists and Mineralogists.

Van Krevelen, D.W. (1961). *Coal: Typology, Chemistry, Physics, Constitution*. Amsterdam: Elsevier.

Vigneresse, J.L. and Cuney, M. (1991). Are granites representative of heat flow provinces. In *Terrestrial Heat Flow and the Lithosphere Structure*, eds. V. Cermák and L. Rybach, pp. 86–110. Berlin: Springer-Verlag.

Vink, G.E., Morgan, W.J. and Vogt, P.R. (1985). The Earth's hot spots. *Scientific American*, 252(4), 32–9.

Von Herzen, R.P. and Maxwell, A.E. (1959). The measurement of thermal conductivity of deep-sea sediments by a needle probe method. *Journal of Geophysical Research*, 64, 1557–63.

Von Herzen, R.P. and Scott, J.H. (1991). Thermal modeling for Hole 735B. *Proceedings of the Ocean Drilling Program, Scientific Results*, 118, 349–56.

Von Herzen, R.P. and Uyeda, S. (1963). Heat flow through the eastern Pacific Ocean floor. *Journal of Geophysical Research*, 68, 4219–50.

Von Herzen, R.P., Detrick, R.S., Crough, S.T., Epp, D. and Fehn, U. (1982). Thermal origin of the Hawaiian Swell; heat flow evidence and thermal models. *Journal of Geophysical Research*, 87(B8), 6711–23.

Von Herzen, R.P., Detrick, R.S., Crough, S.T., Epp, D. and Fehn, U. (1983). Thermal origin of the Hawaiian Swell; heat flow evidence and thermal models. *Eos, Transactions, American Geophysical Union*, 64(18), 310.

Walters, R.D. (1993). Reconstruction of allochthonous salt emplacement from 3-D seismic reflection data, Northern Gulf of Mexico. *American Association of Petroleum Geologists Bulletin*, 77(5), 813–41.

Walther, J.V. and Orville, P.M. (1982). Volatile production and transport in regional metamorphism. *Contributions to Mineralogy and Petrology*, 79(3), 252–7.

Wang, K., Mulder, T., Rogers, G.C. and Hyndman, R.D. (1995). Case for very low coupling stress on the Cascadia subduction fault. *Journal of Geophysical Research*, 100(B7), 12907–18.

Waples, D.W. (1980). Time and temperature in petroleum formation: Application of Lopatin's method to petroleum exploration. *American Association of Petroleum Geologists Bulletin*, 64, 916–26.

Webb, F.J., Wilkinson, K.R. and Wilks, J. (1952). The thermal conductivity of solid helium. *Proceedings of the Royal Society of London, A*, 214, 546–63.

Weller, R.A. and Price, J.F. (1988). Langmuir circulation within the oceanographic mixed layer. *Deep-Sea Research*, 35, 711–47.

Wellman, P. and McDougall, I. (1974). Cainozoic igneous activity in Eastern Australia. *Tectonophysics*, 23, 49–65.

Welte, D.H., Horsfield, B. and Baker, D.R. (eds.) (1997). *Petroleum and Basin Evolution: Insights from Petroleum Geochemistry, Geology and Basin Modeling*. New York: Springer-Verlag.

Wernicke, B. (1985). Uniform-sense normal sense simple-shear of the continental lithosphere. *Canadian Journal of Earth Sciences*, 22, 108–25.

Whitmore, D.H. (1960). Excitation processes in ceramics and anomalous increase in thermal conductivity at elevated temperatures. *Journal of Applied Physics*, 31, 1109–12.

Whitten, D.G.A. and Brooks, J.R.V. (1972). *The Penguin Dictionary of Geology*. London: Penguin Books.

Wignall, P.B. (1994). *Black Shales*. Oxford: Clarendon Press.

Wilkins, R.W.T., Wilmshurst, N.J., Russell, N.J., Hladky, G., Ellacott, M.V. and Buckingham, C.P. (1992). Fluorescence alteration and the suppression of vitrinite reflectance. *Organic Geochemistry*, 18, 629–40.

Wilkins, R.W.T., Buckingham, C.P., Sherwood, N., Russell, N.J., Faiz, M. and Kurusingal, J. (1998). The current status of the FAMM thermal maturity technique for petroleum exploration in Australia. *Australian Petroleum Production and Exploration Association Journal*, 38(1), 421–37.

Willett, S.E. and Chapman, D.S. (1987). Analysis of temperatures and thermal processes in the Uinta Basin. In *Memoir 12. Sedimentary Basins and Basin-Forming Mechanisms*, eds. C. Beaumont and A.J. Tankard, pp. 447–61. Calgary, Alberta: Canadian Society of Petroleum Geologists.

Willett, S. and Pope, D. (1996). Thermo-mechanical modelling of orogenic plateau growth by crustal thickening (abs.). *Eos, Transactions, American Geophysical Union*, 77(46 suppl.), 688.

Williams, C.F. and Anderson, R.N. (1990). Thermophysical properties of the Earth's crust: *In situ* measurements from continental and ocean drilling. *Journal of Geophysical Research*, 95(B6), 9209–36.

Wilson, J.T. (1963). A possible origin of the Hawaiian Islands. *Canadian Journal of Physics*, 41(6), 863–70.

Wisian, K.W., Blackwell, D.D., Bellani, S., Henfling, J.A., Normann, R.A., Lysne, P.C., Förster, A. and Schrötter, J. (1998). Field comparison of conventional and new technology temperature logging systems. *Geothermics*, 27(2), 131–41.

Wyllie, P.J. (1988). Solidus curves, mantle plumes and magma generation beneath Hawaii. *Journal of Geophysical Research*, 93(B5), 4171–81.

Wyrtki, K. (1971). *Oceanographic Atlas of the International Indian Ocean Expedition*. Washington DC: National Science Foundation.

Zhang, Y.K. (1993). The thermal blanketing effect of sediments on the rate and amount of subsidence in sedimentary basins formed by extension. *Tectonophysics*, 218, 297–308.

Ziagos, J.P., Blackwell, D.D. and Mooser, F. (1985). Heat flow in Southern Mexico and the thermal effects of subduction. *Journal of Geophysical Research*, 90(7), 5410–20.

Ziman, J.M. (1960). *Electrons and Phonons*. Oxford: Clarendon Press.

Index

Note: f = figure, n = note, q = question, t = table, x = example

β, *see* stretching factor
κ, *see* thermal diffusivity
λ, *see* thermal conductivity

activation energy, 149–50, 152*f*
Africa, hot spots in, 250
arithmetic mean, 98*f*, 99–100
aromatisation, *see* molecular biomarkers
Arrhenius equation, 149, 182
Australia, heat flow provinces in, 231*f*

back-arc basins, 254*f*, 255–6
Beck, A.E., climate research, 82, 86
BHT correction, *see* bottom-hole temperature
bitumen, 147
Blackwell, D.D.
    heat flow provinces, 28
    thermal conductivity of shale, 142
bottom-hole temperature, 59–67
bottom-water temperature, 74–7, 86
Bullard, Sir Edward, pioneering heat flow
        measurements, 6, 19
Bullard plots, 208*f*
    definition, 210–12
    effect of heat generation on, 213
    reasons for non-linearity, 212–14
Bullard probe, 55
Burnham, A.K. and Sweeney, J.J. (1989), *see*
        EASY%Ro
BWT, *see* bottom-water temperature

CAI, *see* conodont alteration index
calcium, *see* groundwater, geothermometers
Canada, surface heat flow, 85
Carslaw, H.A. and Jaeger, J.C. (1959)
    analytical solutions for heat conduction
        problems, 18
    changes in surface temperature, 78, 83
    one-dimensional heat conduction, 16, 17*q*
    sedimentation and surface heat flow, 216
    transient conductivity probes, 116, 118, 122, 124
Cermák, V.
    climate change, 85*x*
    heat generation in the crust, 38*f*
chiral centres, 182*n*
clay minerals (*see also* shale), 174
compaction models, 131*f*, 213
    Baldwin, B. and Butler, C.O. (1985), 129–30

Falvey, D. and Middleton, M. (1981), 128–9,
        132
    Sclater, J.G. and Christie, P.A.F. (1980),
        127–8, 129, 213*q*
compressional-wave velocity
    by rock type, 27
    relationship to heat flow, 232–4
    relationship to heat generation, 36–9
    relationship to thermal conductivity, 92, 136
conductivity, *see* thermal conductivity
conodont alteration index, 172–4, 198, 199*f*, 201*f*
convection
    in boreholes, 58–9
    in the mantle, 9, 10*f*
    of seawater at spreading centres, 243–4
Cooper, L.R. and Jones, C. (1959), BHT
        correction, 60–2, 65, 86
core, formation of, 6
Cull, J.P.
    Australian heat flow, 20*x*, 37, 233, 270*q*
    climate research, 82*f*
    mantle thermal conductivity, 247, 248*f*
    thermal conductivity probes, 118–19, 121, 123
Curie depth, 70–2, 86
Curie temperature, 70

divided bar apparatus
    cell method, 112–15
    construction, 108–12
drill stem test, 59, 86
DST, *see* drill stem test
dual-needle probe, 119–20

Earth
    age of, 4, 9
    origin, 4–6
    separation of core, 6
EASY%Ro, 164–7
effective wavelength of a temperature cycle, 79
electrical resistivity, upper mantle, 72–3, 86
Ewing probe, 55, 56*f*
exinite, *see* liptinite
extension
    passive margins, 265, 272–3
    pure shear, 257–63
    pure-shear/simple-shear coupled model, 265–8
    simple shear, 263–5

FAMM®, 169–71, 198, 199*f*
FIM, *see* fluid inclusion microthermometry
finite difference modelling, 275–9
    boundary conditions, 284–5, 297–8
    discretisation, 279–84
    errors in, 290–1
    precision, 288–9
    relaxation, 279
Fisher, Reverend Osmond, quotations, ix, 237,
    282
fission track thermochronology, 198, 199*f*
    apatite composition, 195–6
    apparent fission track age, 193–4
    chi-squared statistical test, 194
    confined tracks, 191
    definition, 188–9
    limitations, 195–7
    preparation of samples, 190–1, 192*f*
flexural cantilever model, 266
fluid inclusion microthermometry, 179–81
fluorescence alteration diagram, 169*f*
fluorescence alteration of multiple macerals, *see*
    FAMM®
frequency factor, 150

gamma ray
    activity of elements, 32*t*
    electric well log, 31
    relation to heat generation, 32–4, 33*f*, 35*f*, 36*f*
    spectrometry, 26, 31
geometric mean, 98*f*, 100
global warming, 84–5
groundwater
    effect on heat flow, 219–21, 294
    geothermometers, 67–70, 86

half-space cooling model, 238–41
harmonic mean, 97–9, 98*f*
heat capacity, *see* specific heat
heat flow
    correlation with age, 231, 232*f*
    correlation with electrical conductivity, 234–5
    correlation with heat generation, 229, 230*f*
    correlation with P-wave velocity, 232–4
    definition, 12
    effect of groundwater, 219–21, 294
    effect of salt domes, 226–8
    effect of sedimentation and erosion, 216–19
    effect of surface temperature cycles, 214–16
    global averages, 9, 11*t*
    refraction, 222–4
heat flow equation, derivation of, 14–15
heat flow province, 28, 229, 230
heat generation
    calculation from well logs, 30–6
    correlation with heat flow, 229, 230*f*
    definition, 16
    distribution in crust, 23, 28–9, 38*f*
    effect on Bullard plots, 213
    by faults, 39–41
    by metamorphism, 41–3
    in numerical modelling, 282, 285
    relationship to seismic velocity, 36–9
    by rock type, 27*t*
    in sediments, 30, 31*t*

heat production, *see* heat generation
Himalayan mountains, 256
homogenisation temperature, 180
hopanes, *see* molecular biomarkers
Horner plot, 62–5, 86
hot spots, 248–52
hydrocarbon potential, 150

ice age periodicity, *see* Milankovic cycles
illite crystallinity index, 174, 175*f*
illite–smectite ratio, 174–7
inertinite, 155, 157*f*
island arcs, 254*f*, 255
isomerisation, *see* molecular biomarkers
isostatic subsidence, *see* subsidence, isostatic

Jaeger, J.C., *see* Carslaw, H.A. and Jaeger, J.C.
    (1959)
Jarvis and McKenzie (1980), constant strain
    extension, 261–3, 296
Jones, C., *see* Cooper, L.R. and Jones, C. (1959)

Kelvin, Baron of Largs, 7, 9*n*
kerogen
    definition, 147
    types, 147–8
kinetics
    of fission track thermochronology, 196
    of hydrocarbon generation, 149–53
    of smectite–illite transformation, 175–7
    of vitrinite reflectance, 165–6, 168

Lachenbruch, A.H.
    heat flow during extension, 262, 264
    heat flow provinces, 28, 229, 230, 231*f*
    heating on faults, 39
    Horner correction of BHT, 62
Lees' topographic correction, 224–6
liptinite, 155, 156*f*
Lister probe, 56–7, 125
lithosphere
    continental, 245–8
    definition, 237, 298
    oceanic, 238–45
Lopatin's method, 159

McKenzie (1978), instantaneous pure-shear
    extension, 258, 269, 296
mantle
    convection, 9, 10*f*
    temperature within, 9, 11
    thermal conductivity, 247
metamorphism, *see* heat generation, by
    metamorphism
methyladamantane index, 187
methyldiamantane index, 187
methylphenanthrene index, 185
Milankovic cycles, 85
mixing laws, 97–100
molecular biomarkers, 181–8
moon, formation of, 6

needle probe, 116–19
New England/Central heat flow province, 230*f*
North Sea, 269*x*, 296

numerical modelling, *see* finite difference modelling

palinspastic restoration, 294–5
passive margins, *see* extension, passive margins
Peclet number, 219, 262
phonon conduction, 91–2
plate cooling model, 242
platinum resistance thermometer, 51
point-source probe, 123–5
Pollack, H.N., Hurter, S.J. and Johnson, J.R. (1993), heat flow compilation, 9, 11*t*, 243
porosity
    effect on thermal conductivity, 128*f*
    measurement, 105, 107
potassium (*see also* groundwater, geothermometers)
    abundance in crust, 26*t*, 32*t*, 231
    abundance using electric well logs, 31–2
    gamma activity, 32*t*
    in mantle plumes, 249
    radioactive decay of, 24–6
pre-exponential factor, *see* frequency factor
pure shear, *see* extension, pure shear
pyrolysis, 151, 178–9, 198, 199*f*

radiative conduction, 94–6
radioactive decay
    contribution to surface heat flow, 9, 24, 237
    discovery, 7
    modes, 25*f*
ridge-push force, 252*f*
RockEval, *see* pyrolysis
Roy, R.F.
    heat flow provinces, 28, 29*q*
    heating on faults, 39
Rutherford, Ernest, 8–9*n*

salt domes, effect on heat flow, 226–8
sandwich probe, 121*f*
Sass, J.H.
    Australian heat flow, 20*x*, 230, 231*f*
    heating on faults, 39
    thermal conductivity anisotropy, 115, 143
sediments, blanketing effect, 269, 296
seismic velocity, *see* compressional-wave velocity; shear-wave velocity
shale
    drying, 107
    thermal conductivity of, 142–4
shear-wave velocity
    relationship to surface heat flow, 11
    relationship to thermal conductivity, 92, 136
Sierra Nevada heat flow province, 230*f*
silica, *see* groundwater, geothermometers
simple shear, *see* extension, simple shear
slab-pull force, 253*f*
smectite, *see* illite–smectite ratio
sodium, *see* groundwater, geothermometers
South Africa, geotherms deduced from xenoliths, 72*f*
specific heat, 13
square-root mean, 98*f*, 100
steranes, *see* molecular biomarkers
stretching factor, 257–8, 271, 272*f*, 295–7

subduction zones, 252–7
subsidence
    isostatic, 260, 295
    thermal, 260–1
sulphur, effect on oil generation, 151*t*
Sweeney, J.J. and Burnham, A.K. (1990), *see* EASY%Ro

TAI, *see* thermal alteration index
temperature
    cycles at Earth's surface, 78–82, 214–16
    definition, 12
    at Earth's surface, 77–8
    effect on thermal conductivity, 134–6
    flowchart for determining, 49*f*
    of homogenisation, 180
    logging tools, 50–3
    precision logs, 86, 210
    relationship to electrical resistivity, 72–3, 86
    of the sea floor, *see* bottom-water temperature
    of the sea surface, 77*f*
thermal alteration index, 171–2, 198, 199*f*, 201*f*
thermal conductivity (*see also* phonon conduction; radiative conduction)
    anisotropy, 114–15, 282
    definition, 13, 90
    effect of porosity, 128*f*
    effect of pressure, 93
    flowchart for determining, 103*f*
    of mantle, 247
    measurement, *see* divided bar apparatus; needle probe
    of middle and lower crust, 246
    by mineral type, 101*t*
    in numerical modelling, 283–4
    of olivine, 96*f*
    of pore fluids, 112, 133*t*
    relationship to thermal diffusivity, 15*n*
    by rock type, 93*f*, 104*t*
    of shale, 142–4
    temperature correction, 134–6
    of water, 112
thermal diffusivity
    definition, 15
    relationship to thermal conductivity, 15*n*
thermal equilibrium in boreholes, 50
thermal gradient
    definition, 12
    at Earth's surface, 80–1, 84
    precision log, 53*n*, 54*f*
thermal maturity, 146
thermal resistance, 210
Thompson, William (*see also* Kelvin, Baron of Largs)
    heat flow pioneer, 6, 7*n*
    quotation, 90, 146, 207
thorium
    abundance in crust, 26*t*, 231
    abundance using electric well logs, 31–2
    gamma activity, 32*t*
    radioactive decay of, 24–6
Tibetan Plateau, 256, 257
time–temperature index, 159, 162*f*, 201*f*
Tmax, 178*f*, 179, 201*f*
topography, *see* Lees' topographic correction

Touloukian, Y.S., thermal conductivity
    compilations, 21, 101*t*, 133*t*
transformation ratio, 179
trenches, 254*f*, 255

Uinta Basin, Utah, 66
Umklapp processes, 92
underplating, magmatic, 270–2, 296
units
    conversion between systems, 19*t*
    SI, Système International d'Unités, 18, 19*t*
uranium
    abundance in crust, 26*t*, 231
    abundance in mantle, 8
    abundance using electric well logs, 31–2
    gamma activity, 32*t*
    radioactive decay of, 24–6
    spontaneous nuclear fission, 189
Ussher, Archbishop James
    origin of the Earth, 4
    quotation, 3
Uyeda, S.
    effect of sedimentation on surface heat flow,
        216, 217*f*, 218*f*
    heat refraction in basins, 223

Van Krevelen diagram, 147–9
Verne, Jules, quotations, 23, 47

violin-bow probe, *see* Lister probe
virgin rock temperature, 59–60
vitrinite, 155, 157*f*, 158
vitrinite reflectance
    comparison with other maturity indicators,
        199*f*, 201*f*
    EASY%Ro, 164–7
    limitations, 167–8
    marine influenced material, 168
    and maximum palaeotemperature, 163
    measurement, 158–9
    R_o versus R_{vmax}, 159
    suppression, 168, 169
    when to use, 198, 199*f*
Von Herzen, R.P.
    correcting borehole temperature logs, 53
    effect of sedimentation on surface heat flow,
        216, 217*f*, 218*f*
    heat refraction in basins, 223
    needle probe, 116

Waples, D.W.
    Lopatin's method, 159
    time–temperature index, 161, 162*f*
well log analysis, 136–42

xenoliths, 71–2, 86